# Concrete Construction Manual

KIND-BARKAUSKAS
KAUHSEN
POLONYI
BRANDT

BIRKHÄUSER – PUBLISHERS FOR ARCHITECTURE
BASEL · BOSTON · BERLIN

EDITION DETAIL
MUNICH

The original German edition of this book was conceived and developed by
**DETAIL**, Review of Architecture

Editor:
Bundesverband der Deutschen Zementindustrie e.V., Cologne

Authors:

Friedbert Kind-Barkauskas, Dr.-Ing., architect
Bundesverband der deutschen Zementindustrie e.V., Cologne (parts 1, 2, 4 and Glossary)

Bruno Kauhsen, Prof. Dr.-Ing., architect
Department of Architecture, North-East Lower Saxony Polytechnic (parts 4 and 5)

Stefan Polónyi, Emer. Prof. Dr.-Ing. E.h. Dr. h.c. Dr.-Ing. E.h.
Claudia Austermann, Dipl.-Ing.
Faculty of Building, Dortmund University (parts 3 and 4)

Jörg Brandt, Dr. sc. agr.
Bundesverband der Deutschen Zementindustrie e.V., Cologne (parts 2 and 4)

Assistants:
János Brenner, Prof. Dr. techn. Dr. sc. techn., architect, Budapest (part 1)
Martin Hennrich, Dipl.-Ing., Aachen (part 5)
E. A. Kleinschmidt, Dipl.-Ing., Dorsten (part 3)
Martin Peck, Dipl.-Ing., Bundesverband der Deutschen Zementindustrie e.V., Cologne (part 2)
Olaf Sänger, Dipl.-Ing., Dortmund (part 3)

Editorial services:
Sabine Drey, Dipl.-Ing.
Christian Schittich, Dipl.-Ing., architect
Institut für internationale Architektur-Dokumentation GmbH & Co. KG, Munich

Translators (German/English):
Gerd Söffker, Philip Thrift, Hanover

A CIP catalogue record for this book is available from the Library of Congress,
Washington, D.C., USA

Deutsche Bibliothek – Cataloging-in-Publication Data

Concrete construction manual / [publ. by: Institut für Internationale Architektur-Dokumentation GmbH, Munich].
Jörg Brandt ... [Transl. from German to Engl. Gerd Söffker, Philip Thrift]. – Basel; Boston; Berlin: Birkhäuser;
München: Ed. Detail, 2002
ISBN 3-7643-6724-5

This work is subject to copyright. All rights are reserved, whether the whole or part of the material is concerned, specifically the right of translation, reprinting, re-use of illustrations, recitation, broadcasting, reproduction on microfilms or in other ways, and storage in databases. For any kind of use, permission of the copyright owner must be obtained.

This book is also available in a German language edition (ISBN 3-7643-6685-0).

© 2002 Birkhäuser – Publishers for Architecture, P.O. Box 133, CH-4010 Basel, Switzerland
Member of the BSpringer Publishing Group.

Printed on acid-free paper produced from chlorine-free pulp. TCF ∞

Printed in Germany

ISBN 3-7643-6724-5

9 8 7 6 5 4 3 2 1     http://www.birkhauser.ch

# Contents

| | | | | | |
|---|---|---|---|---|---|
| **Part 1 · Concrete in architecture** | 8 | The concrete surface | 65 | **Part 4 · Construction details** | 178 |
| Friedbert Kind-Barkauskas | | Design concepts | 65 | Friedbert Kind-Barkauskas | |
| | | The constituents of the concrete mix | 65 | Bruno Kauhsen | |
| The development of concrete technology | 9 | The effects of formwork | 66 | Stefan Polónyi | |
| Lime mortar applications in antiquity | 9 | Finishing the concrete surface | 68 | Jörg Brandt | |
| Roman buildings with opus caementitium | 10 | The use of coatings | 73 | | |
| | | The effects of the weather | 74 | Loadbearing external wall | 180 |
| The production of cement and concrete | 11 | References | 77 | Loadbearing internal wall | 183 |
| First trials with reinforced concrete | 13 | | | Non-loadbearing facade | 184 |
| The development of prestressed concrete | 17 | Building science | 78 | Flat roof of impermeable concrete | 185 |
| | | General | 78 | Solid roof | 186 |
| Reinforced concrete in the Modern Movement | 18 | Basic requirements | 78 | Joints | 187 |
| | | The interior climate | 78 | Fixing of facade panels | 189 |
| Early industrial and commercial structures | 18 | Energy economy and thermal insulation | 87 | Stairs | 191 |
| The early 20th century | 20 | The effect of noise and sound insulation | 93 | | |
| The development of stressed skin structures | 22 | Behaviour in fire and fire protection | 94 | **Part 5 · Built examples** | 194 |
| Expressionism | 24 | Building science requirements – overview | 96 | Bruno Kauhsen | |
| The early Modern Movement | 25 | The properties of components | 100 | Built examples in detail – overview | |
| Frank Lloyd Wright | 27 | References | 104 | Examples 1 – 33 | 195 |
| Le Corbusier | 28 | | | | |
| Shell structures | 30 | | | **Appendix** | 280 |
| Pier Luigi Nervi | 31 | **Part 3 · Reinforced concrete in buildings** | 106 | | |
| Architecture at the height of the Modern Movement | 32 | Stefan Polónyi | | Index and glossary | 280 |
| | | Claudia Austermann | | Index of names | 293 |
| Late-Modern and Postmodern | 36 | | | Picture credits | 294 |
| Dutch structuralism | 38 | Multistorey structures | 108 | | |
| The use of precast concrete | 39 | Aspects of industrial production | 108 | | |
| Contemporary architecture | 41 | Joints in structures | 110 | | |
| References | 44 | Roofs | 112 | | |
| | | Floors | 113 | | |
| | | Walls | 128 | | |
| | | Columns | 129 | | |
| **Part 2 · Fundamentals** | 46 | Service cores, stairs | 134 | | |
| Friedbert Kind-Barkauskas | | Facades | 135 | | |
| Jörg Brandt | | High-rise buildings | 138 | | |
| | | Suspended high-rise buildings | 140 | | |
| Concrete – the material | 47 | Single-storey sheds | 142 | | |
| The composition of concrete | 47 | Single-storey sheds of trusses and frames | 143 | | |
| Types of concrete | 50 | Stressed skin structures | 151 | | |
| Properties of concrete | 51 | Suspended roofs | 165 | | |
| Exposure classes | 51 | Foundations | | | |
| Reinforcement | 60 | Shallow foundations | 168 | | |
| Concrete masonry | 61 | Deep foundations, pile foundations | 172 | | |
| Application | 61 | Securing excavations | 173 | | |
| References | 64 | Building below the water table | 174 | | |
| | | References | 175 | | |

# Preface

"To gain some kind of foothold in the broad field of the architecture of our age, where the confusion or complete lack of principles in relation to style is on the increase and criticism of its application is very severe among the infinite mass of structures that have been built on this Earth in various eras, I would like to express one main principle: architecture is construction! I derive a second main principle for good architecture from the following observation: every completed construction in a certain material has its very particular character and could not sensibly be built in the same way using any other material."

Karl Friedrich Schinkel

Building is an elementary human activity. It corresponds to mankind's need to shape the environment sensibly for its own benefit. Architecture is the result of the synthesis of different utilisation requirements, constructional options and artistic concepts. The building customs of a period, which also leave their impression on society, develop according to the respective economic circumstances and political notions. If we delve into the role of concrete in architecture, it becomes clear that a very long period of evolution was necessary – and is still not over.

The use of limestone as a building material has been proved through finds of mortar dating from 12,000 BC. This indicates that opus caementitium – Roman concrete – was produced and used from the 2nd century BC onwards. This became the embodiment of imperial building and enabled great achievements in architecture and engineering. But the knowledge of how to produce concrete was lost with the fall of the Roman Empire. The basics of its production were not rediscovered until 1,500 years later.

The actual evolution of concrete in modern history began with the invention of Portland cement in England in 1824. Trials with iron reinforcement in the concrete, which were undertaken around 1850 in England and France simultaneously, soon permitted longer spans and finally resulted in a totally new type of architecture. Initially, the early, really quite idiosyncratic, efforts resulted in reinforced concrete being regarded as a substitute material for timber, iron and stone. However, the consummate and independent use of the material and a sound theory for the accurate analysis of its structural performance quickly developed.

At first, traditional concepts of building and artistic expression prevailed, but these gradually became more adventurous as previous ideas were cast aside. This led to a new architectural style, the Modern Movement, which developed in Europe after 1900. Reinforced concrete had a crucial influence on this. However, the use of this versatile material for linear or planar elements is still a major factor in building today.

This second edition of the Concrete Construction Manual has been fully revised. Furthermore, the appearance of new cement and concrete standards at European level demanded a complete reworking of the chapter on the fundamentals of concrete. Up-to-date examples illustrate changes in architectural fashions and further outstanding applications of reinforced concrete in modern structures. The subtitle "Designing with reinforced concrete in building works" makes it clear that this specialist publication in the series of construction manuals from Edition DETAIL is primarily aimed at architects and engineers. However, this book is also intended to show teachers and students the diverse range of applications of concrete and so contribute to avoiding errors of planning and workmanship in practice.

The examples of architecture and construction are intended to provide inspiration for our own designs and for building with reinforced concrete in a way suited to that material. Furthermore, they demonstrate the chance we have to shape our built environment in a way that is varied, stimulating, well proportioned and hence human.

The publisher and the authors would like to express their sincere gratitude to the architects, engineers and companies named in this book for the loan of their material in the preparation of this book. Thanks are also due to Christian Schittich and the other assistants for coordinating the production.

On behalf of the publisher and the authors
Friedbert Kind-Barkauskas
December 2001

# Part 1 · Concrete in architecture

**Friedbert Kind-Barkauskas**

**The development of concrete technology**

    Lime mortar applications in antiquity
    Roman buildings with opus caementitium
    The production of cement and concrete
    First trials with reinforced concrete
    The development of prestressed concrete

**Reinforced concrete in the Modern Movement**

    Early industrial and commercial buildings
    The early 20th century
    The development of stressed skin
    The early Modern Movement
    Frank Lloyd Wright
    Le Corbusier
    Shell structures
    Pier Luigi Nervi
    Architecture at the height of the Modern Movement
    Late-Modern and Postmodern
    Dutch structuralism
    The use of precast concrete
    Contemporary architecture

**References**

# The development of concrete technology

The use of cement and concrete in modern construction work is the result of many pioneering engineering and scientific discoveries. Some of these stem from ancient times and even today have lost none of their validity. Since the purely fortuitous use of limestone in ancient times, the evolutionary path leads via the specific preparation of this material and its combination with other materials to form a new material – lime mortar. The recognition of the hydraulic properties of certain combinations of materials enabled the development of opus caementitium – Roman concrete – and its use in building work. Today, we are still able to witness the great architectural achievements of that period that made use of this material.

During the Middle Ages the material was more or less forgotten. It was not until the middle of the 18th century that it was rediscovered in conjunction with investigations into the hydraulicity of hydraulic lime and the production of cement. One hundred years later, the tamped concrete technique evolved based on the concept of loam construction. That was followed by diverse experiments to improve the tensile strength of components made from concrete by including iron inserts, and thereafter the perfection of reinforced concrete using iron. The development of the fundamental methods of analysis finally led to a generally applicable theory of iron-reinforced concrete and hence also to our modern reinforced concrete technology employing steel.

Further developments in the field of prestressed concrete have rendered possible ever longer spans while at the same time saving weight. For a number of years high-strength concretes for special applications have been produced for civil and structural engineering projects. Alongside this there are experiments with glass fibre reinforcement and self-compacting concrete. Still newer methods for finishing concrete surfaces promote the development towards a more distinctive use of concrete in architecture.

## Lime mortar applications in antiquity

It is now no longer possible to determine whether the earliest applications of lime mortar known to us from ancient times were deliberate or merely chance results of the use of the materials available in a particular local situation. But without doubt, the observance of chance reactions played an important role in the development of mortar. For instance, the disintegration in the rain of certain limestones after being well heated, and the subsequent renewed solidification of the remoulded material, as well as the repetition and testing of such processes.

Finds in eastern Turkey dating from about 12,000 BC represent the oldest known use of lime mortar as a building material. Then about 6000 years later lime was used as a binder in the mortar in the construction of clay brick structures in the Jericho culture of Palestine. Among the remains of old buildings dating from about 5500 BC excavated near Lepenski Vir in the Carpathian Mountains are ground floor slabs with a concrete-like composition made up of fired lime, sand and loam. The use of lime mortar in important religious buildings has been proved in ancient Egypt, Troy and Pergamum. This method of building is mentioned several times in the Old Testament, which was written around 1200 BC.

The Phoenicians realised that the volcanic rock from the island of Santorini – ground and mixed with lime, sand and water – produced a waterproof mortar. They used this material for building their irrigation systems, among other things, and disseminated this knowledge around the whole of the Mediterranean. The waterproof plaster made from lime and loam used in the cisterns of Jerusalem, proclaimed by King David as his residence around 1000 BC, is still intact today. Fired lime was also known to the Greeks; they had been using this material since the 7th century BC, mixed with crushed marble to form a lime plaster. The lime content of the cisterns on Santorini was, for example, about 43%. However, the masonry was built of dry ashlar walling. In contrast, it has been proved that lime mortar was used in the final structures of Nebuchadnezzar in Kasr and Babylon (around 600 BC) and in the so-called "Long Wall" of Athens (about 450 BC). This also applies to sections of the Great Wall of China built around 300 BC. In this case it has also been established that the subsoil was consolidated with lime to improve its load-bearing capacity.

The development of concrete technology

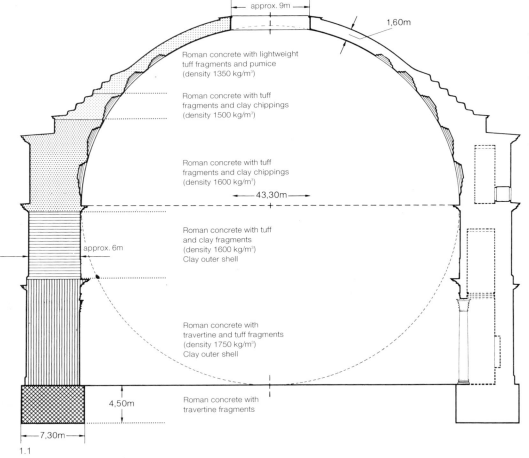

## Roman buildings with opus caementitium

The 3rd century BC witnessed a new method of wall construction in Lower Italy: lime mortar and rubble stone tipped into the cavity between leaves of jointed ashlar stone and compacted by tamping. Anchor stones (diatonoi) linked the two masonry leaves and ensured the necessary stability until the emplekton ("tamped material") had hardened. In his work De re rustica, published in 184 BC, Marcus Porcius Cato describes a lime mortar; he recommends a mixture of one part slaked lime to two parts sand.

It is at this point that the Romans take up the idea of Greek cast-in-situ masonry. Among other materials they used tuff, marble rubble and clay bats as the aggregate. In his work De architectura libri decem, published in 13 BC, Vitruvius describes for the first time the production of hydraulic mortar and concrete comprising hydraulic mortar and stone chippings. This opus caementitium is without doubt comparable with today's concretes of equal compressive strength.

Opus caementitium is waterproof under certain conditions and contains sufficient quantities of coarse aggregate, up to about 70 mm in size, as well as gravel and sand as the fine aggregate. The hydraulic lime mortar and the aggregates were mixed together prior to being placed and then compacted mechanically by means of tamping. The outer leaves of the masonry were mainly of stone and clay, often richly decorated. Concrete surfaces with an "as struck" finish are found almost exclusively in utility buildings such as cisterns and thermal baths. Large foundations, such as those to the Colosseum in Rome (completed in 80 AD), and port structures, e.g. the breakwater at Naples, were built in this material; the impressions left by the formwork boards can still be seen. The foundations to the breakwater at Pozzuoli were built by Caligula using large precast concrete blocks subsequently sunk into the water.

Agrippa started building what is probably the most spectacular structure in ancient Rome – the Pantheon – in 27 BC. The cylindrical wall, measuring 43.4 m in diameter, is crowned by a self-supporting monolithic dome construction made of concrete. The cross-section matches the force diagram so exactly that the thrust of the dome can be carried by the walls without the need for buttresses. As investigations have shown, the walls and the waffle-type dome construction employ concretes of different densities so that the weight decreases considerably towards the apex of the dome with its 9 m dia. rooflight.

The dome of the Hagia Sophia in Constantinople (now Istanbul), erected between 532 and

1.1

1.2

## The production of cement and concrete

537 AD under the direction of the architects Anthemios of Tralles and Isidoros of Milet in the reign of Justinian, measures 32 m in diameter. It was not until 1570 that a diameter of 43 m was realised again – in Sinan's Selimiye Mosque in Edirne. The vaulting caps were built on clay masonry acting as permanent formwork. Four arches enclose the central dome. These are supported by the vaulting caps of the apses and in this way are relieved of the thrust from the great dome.

The Leaning Tower of Pisa is regarded as one of the last great concrete structures in the Roman tradition in Italy. Construction started on the 58 m high tower in 1173. Despite being stabilised, which reduced the angle of tilt by about 10%, the tower still leans at an angle of about 5° from the vertical. A monolithic concrete cylinder with walls 2.7 m thick contains the helical stairs and is clad on both sides with hard marble.

### The production of cement and concrete

The knowledge about the correct composition of opus caementitium was lost in the Middle Ages. The mortar for fortifications and secular buildings was a mixture of pure loam and lime or sand. Occasionally, gypsum and clay brick dust were mixed in as additives. Different organic substances were added in order to try to improve the strength of the mortar. Various sources have reported on the addition of vinegar, milk, or – in 1450 in Vienna during the building of St Stephen's Cathedral – wine of the latest vintage! In the 16th century the Dutch discovered the hydraulic effectiveness of ground tuff (i.e. trass), without knowing the underlying chemical reasons for this. Trade with this valuable building material was soon to become an important economic factor throughout Europe. All attempts to produce an effective binder continued without any real plan until well into the 18th century – and for the most part devoid of any noteworthy success.

A technical dictionary published in 1710 describes crushed clay bricks for the first time as "cement". Bernard Forest de Bélidor, an engineering officer in the French army, published a manual entitled La Science des Ingénieurs in 1729. Its topics included the production of mortar from different types of limestone, the use of various hydraulic additives and the production of "cast-in-situ vaulting" from hydraulic lime. The term "Béton" appears

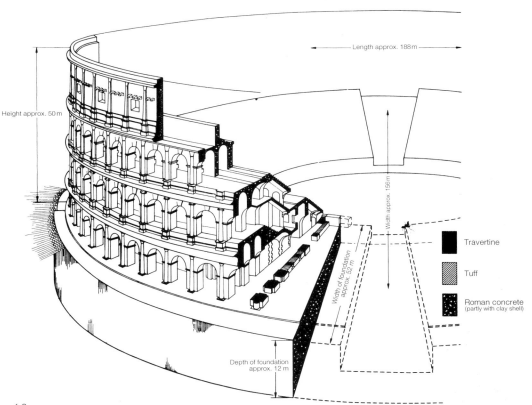

1.3

1.1 Pantheon, Rome, section illustrating various types of concrete
1.2 Pantheon, Rome, begun in 27 BC
1.3 Colosseum, Rome, completed in 80 AD
1.4 Leaning Tower of Pisa, begun in 1173

1.4

# The development of concrete technology

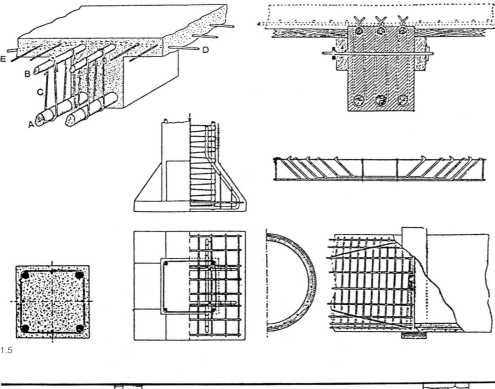

1.5

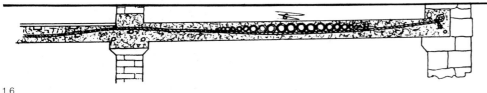

1.6

1.7

in his work of 1753 Architecture hydraulique. This designated a mixture of water-resistant mortar and coarse aggregates and borrowed the old French terms for masonry "Bethyn" or "Becton".

The Englishman John Smeaton established the principles of hydraulicity in 1755. He discovered that a certain clay content in cement was the reason for the setting properties underwater and the later waterproof properties of the mortar. Thereupon, he was appointed to rebuild the recently destroyed Eddystone lighthouse off the English coast near Plymouth. For his mortar, which Smeaton claimed would produce a cement equal to the best Portland stone in terms of strength and durability, he mixed equal parts of the local Aberthaw lime and pozzolana from Civitavecchia in Italy. In 1796 the Englishman James Parker succeeded in producing Roman cement, a substitute for pozzolana and trass, which hardens without the addition of lime. The name of this cement was derived from its colour, which was very similar to that of Roman pozzolana.

However, the chemical relationships to explain the durabilities of different combinations of materials remained unclear. It was not until 1815 that the Berlin chemist Johann Friedrich John described the reasons why mortar made from limestone is more durable than mortar made from seashells. To come to this result he had examined samples of mortar from historical structures and established that it is the chemical relationship between silicic acid, alumina and lime – simultaneously subjected to high temperatures – that produces the bonding force. John was awarded a prize for his work by the Dutch Academy of Science.

In 1824 the English master bricklayer Joseph Aspdin developed a mixture of clay and limestone that he called "Portland cement". He described it as a method for "an improvement in the modes of producing an artificial stone". He had based the name of his "cement" and the artificial stone for "stuccoing buildings, waterworks and cisterns" or other suitable building works on Smeaton's strength comparison with Portland stone. This name for this type of cement has been retained to this day. In 1825 a small factory in the English town of Wakefield was the first to produce Portland cement on a commercial basis. The next development was in 1844 when the Englishman Isaac Charles Johnson introduced firing up to sintering, instead of the underfiring used hitherto, into the production of cement. This brought about considerable improvements to the material's properties.

## First trials with reinforced concrete

The first structures built entirely of concrete appeared in France and England at the start of the 19th century. The English stucco master William Boutland Wilkinson succeeded in building the first floor slab reinforced with wire ropes in 1852. And in 1854 he applied for a patent for an iron-reinforced concrete composite floor slab. The text of the patent describes it thus: "The invention concerns fire-resistant structures with concrete floors reinforced by means of wire ropes and thin iron bars embedded below the middle axis of the concrete." [20] He built a two-storey house for himself in 1865 in Newcastle, England, completely of concrete with waffle floors and precast stairs. As the reconstruction during demolition later revealed, he had placed the iron reinforcement in the tension zone and for multi-bay slabs had transferred this to the top over intermediate and end supports.

The application of iron inserts for stabilising structures and components was also the subject of an investigation by T.E. Tyerman. His work was patented in 1854. He also pointed out the need to bend the iron inserts to produce a better bond within the mortar. Just one year later the French building contractor François Coignet developed a tamped concrete method – imitating loam construction – for erecting structures and components of all kinds; he called it "Béton aggloméré". At the same time, he applied for a patent in England for his use of criss-crossing iron bar reinforcement in concrete floor slabs. He built a three-storey house of concrete in St Denis. In the same year, 1855, the German engineer Max von Pettenkofer – based on his own research – published the method of production for Portland cement, which until then had been kept secret, and hence opened up the way for the start of cement production in Germany.

During the same period, the Frenchman Josef Louis Lambot was working on the problem of how to use concrete reinforced with iron for tension members. Concerning the use of reinforced concrete as a substitute for timber in the construction of water tanks, plant pots and in shipbuilding, he wrote: "My invention has a new product as its object, which serves to replace the timber in shipbuilding and wherever it is at risk of damage through moisture, like in wooden floors, water tanks, plant pots, etc.

1.5 François Coignet, drawings from his patent for the reinforcement of concrete slabs, 1854
1.6 William Boutland Wilkinson, drawing from his patent for iron-reinforced composite slabs, 1854
1.7 Concrete mixer, 19th century
1.8 Josef Louis Lambot, facsimile of his patent for reinforced concrete as a substitute for timber, 1855

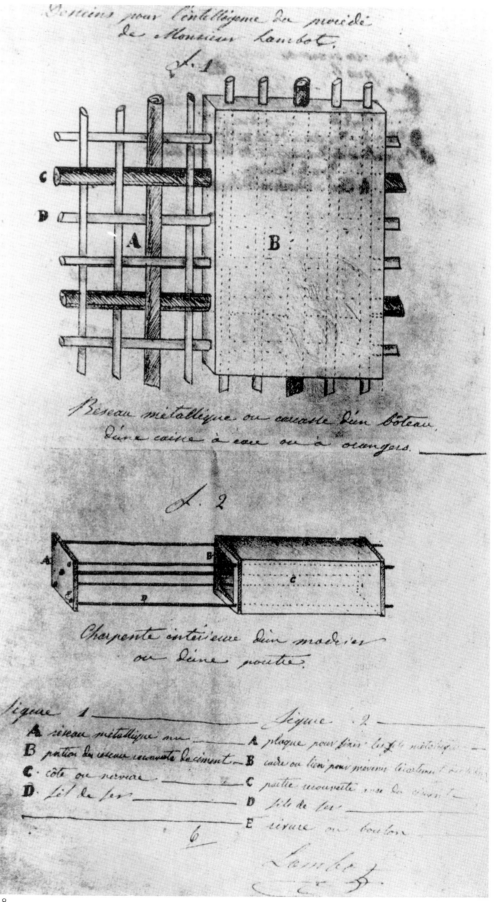

1.8

# The development of concrete technology

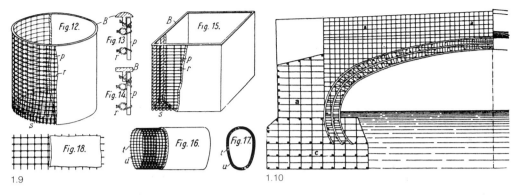

1.9   1.10

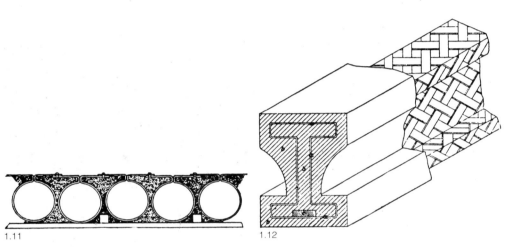

1.11   1.12

1.13

The new substitute material consists of a metal mesh of wires and struts connected or woven together in any form. I give this mesh a form that best fits the object I wish to produce and subsequently embed it in hydraulic cement and thus also eliminate any joints." [20] Lambot called his material of the 1855 patent "Ferciment".

Another Frenchman, Joseph Monier, carried out similar experiments. As he was a gardener, he had the idea of producing flowerpots from a plaited mesh and enclosing this in cement. As he developed his idea, he was able to develop a "method for producing objects of various kinds from a combination of metal framing and cement", for which he was granted a patent. He described the purpose of his invention thus: "to improve the durability of cement while saving on cement and work" [20]. He also applied his principle to building bridges of iron-reinforced concrete. However, as the drawings show, the arrangement of his reinforcing bars reveal that he was totally unaware of the flow of forces within a component or structure. The drawings also reveal that the internal structural relationship of the composite action of the materials concrete and iron was not yet known. Nevertheless, "strain tables" for concrete were published in the "Journal des Ponts et Chaussées" in the same year.

The first standards for cement were introduced in Germany in 1877. The oldest concrete structures in Germany still standing today were erected between 1871 and 1875 in Berlin (Victoriastadt). Six of the original 60 multistorey residential blocks still exist. Other structures built of unreinforced (plain) concrete were erected in 1879 on the occasion of a horticultural exhibition in Dreieichenpark in Offenbach near Frankfurt am Main. It was about this time that the lawyer Thaddeus Hyatt and others in America recognised the structural relationships in iron-reinforced concrete. Numerous components reinforced with iron were developed, e.g. for loadbearing concrete structures, with stirrups and gusset plates, and slabs of concrete and glass. In Hyatt's all-embracing patent of 1878 it states: "Cement concrete is combined with strips and bars of iron to form slabs, beams or vaulting such that the iron is only used on the tension side." [20]

Hyatt also discovered the fire resistance of the material when the iron is completely encased in the concrete. (His fire test with a concrete building erected specially for this purpose in London became famous.) In addition, he investigated the durability of the bond effect between concrete and iron reinforcement, the sufficiently even thermal expansion of the two materials and their different elasticities. And last but not least, it was also Hyatt who advocated the particularly effective structural form of the T-beam. He stressed the suitability of this composite building material not only for loadbearing members in structures but indeed – owing to its weathering resistance and low maintenance costs – for bridge-building, too.

The Coignet company achieved great economic success in France around 1880 with the improved technology of iron-reinforced concrete construction. Edmond Coignet came to the conclusion that a favourable water/cement ratio, continuous mixing and careful compaction would produce concrete of good quality. The speciality of his company was the construction of buildings entirely of concrete, erected with the help of reusable formwork. Meanwhile in Germany, the building contractors Conrad Freytag and Carl Heidschuh purchased the Monier patent, but sold their rights to its use in the Berlin region one year later to Gustaf Adolf Wayss. The purchase of the Monier patent marked the beginning of reinforced concrete construction in Germany on a larger scale. For example, the construction of the new Reichstag building in Berlin to a design by the architect Paul Wallot included for lightweight partitions. However, a patent application was initially refused because the Berlin master bricklayer Carl Rabitz had already built similar walls in Schinkel's "Altes Museum" and had been granted a patent for them. Only after the site manager on the Reichstag project, government master-builder Mathias Koenen, had examined the new partitions was it conceded that they indeed represented a new type of construction.

Koenen published his method of analysis for the moment of resistance of a rigid slab with iron bars positioned near the soffit in the *Zentralblatt der Bauverwaltung*. In tests carried out jointly with Wayss he had established that a concrete slab with iron reinforcement can carry many times the load possible with an unreinforced slab of the same dimensions. Wayss therefore also took over the construction of loadbearing floor slabs and vaulting for the Reichstag project in Berlin. In 1887 Wayss published The Monier System (Iron Skeleton with Concrete Filling) in its application to Building, in which Koenen presented the results of his experiments and his method of design for rigid slabs. This was the first manual for iron-reinforced concrete, and was a major influence in disseminating the knowledge about this new type of construction throughout Germany. In Prussia in the same year, DIN 1164 became the first industrial standard.

All previous trials with iron reinforcement in concrete had indicated that the loadbearing effect was still limited by the potential formation of cracks at higher loads. Therefore, the American Jackson presented an invention in 1886 in which concrete elements could accommodate more tension by including iron bars stressed by means of threads and nuts. His method, however, enabled only a very small prestress to be achieved; tensile stresses and horizontal thrust still presented a problem. The German engineer Doehring took up Jackson's idea and developed a method for prestressing iron-reinforcement in concrete components in his own tests into how to prevent cracks in iron-reinforced concrete construction. At the same time, Ernest Leslie Ransome from England was also working on iron reinforcement for concrete. The result of his research was a profiled reinforcing bar, patented in 1893.

But two years prior to this Edmond Coignet had already produced precast concrete elements in an on-site factory while building the Biarritz Casino. Since then the Casino has been regarded as the starting point for modern construction with precast concrete components. In the following years Coignet used large-format prefabricated components of iron-reinforced concrete in other construction projects. In the meantime, the French stonemason François Hennebique was working on a method for producing iron-reinforced concrete composite construction. In a series of tests he perfected the construction of the beam-and-slab floor connected monolithically with iron-reinforced concrete columns. So he succeeded in producing what is for reinforced concrete probably the most typical type of construction. Even at this early stage, the arrangement of the reinforcement in his very economical building system corresponded exactly to the flow of forces in the structural analysis. Numerous factories and warehouses with relatively high imposed loads, e.g. in Lille (1892) were therefore built using the "Système Hennebique". In 1896 Hennebique's platelayers' cabin became the first mass-produced building. This, the first portable building unit, consisted of 50 mm thick iron-reinforced concrete panels. Furthermore, in 1904 he demonstrated all the potential applications of his iron-reinforced concrete in the construction of his own house at Bourg-la-Rheine.

In the meantime, the world's first iron-reinforced high-rise building had been built in America. The 16-storey "Ingall's Building" (1902) in Cincinnati made use of an iron-

1.9 Joseph Monier, drawing from his German patent of 1871
1.10 Joseph Monier, drawings from his bridge patent of 1873
1.11 Thaddeus Hyatt, drawings from his first iron-reinforced concrete patent of 1871
1.12 Thaddeus Hyatt, drawing from his patent of 1874
1.13 Building at the garden exhibition, Offenbach, 1879
1.14 and 1.15 Viewing platform, Helsingborg, Sweden, during construction and after completion (1903)

1.14

1.15

# The development of concrete technology

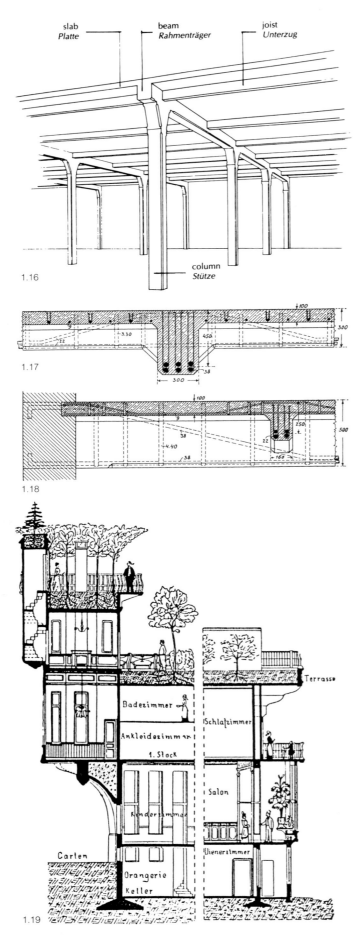

1.16
1.17
1.18
1.19

reinforced frame developed by Ransome on the basis of Hennebique's work. Experience with the new composite building material iron-reinforced concrete increased and it was realised that the formation of cracks also provided an indication of changes in length of the concrete caused by various influences. Even before the start of the 20th century the German engineer Schumann had discovered in five years of tests that the "swelling" of concrete due to water absorption, and "shrinkage" due to drying out, were the causes of hairline cracks in exposed concrete elements. In 1893 he discovered that the setting of concrete was linked to small changes in volume but that they were less than in other materials. Shortly afterwards the Austrian engineer Fritz Edler von Emperger published his fundamental work on the theory of iron-reinforced slabs according to the theory of elasticity. This was followed by the *Handbuch für Eisenbetonbau* (Iron-reinforced Concrete Construction Manual) and the *Betonkalender* (Concrete Yearbook); the latter is an important work of reference that still appears regularly today. However, this new knowledge spread only very gradually.

The World Exposition took place in Paris in 1900. One great attraction for the public was Edmond Coignet's "Château d'Eau". This fantastic design was built entirely of iron-reinforced concrete. Its baroque diversity of form demonstrated impressively the many possible applications of the new material. This structure – like the Crystal Palace of the Great Exhibition in London in 1851 and the Eiffel Tower in Paris in 1889 – was intended to show that great progress had been made in the field of building technology. In 1902 Mathias Koenen published *Grundzüge der statischen Berechnung der Beton- und Eisenbetonbauten* (Principles for the Structural Analysis of Concrete and Iron-Reinforced Concrete Structures). Emil Mörsch later developed this into a general theory of reinforced concrete construction.

At the same time, the engineer and building contractor Robert Maillart began tests connected with the design of heavily loaded floor slabs at his works in Switzerland. His idea of a total loadbearing structure carried by columns alone and without additional beams represented the antithesis to Hennebique's well-known loadbearing system. In 1909 Maillart developed his method of analysis for flat slabs with "mushroom" columns and just one year later built a warehouse in Zurich according to his proposals. This was followed by the so-called filter house in Rorschach.

## The development of prestressed concrete

After the initial attempts to produce prestressed concrete by Jackson and Doehring, Koenen in Germany, Sacrez in Belgium, Lund in Sweden and Steiner in America took up this idea again. Worthwhile improvements to the method failed at first due to the fact that the creep and shrinkage of concrete were not yet generally appreciated. But in their patents of 1907 and 1908, Sacrez and Steiner were already calling for the embedded reinforcement to be stressed to such an extent that upon inducing the load in the component concerned, the compressive forces in the concrete must first be neutralised by stressing the reinforcement before tensile stresses and hence cracks can occur in the concrete.

The French engineer Eugène Freyssinet investigated the phenomenon of creep in concrete and from this derived his own method for producing prestressed concrete in 1911. He realised that the creep of concrete decreased as the compressive strength and density of the concrete increased. The smaller the creep, the smaller the loss of prestress due to creep is. On the other hand, the larger the remaining prestress after losses have taken place, the less significant the loss of prestress is. It is therefore necessary to prestress steel with a high tensile strength. Some years later Wilson in England took up the idea of prestressed iron reinforcement in concrete again and developed a beam with both conventional reinforcement and wire ropes highly prestressed before placing the concrete.

Walter Bauersfeld was granted a patent in 1922 for his method for producing domes and similar curved surfaces of reinforced concrete. Freyssinet and J. Séailles also continued working on prestressed concrete technology. They recognised the importance of low-mortar, high-strength concrete and developed the method of compaction by vibration which today is an essential requirement when employing prestressed concrete. The Wayss & Freytag company introduced the term "Spannbeton" (prestressed concrete) into Germany in 1935. The prestressed concrete components they produced under licence using the Freyssinet method were characterised by the fact that the concrete was prestressed in such a way by inducing certain forces that it was not subjected to tension – or only to a limited extent – under service loads. Just one year later Franz Dischinger succeeded in proving that beam bridges in prestressed concrete could be built to span up to about 150 m. At that time the maximum span for reinforced concrete construction was merely 70 m.

The first book on prestressed concrete was published in 1943 by Emil Mörsch. He explained the method of analysis for the material and described a number of projects, including the first prestressed concrete structure by Wayss & Freytag in Germany – a bridge over the motorway in Oelde, which still stands today.

In the meantime, prestressed concrete had become established in almost all areas of construction. Throughout the world, large uninterrupted spans were being constructed using this technology. Besides bridge-building with its various structural systems and precast concrete systems for civil and structural engineering projects, prestressed concrete is particularly suitable for the roof members over single-storey sheds as well as northlight and barrel vault roofs.

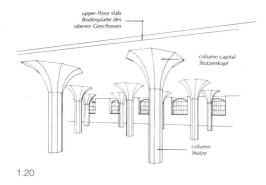

1.20

1.16 François Hennebique, concrete frame, 1904
1.17 and 1.18 François Hennebique, drawings from his patent for reinforced concrete composite construction, 1892
1.19 François Hennebique, own house in Bourg-la-Reine, 1904
1.20 Robert Maillart, warehouse, Zurich, 1910
1.21 Elzner and Anderson, Ingalls Building, Cincinnati, 1902

1.21

# Reinforced concrete in the Modern Movement

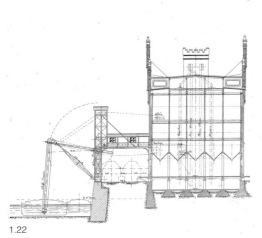

1.22

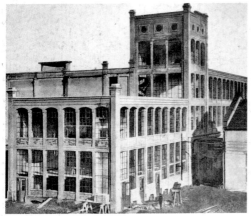

1.23

1.24

The significance of reinforced concrete in architecture must be evaluated according to various aspects. One of these is the material itself. For thousands of years, structures had mainly been constructed from primary, i.e. from the technological viewpoint, homogeneous materials, e.g. stone, timber, loam, fired lime, straw or reeds. It was not until the mid-18th century that the Industrial Revolution added iron to this list, and steel production first began in the 19th century. Both these materials were new, and although not primary, they were still homogeneous materials. Iron and steel were ideally suited to fulfilling the new demands society was placing on the built environment. These included, for example, stations and other structures for the railways, bridges, factories, exhibition buildings and warehouses. The new building materials iron and steel came into their own because they were equally useful in tension and compression, an aspect that was barely possible with the traditional materials, with the exception of timber.

The development of reinforced concrete marked the dawn of a new age. Reinforced concrete was the first heterogeneous building material – in this case using steel, cement, sand, gravel and water – whose artificial composition, possessed much better properties than its individual components. The strength of steel in tension and that of concrete in compression were united in one material. Essentially, any reinforced concrete cross-section can be given the necessary tensile and compressive strength to accommodate the intended range of stresses. No primary building material can carry so much load. Reinforced concrete is also more fire-resistant than most homogeneous building materials. And it represents an economic advantage because it comprises generally inexpensive, readily available materials.

Reinforced concrete rendered possible a completely new approach to construction; the monolithic frame on a preset column grid and its further evolution into plan layouts that no longer relied on loadbearing walls were early accomplishments. Soon, designs inconceivable in steel and timber could be implemented. These included the flat slab with flared column heads, representing a seamless transition between floor slab and column, where indeed the slab requires no intersecting supporting beams.

But constructions most typical of the material are those in which the distinction between active and passive, i.e. loadbearing and non-loadbearing, elements is eliminated. This is the advantage of reinforced concrete over other building materials. However, its malleability and untreated, exposed surface offered new, hitherto unknown opportunities.

Progressive architects were excited by the new material. For example, in 1922 the Dutch architect Hendrik Petrus Berlage wrote: "After iron, reinforced concrete is probably the most important invention in the realm of materials, perhaps the most important of all because reinforced concrete possesses all the properties that are missing in iron – and because the properties of stone and iron are united in this building material. For what has now become possible in principle? No more and no less than the construction of surfaces without seams, walls without joints. On a stone wall that was only possible after plastering it, not before, and moreover this enabled the straight span between two supports spaced at practically any distance. So it has become technically feasible to construct the two most important elements of architecture – the wall and the member spanning between supports – with almost any dimensions. In addition, there is the unification of floor and ceiling, also as a complete unit, and again in all possible sizes. This building material is a technical triumph over the difficulties caused by all the building materials produced up to now." [5]

### Early industrial and commercial structures

The earliest and most decisive successes for this new engineering material were in the field of industrial and commercial buildings, where its low cost and the new structural opportunities of spanning large areas were appreciated. Architects and engineers had been liberated from the constraints of traditional building methods. The close of the 19th century saw a number of factories built using the Hennebique system, e.g. the Charles Six spinning mill in Tourcoing, the flour warehouse in Nantes, the Barrois spinning mill in Lille, the grain silo in Strasbourg, and the La Cité spinning mill in Mulhouse.

Early industrial and commercial buildings

Ransome's structural system, which took into account formwork procedures and the casting of the entire loadbearing structure as well as reinforcement, was first used by the United Shoe Machinery Company in 1903. The reinforced concrete frame of the plain cubist structure is readily discernible on the facade, similar to Albert Kahn's Winchester Gun Factory (completed in 1906). Kahn was also responsible for the Ford Highland Park Plant in Detroit (1910). This building was based on a 6.0 x 4.5 m grid and measures 288 x 22.5 m; it was the first factory for the industrial production of cars. The plan layout of the clearly arranged frame building could be subdivided as required. All the access and sanitary facilities were housed together in an external core.

In Germany, around the start of the 20th century, reinforced concrete started replacing steel as the material for many large market halls. In addition to better fire resistance, decisions were based on economic and aesthetic considerations. This is why there is frequently a great discrepancy between the external appearance and the interior of these buildings. While the facades were often provided with a historic cladding, the reinforced concrete construction was left exposed internally. For instance, the new market hall in Breslau (now Wrocław, Poland) completed in 1908 was given a neo-Gothic facade. We can see here that, being a new material, concrete still had to find its own character. The loadbearing structure was merely a steel frame encased in concrete. It even attempted to imitate a riveted iron construction by painting.

In contrast, the market hall by Richard Schachner on Munich's Thalkirchnerstrasse was a much more consistent design. Its reinforced concrete structure was unencumbered by traditional forms and sentiments and, both inside and outside, it made no attempt to hide its modern style.

1.25

1.26

1.27

1.22  Eduard Züblin, design for a grain silo in Strasbourg, 1898
1.23  Eduard Züblin, "La Cité" spinning mill, Mulhouse, 1900
1.24  Albert Kahn, Winchester Gun Factory, 1906
1.25  Plüdemann and Küster, market hall, Breslau (now Wrocław), 1908
1.26  Richard Schachner, market hall, Munich, 1911
1.27  Auguste Perret, Esders Clothing Factory, Paris, 1919

# Reinforced concrete in the Modern Movement

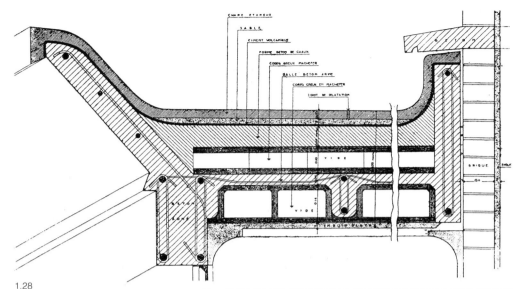

1.28

### The early 20th century

The first residential building in which reinforced concrete was used in a characteristic form in the make-up of its facade is acknowledged to be Auguste Perret's apartment block at 25 Rue Franklin, Paris. The architect allowed the reinforced concrete frame of the building to remain clearly visible on the facade. The difference between loadbearing structure and decorated, masonry infill panels is thus clearly identifiable. The consistent use of reinforced concrete, which, however, remained completely hidden everywhere behind a cladding of ceramic tiles, was also connected with the architect's stylistic concepts; the confined site and the wish to obtain better views led to the oriels, which were possible only in concrete or steel.

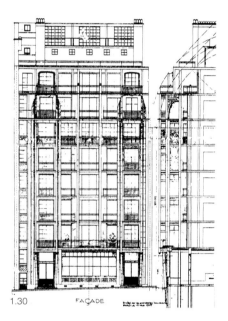

1.30

Perret pioneered the use of the new concrete architecture. He was able to develop his own architectural language out of a combination of neoclassical tradition and the use of the material in a suitable way. This was reflected in other important buildings by Perret, like the garage on the Rue Ponthieu in Paris (1906), in which he added ornamental glass panels to the loadbearing frame of fair-face concrete.

Reinforced concrete was also quickly adopted for religious buildings. But Anatole de Baudot's Church of St Jean de Montmartre in Paris (1894-97) still showed no typical concrete characteristics.

Built about 10 years later, the Unity Church in Oak Park by Frank Lloyd Wright was described by the architect himself as "the world's first concrete monolith". The search for an inexpensive method of building led him to use reinforced concrete. For a long time this church

1.29

The early 20th century

was regarded as the most modern church in the USA. Besides special structural features like the grid of beams supporting the roof, its modern design is also revealed in Wright's handling of the material; in order to influence the texture and colour of the concrete, he asked for a special gravel to be added to the mix. This technique, since improved, is still used today to extend the range of artistic options. The first fair-face concrete church in Europe was built between 1910 and 1913 in Vienna – the Church of the Holy Ghost by Josef

1.31

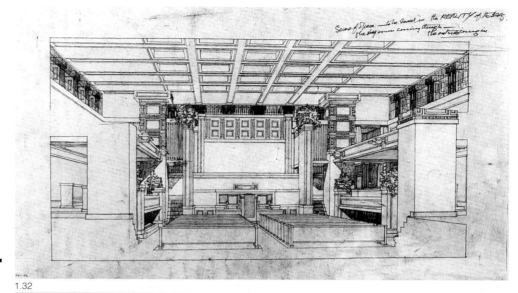
1.32

Plečnik. Reinforced concrete gave the architect the freedom to quote the most diverse classical forms and styles both on the facade and within the interior. In the subterranean crypt he experimented with the expressive possibilities of the material. The sculpted capitals and bases of the columns remind the observer of early Christian architecture.

In 1922 Perret designed the Notre Dame Church at Le Raincy near Paris. As the load-bearing structure had been moved to the inside, he was able to reduce the wall to a transparent lattice. This building had a lasting effect on Karl Moser's St Antonius Church in Basel (1927). He said of this church: "In attempting to achieve a harmonious, capacious and bright interior but still build the church with the most economic means, it was decided to opt for reinforced concrete."

The School of Anatomy at Munich University was built almost at the same time as Frank Lloyd Wright's Unity Church. The use of the material is just as consistent as his but the architectural language explores completely different paths. This building, to a design by the architect Max Littman, is quite rightly regarded as one of the first important structures

1.33  1.34

1.28  Auguste Perret, Hotel Particulier, Paris, detail of junction at top of projecting bay
1.29  Auguste Perret, church at Le Raincy near Paris, 1922
1.30  Auguste Perret, apartment block in Rue Franklin, Paris, 1903, elevation and section
1.31 and 1.32  Frank Lloyd Wright, Unity Church, Oak Park, Illinois, 1904-06
1.33  Josef Plečnik, Church of the Holy Ghost, Vienna, 1910-13, crypt
1.34  Max Littman, School of Anatomy, Munich University, 1906
1.35  Robert Maillart, warehouse, Zurich, 1910

1.35

# Reinforced concrete in the Modern Movement

1.36

1.37

erected completely in reinforced concrete in Germany because it is a credible demonstration of all the structural and aesthetic qualities of this material. In 1908 the *Süddeutsche Bauzeitung* expressed it thus: "The new building is unique in more than one sense. First, Germany has no other building of this type, totally new and satisfying all modern requirements so magnificently and thoroughly; second, the solution to the practical task is certainly new; and third, the use of the material, reinforced concrete, to an extent never dared before on the exterior. One can say that the whole main building is actually made from one piece of concrete; ... The walls stand there smooth and tranquil in the fine matt grey of the concrete, and virtually their only decoration is the lines resulting from the construction of the reinforcing framework which determines the panels of the ceilings or the simple subdivision of the walls internally as well." [52]

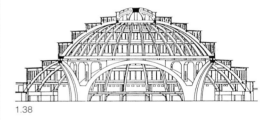

1.38

## The development of stressed skin structures

One of the greatest concrete structures from this period is Max Berg's Centennial Hall (now Hala Ludowa) in Breslau (now Wrocław) (1911-13). This was intended to demonstrate the latest accomplishments in construction at a great exhibition. It was the first prestige building of this order of magnitude dominated internally and externally by fair-face concrete. The

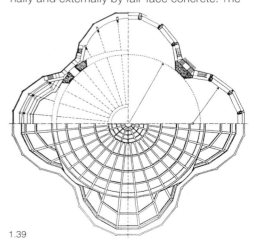

1.39

# The development of stressed skin

structural calculations for the building represent an incredible engineering achievement; after almost 1800 years the span of the Pantheon in Rome had at last been exceeded. The supporting structure to the 65 m span dome consists of 32 ribs resting on a ring beam (in tension) of steel sections at the base and another (in compression) at the apex (clear diameter 14.4 m). Surmounting the dome is a lantern 5.75 m high, which is supported on the upper ring beam. Four huge reinforced concrete arches curved on both plan and elevation support the dome. On the outside these arches are propped by flying buttresses, interconnected by stiffening ribs, which at the same time enlarge the hall. The arches rise 16.7 m and span 41.2 m.

Despite its modernity and boldness, the historical references in the construction are quite apparent. For example, the plan layout of the Centennial Hall with its four apses re-create the Italian Renaissance churches in Todi and Milan, while the loadbearing structure with the circumferential ring of flying buttresses reveals Gothic parentage. Comments of Berg in which he refers to the monumentality of medieval cathedrals prove that these similarities are not coincidental. Even though the Centennial Hall was a phenomenal achievement, the possibilities that reinforced concrete brings to stressed skin structures were not fully exploited. That was left to the airship hangar in Orly completed by Eugène Freyssinet three years later. This single-storey shed is a folded plate structure in which the non-loadbearing surfaces also have structural functions. In section the construction follows the shape of the thrust line and there is no longer a distinction between structure and

1.41

1.42

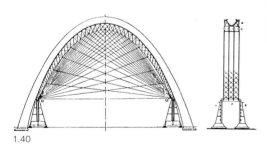
1.40

1.36 to 1.39 Max Berg, Centennial Hall, Breslau (now Wrocław), 1911-13
1.36 Under construction
1.37 Interior view
1.38 Section
1.39 Plan
1.40 Eugène Freyssinet, airship hangar with travelling concreting scaffold, Orly, 1916
1.41 Walter Bauersfeld, Dyckerhoff & Widmann, planetarium, Jena, 1924-25
Reinforcing mesh for shell dome
1.42 Walter Bauersfeld, Dyckerhoff & Widmann, planetarium, Jena, 1924-25
General view
1.43 Eugène Freyssinet, airship hangar, Orly, 1916

1.43

1.44

envelope. The reinforced concrete slabs are just 90 mm thick for a span of 75 m and cross-sectional depths of between 3.0 and 5.4 m, which led to an enormous saving in material.

The revolutionary development of shell structures was promoted in the following years above all through the experiments of Franz Dischinger, Walter Bauersfeld and Ulrich Finsterwalder, in collaboration with the Dyckerhoff & Widmann company. Early examples are an experimental structure in Jena (1922), the production building for the Schott company (1924) and the planetarium for the Carl Zeiss company (1925), both also in Jena. The reinforced concrete shells were based on a precisely calculated mesh of reinforcement whose actual stresses essentially match the values calculated for them beforehand. This mesh was then encased in sprayed concrete in distinct horizontal rings starting at the bottom.

1.46

### Expressionism

The architects of the Expressionist period recognised the sculptural potential of concrete and at first tried to employ this in their structures as an artistic element. At the outset, they met with a number of problems, large and small. For example, the irregularly curved surfaces of Erich Mendelsohn's Einstein Tower in Potsdam (1920) led to formwork difficulties, which is why the upper part of the structure was built in traditional masonry and subsequently clad with a thin layer of shaped concrete. Rudolf Steiner's Goetheanum in Dornach near Basel was built between 1925 and 1928 after the previous timber building burned down. Steiner managed here to create a supplely moulded, monolithic concrete construction, which could not have been built with such a power of expression in any other material. According to Steiner, the Goetheanum was intended to express "the substance of organic design". However, this concept required elaborate and expensive formwork, without which the numerous curved surfaces and sharp edges could not have been realised. This formwork, the responsibility of the carpenter Heinrich Liedvogel, was made from thin strips, which were bent while wet and nailed over ribs.

1.45

## The early Modern Movement

Up until the outbreak of World War I there were a few avant-garde architects who tried their hand at the cubist-rational architecture of the "New Architecture". We cannot speak of a uniform style here, also in the light of the currents of historicism, traditionalism and Art Nouveau still strong at this time. Little was built during World War I, but in the first post-war years, after the collapse of the old order, a fertile soil for new ideas ensued. Many developers and architects were now of the opinion that new social and political structures should be reflected in a new architecture. The influences of painting, from which the fundamental impulses for renewal emanated, made decisive contributions to establishing the International Style.

The early Modern Movement was primarily interested in formal, spatial and social issues. Characteristic reinforced concrete constructions were initially not the prime objective and therefore appear mainly in projects that were never built. One example of this is Tony Garnier's plan for the ideal city, Cité Industrielle, in the Rhône valley, for 35,000 industrial workers. The designs, which were drawn between 1901 and 1917, proposed forms that would exhaust the structural and architectural potential of reinforced concrete in columns, pilotis, cantilevers, ribbon windows and glass walls. The long, thin cantilever of the railway station, supported by a slender column with flared head is especially remarkable and foreshadows later developments. The idea of using the roof as a terrace is quite revealing among Garnier's proposals.

Le Corbusier developed his Maison Domino units made from prefabricated standard elements in 1914. Only the columns, floor slabs and stairs of reinforced concrete were specified, the rest could be filled in according to the user's own ideas. Ludwig Mies van der Rohe's project for an office building in Berlin (1922) was pioneering in its use of a reinforced concrete frame coupled with ribbon windows, although the structure does not necessarily have to be realised in concrete.

1.47

1.48

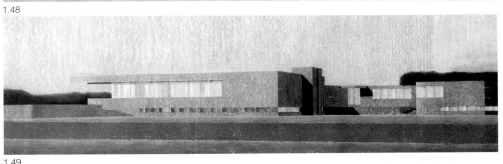
1.49

1.50

1.44 Erich Mendelsohn, Einstein Tower, observatory and astrophysical institute, Potsdam, 1920-21
1.45 Rudolf Steiner, Goetheanum, Dornach, 1928
1.46 Erich Mendelsohn, design sketch for Einstein Tower, Potsdam, 1920-21
1.47 Tony Garnier, Cité Industrielle, housing development project, 1901-17
1.48 Ludwig Mies van der Rohe, country house of reinforced concrete, 1923, perspective view
1.49 Ludwig Mies van der Rohe, country house of clay brickwork, 1924, perspective view
1.50 Ludwig Mies van der Rohe, project for an office block of reinforced concrete, 1922

# Reinforced concrete in the Modern Movement

1.51

1.52

1.53

According to the view of the architects influenced by Cubism in the 1920s and 1930s, the appearance of a building should be determined by good proportions and smooth wall surfaces. The search for forms that do justice to the construction or the material thus came to the fore. The country house project (1923) of Ludwig Mies van der Rohe is typical. Here, the architect designed – in terms of proportions and fenestration – absolutely identical alternative facades in reinforced concrete and facing brickwork.

The sober, plain structures of Adolf Loos or the purist villas of Le Corbusier in Stuttgart's Weissenhof Estate, in Garches or Poissy employ either traditional masonry or are supported by steel columns behind cladding. Even a form so typical of concrete such as the "levitating" roof of Mies van der Rohe's Barcelona Pavilion, which illustrates impressively the flowing space concept for the first time, was probably made from steel and merely encased in plaster.

In Gerrit Thomas Rietveld's Schröder house in Utrecht (1924), the most consistent and best known structure of the Dutch De Stijl movement, reinforced concrete construction was imitated only in the various masonry facade elements or cantilevering balconies.

Lovell Beach House is, by contrast, a true non-plastered reinforced concrete construction. The Austrian architect Rudolf Schindler, who had worked previously with Frank Lloyd Wright, designed this house on the Californian coast in 1926. In the holiday complex "El Pueblo Ribera" (1923-25), Schindler experimented with the "slab-cast" method he had developed. In this method two 400 mm high horizontal planks, lined with building paper, form the formwork which is slid into vertical guiding timbers. Triangular, horizontal fillets hold the paper in place. The groove between two "lifts" of concrete remains visible in the finished wall.

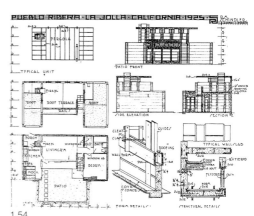

1.54

## Frank Lloyd Wright

It was not until the 1930s, among the architects of the Classical Modern Movement, that the typical material properties of concrete became prominent. Frank Lloyd Wright demonstrated the different options this material provided impressively on various buildings. In 1930 he wrote: "It is not easy to see in this conglomerate a high aesthetic property, because, in itself it is amalgam, aggregate, compound ... The net result is, usually, an artificial stone at best, or a petrified sand heap at worst ... Surely, here, to the creative mind, is temptation. Temptation to rescue so honest a material from degradation. Because here in a conglomerate named "concrete" we find a plastic material, that as yet has found no medium of expression that will allow it to take plastic form." [16]

The house completed in 1939 for the Kaufmann family at Bear Run, Pennsylvania, is also known as "Falling Water" owing to its position overlooking a waterfall. It became typical of the concrete structures of the Modern Movement, despite its masonry walls. The decisive elements, however, are the reinforced concrete slabs cantilevering out on all sides. This device exploits the structural possibilities of the material downright radically and achieves a clever interweaving of the living space with the natural surroundings.

The administration building of the Johnson Wax Company in Racine, Wisconsin, was built in the same year. In this introspective design, Wright punctuated the open-plan office space using free-standing reinforced concrete columns with "mushroom" heads. However, these do not support a solid slab but rather glazed rooflights. In other areas the same "mushroom" heads are employed below but separated from reinforced concrete slabs.

Another, again completely different but nevertheless typical, concrete structure can be seen at the Solomon Guggenheim Museum in New York. Wright's design for the building stems from the mid-1940s, but it was not built until the late 1950s. The helical form, slightly reminiscent of a hopper, could not have been realised in any other material other than reinforced concrete.

1.51 Ludwig Mies van der Rohe, Barcelona Pavilion, 1928-29
1.52 Gerrit Thomas Rietveld, Schröder House, Utrecht, 1924
1.53 Rudolph M. Schindler, Lovell Beach House, Newport Beach, California, 1925
1.54 Rudolph M. Schindler, "El Pueblo Ribera" holiday complex, La Jolla, California, 1925
1.55 Frank Lloyd Wright, "Falling Water", Bear Run, Pennsylvania, 1935-39
1.56 Frank Lloyd Wright, administration building, Johnson Wax Company, Racine, Wisconsin, 1936-39

1.55

1.56

1.57

1.58

## Le Corbusier

Le Corbusier is a key figure in building with concrete. A clear change in his work is evident around 1930. Up to this time almost all of his buildings had been rendered and painted white. But on his Swiss Pavilion for the City University in Paris he omitted this coating for the first time and allowed the reinforced concrete to remain exposed with its untreated "as struck" finish (revealing the textures and

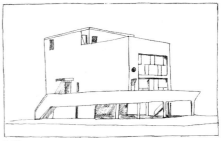
1.60

impressions of the formwork). Le Corbusier's thorough knowledge of reinforced concrete, which was now to become his most important material, came from working in the office of Auguste Perret in 1908 and 1909. Almost all of his structures realised after 1930 make use of the sculptural potential of concrete, e.g. in the form of the oversized large, shady "brise-soleil" on the La Cité d'Affaires high-rise building in Algiers (1939), or in his later residential complexes in which he combined several hundred housing units imaginatively and without monotony.

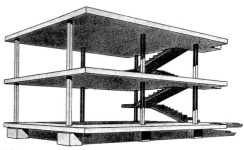
1.61

The first and probably best known of these structures was built between 1945 and 1953 in Marseille. In the shape of massive pilotis and a supporting framework of reinforced concrete and steel, he was able to achieve a style so typical of concrete. The "brise-soleil", partly precast, partly cast in situ, combined with loggias and spandrel panels in front of the facade gives the building a monumental plasticity. Finally, Le Corbusier succeeded in also exploiting the potential of cast concrete by placing positive-negative shapes in the formwork for decorative purposes. His bizarre, sculpted roof structures are striking examples of the unconstrained artistic use of the material.

1.59

Le Corbusier turned his back on the Modern Movement with his sculptural design for the Notre-Dame-du-Haut pilgrimage chapel at Ronchamp (1950-54). He had to contend with being labelled as a "popularist" by a number of doctrinaire advocates of the Modern Movement. The unconstrained, sculpted shape of the structure with its non-linear form consists of a concrete frame with rubble stone infill panels. The roof is formed by two unfurling and curving reinforced concrete slabs whose massiveness – coupled with an almost autonomous monolithic concrete bell-tower – dominates the external appearance. The irregular openings in the internal and external walls seem less dependent on construction or material, like the fat lime facades coated with cement mortar. But the irregularly curved roof was only possible with elaborate formwork.

However, Le Corbusier's other religious building dating from this period – La Tourette, the

1.63

1.62

Dominican Noviciate, at Eveux near Lyon, completed in 1960 – does exhibit a form characteristic of the material. Situated on a steep slope, the square, apparently regular structure has an enclosed inner courtyard which at first suggests strict isolation but upon closer inspection reveals gently sloping external walls and frivolous additions which relieve the austerity. La Tourette demonstrates Le Corbusier's preference for the rough and raw, the "Béton Brut". The building lives from the interplay of the light: in sunshine it appears lucid and full of contrast, but during rainfall sombre and almost melancholy.

1.57 Le Corbusier, Villa Savoie, Poissy, 1929
1.58 Le Corbusier, pilgrimage church, Ronchamp, 1954
1.59 Le Corbusier, Unité d'Habitation, Firminy-Vert, 1962-68
1.60 Le Corbusier, design sketch for "Citrohahn" house, project, 1921
1.61 Le Corbusier, "Maison Domino" system, project, 1914
1.62 Le Corbusier, sketches comparing conventional and modern methods of building, 1929
1.63 Le Corbusier, La Tourette Monastery, Eveux, 1957
1.64 Le Corbusier, La Tourette Monastery, Eveux, 1957, interior of chapel

1.64

# Reinforced concrete in the Modern Movement

1.65

1.66

## Shell structures

Since the 1930s reinforced concrete shell structures have permitted vast areas to be roofed over without intermediate supports. Their forms had been unknown before that time. Such shells always consist of curved surfaces, which cannot carry concentrated loads. As a result, roof-mounted additions are impossible. The unequivocal consequence of material and construction is therefore always a clearly perceptible, geometrically defined shape, identical inside and outside. One engineer who made a great contribution to promoting shell construction was Eduardo Torroja. His dome over the market hall in Algeciras, the La Zarzuela racecourse grandstand in Madrid (1935) and the ball games hall, also in Madrid, helped this new form of construction gain general recognition.

Robert Maillart designed the Cement Industry Pavilion for the Swiss state exhibition of 1939. This was a barrel vault with an unusual shape and incredible effect. It measured 11.7 m high but was just 60 mm thick. This slender construction was subjected to a number of loading tests that confirmed the stability of the thin shell. It was demolished one year later as planned.

The design of the Church of Our Lady of Miracles in Mexico City (1955-57) signalled a completely new departure. Wafer-thin membranes in the form of hyperbolic paraboloids span the interior. The Spanish engineer Felix Candela and the architect Enrique de la Mora employed reinforced concrete here for their extravagant artistic intentions. They demonstrated new, previously unconceived traits of the material. Candela developed the shells not only based on his sound knowledge of structural engineering and building, but also with unique intuition. In 1958 he, together with the brothers Joaquín and Fernando Alvarez Ordoñez, designed the Mantiales restaurant in Xochimilcho. This has become famous owing to its extraordinary shape – an eight-petalled flower of parabolic concrete shells.

1.67

## Pier Luigi Nervi

The Italian engineer and architect Pier Luigi Nervi found a language that was all his own. He exhausted the sculptural potential of reinforced concrete and gave it a hitherto unknown countenance. Nervi's view was that the loadbearing structure was the most important design element of a building. In 1935 he designed an aircraft hangar to cover an area of 400 m² with a span of about 40 m. His design was realised in Orvieto in 1936 and later in a number of other cities. The simple structure of the intersecting diagonal loadbearing ribs (the edge beams are relieved), exposed on the soffit, are supported by just six raking columns. With such a simple construction, their effect is especially eloquent. One of the most consistent structures from Pier Luigi Nervi is the Palazetto dello Sport in Rome (1957). The complex ribbing of the relatively shallow, round dome was built using Ferro-Cemento, a variation on reinforced concrete developed by Nervi himself. This material is cast in place in multiple reusable and transportable formwork boxes.

As an engineer and architect, Pier Luigi Nervi succeeded in discovering completely new paths in the design of reinforced concrete buildings. He abandoned the use of large supporting structures to cover single-storey sheds and instead constructed the loadbearing system out of a multitude of pieces that each followed the flow of forces exactly. Prefabrication provided the extremely precise and slender elements that he required. The resulting structures, with their lightweight, elegant appearance, achieve surprising spatial effects.

Nervi did not apply any fundamentally new structural principles but helped to express a creative stance looking for new forms in architecture and also developed the necessary engineering means to do this.

1.68

1.69

1.70

1.65  Robert Maillart, Cement Industry Pavilion, Zurich, 1939
1.66 and 1.67  Felix Candela, Mantiales Restaurant, Xochimilcho, Mexico, 1958
1.68  Pier Luigi Nervi, aircraft hangar, Palaschetto, 1939, under construction
1.69 and 1.70  Pier Luigi Nervi, Palazetto dello Sport, Rome, 1957
1.69  General view
1.70  View of underside of ribbed roof construction

1.71

1.72

1.73

## Architecture at the height of the Modern Movement

The exponents of the early Modern Movement in the 1920s and 1930s did not enjoy full acceptance in wider architectural circles nor among the general public. They remained somewhat elitist and avant-garde. Their surfaces and cubes reduced to the simplest geometrical forms of the rectangle, as well as the buildings made from these components, were ironically designated as concrete cubes, concrete boxes and so on. Precisely because of their apparent simplicity, such structures were often completely misunderstood, particularly since it appeared easy to copy them. This criticism cannot be simply portrayed as an insignificant phenomenon accompanying a new architectural fashion and disregarded because virtually all innovative attempts are caricatured in this way, for it left a deep impression which gave the material and the word concrete a bad press for a long time. Of course, this negative assessment also had its origins in the well-known shortcomings in the buildings of the early Modern Movement.

After World War II the political obstacles to the general spread of the Modern Movement across Europe had been removed. A greater emphasis on the "honest" use of materials, influenced by Le Corbusier in particular, became fashionable again. Concrete, and above all the as struck fair-face variety, which Le Corbusier honoured by calling it "a face covered in wrinkles", evolved into a mass product. The properties specific to reinforced concrete were increasingly exploited. There emerged an appreciation of the unrestricted mouldability of the material plus the opportunity to erect buildings "from one mould" and create seamless transitions between walls, floors and ceilings. Likewise, the fact that surfaces no longer needed subsequent treatment was also received positively. There was also the economic factor that encouraged building with this comparatively cheap material. And then there was the variety of designs with which concrete could be used as a loadbearing structure over long spans. The options ranged from simple or cranked surfaces to curved shells, from heavy vaulting to wafer-thin arches, and from rough rock imitations to sylphlike skeletons. However, in many places the formal language was overrated and led to – sometimes embarrassing – imitations.

The Halen Estate near Bern (1955-61, architects: Atelier 5) represents a dense development of houses with a strict composition. Their position on a slope gave them a great exemplary character for the urban development and architecture of that period. The architects succeeded in taking Le Corbusier's principles of simple mass housing one step further and applying them to more upmarket requirements.

Louis Kahn was also influenced by Le Corbusier. In his opinion, structural steelwork did not permit true walls and true columns, with the actual loadbearing structure usually concealed behind the necessary fire-resistant cladding; the majority of his designs therefore made use of concrete. The Jonas Salk Institute in La Jolla, California, (1959-65), is almost exclusively fair-face concrete internally and externally. Kahn preferred his concrete surfaces to be as smooth as possible, and distanced himself from Le Corbusier, who always went to great lengths to achieve a rough, almost rustic texture by employing selected boards. Kahn subdivided the surfaces with a network of deep rebates to indicate the individual formwork panels. And the holes of the formwork ties, carefully arranged to create a picture, remained visible for the first time. This device has once again become topical through the work of Tadao Ando.

The originally simple geometrical forms continued to develop in the construction of formwork. The new, expressive and dynamic roofs whetted the appetites of a new generation of architects. These included Eero Saarinen, who first realised his ideas with the roof over the ice hockey stadium in Yale. At the TWA Terminal at the John F. Kennedy Airport in New York (1956-62), he succeeded, in a much more impulsive way than at Yale, in combining the functional requirements of an airport terminal with an experiment in form. The result was a

1.74

1.75

1.76

1.77

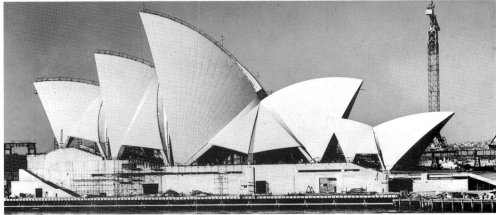
1.78

1.71  Paul Rudolph, house for Arthur W. Milan, Jacksonville, Florida, 1960-62
1.72  Arthur Erikson, Anthropology Museum, Vancouver, 1971-77
1.73  Louis Kahn, Jonas Salk Institute, La Jolla, California, 1959-65
1.74  Jørn Utzon, Sydney Opera House, 1956-76, details of arched rib and precast concrete elements
1.75  Eero Saarinen, TWA Terminal, New York, 1956-62
1.76  Kenzo Tange, National Gymnasium for Tokyo Olympic Games, 1964
1.77  Jørn Utzon, Sydney Opera House, 1956-76, under construction
1.78  Jørn Utzon, Sydney Opera House, 1956-76

futuristic-looking structure, which appears to defy the force of gravity and thus symbolise flight. In an age in which air travel was not yet an everyday reality, this concept for the design of an airport terminal building was certainly understandable.

Sydney Opera House was designed and built between 1956 and 1973 by Jørn Utzon in cooperation with Ove Arup. It is based on the winning competition design by the Danish architect, which provides two theatres and the associated facilities beneath a series of interleaved, white, reinforced concrete shells. The structural analysis and construction of these shells, which are intended to lend the building the character of a sailing ship, were possible only after several redesigns and elaborate experiments. The unusually complex geometry meant it took the architects and engineers several years to find a solution that would allow them to build arch segments of varying curvature from the same set of modular precast concrete units. Finally, a three-sided segment cut from a sphere was used for constructing the concrete shells. The regularly shaped, curved surfaces were then taken from this segment. Although the structural supports to the shells impair the interior and were not part of the original design, they proved indispensable. Despite having taken many years to design and construct and despite all the compromises that were necessary, Sydney Opera House is a unique structure. With the white ceramic cladding to its sail-like shells reflecting the light, it has become the symbol of a whole continent.

Kenzo Tange from Japan also used curved forms, without doubt typical of concrete, for his National Gymnasium for the Tokyo Olympic Games in 1964 (with Yoshokatsu Tsubui). Tange is convinced that "only beauty can be functional" and has repeatedly polemicised against the "monotony of the International Style". And indeed, during this period Japan, which had approached the Modern Movement with much less prejudice, was providing plenty of impulses that European architecture gratefully accepted.

In Canada it was Arthur Erikson, influenced by Le Corbusier, Louis Kahn and Paul Rudolph, who stimulated the progress of concrete architecture decisively. His large projects were almost exclusively of fair-face concrete. The Simon Fraser University in Vancouver is interesting; it is a self-assured, modern building that plays a key role in Erikson's output. Other important buildings by this architect are the Anthropology Museum and the courthouse complex in the same city.

Fair-face concrete became a characteristic element in German architecture during the

1.79

1.80

Architecture at the height of the Modern Movement

1950s and 1960s. Many structures of this period are based on international models. One architect who developed a style all of his own was Gottfried Böhm. In his expressionistic works he skilfully used the sculptural opportunities of the monolithic material. The pilgrimage church in Neviges, completed in 1968, represents a climax in his creative abilities. Together with its ancillary buildings (shops, restaurants, communal facilities), this religious edifice, which rises as a folded plate structure of crystalline forms to create a series of jagged peaks, forms a whole complex. This is intended to assist the movements of larger groups of pilgrims and visitors. The church interior itself is devoid of ornamentation and achieves its effect through the towering forms. The coloured glass windows create richly contrasting accents within the barrenness of the interior space.

Böhm's town hall in Bensberg was built at about the same time. Here, the architect succeeded in integrating, sensitively but not compromisingly, a modern local government administration centre into the historic fabric of the town. The light-coloured concrete structure with its small-scale units and impressive depth forms an attractive composition alongside the old castle. The two towers and the exciting juxtaposition of old and new has proved to be a successful combination.

The Protestant Church of Reconciliation designed by Helmut Striffler was built between 1965 and 1967 in the former Nazi concentration camp at Dachau. This memorial, sunk into the ground, is unobtrusive and restrained, and seems to express a wish for reconciliation and salvation. The desolate stone "desert" of the forecourt reminds us of the human suffering and horrors that took place here during the years of the Nazi dictator.

The Keramion in Frechen, a gallery for contemporary ceramics, by Peter Neufert in collaboration with Stefan Polónyi (1970), consists of a membrane shell in double curvature just 80 mm thick. This structure with its long cantilever measures 32 m in diameter. The roof, with its upturned cone apparently mounted in the middle, has a prestressed edge member and, like the five cone-like columns founded 8 m below ground, traces the direction of the forces to be carried. The circular facade 5 m high consists of frameless panes of glass bonded together along their edges. This succeeds in fusing the structure and its contents into an interesting architectural entity.

1.79  Gottfried Böhm, Bensberg Town Hall, 1968
1.80  Gottfried Böhm, pilgrimage church, Neviges, 1968
1.81  Atelier 5, Halen Estate, Bern, 1961
1.82  Helmut Striffler, Church of Reconciliation, former Nazi concentration camp, Dachau, 1965-67
1.83  Peter Neufert, Stefan Polónyi, Keramion, Frechen, 1970

1.81

1.82

1.83

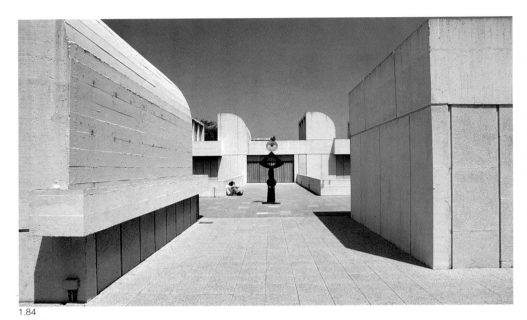
1.84

1.85

## Late-Modern and Postmodern

The International Style became generally widespread during the 1950s and 1960s. It became *the* architectural credo. There was no other significant orientation in design ideas and no other noteworthy interpretation of the interaction between function, design and form. Architecture and functionalism virtually became synonyms. However, by the end of the 1960s there was growing criticism of the building concepts, which had become entrenched in stereotypes and repetitive solutions.

The Italian architect Carlo Scarpa was an individualist whose work cannot be placed in any stylistic compartments of that period. He rose above the sobering functionalism of his age and chose an allegorical, semantic form of expression. His forms are varied and imaginative, and most have a symbolic content. He frequently used concrete. One of Scarpa's main works is the cemetery for the Brion family at San Vito d'Altivole. His masterful use of the design opportunities presented by the material created one of the most eloquent concrete structures of those years.

The angular complex, resembling a little village cemetery, covers about 2000 $m^2$ and essentially consists of five striking elements: the covered gateway, the chapel with adjoining cloister, the arch with the two sarcophagi of the owners beneath, the tombstone honouring the family members and the Pavilion of Meditation set in a water garden. Conspicuous here are the untreated, as struck surfaces, which – like natural stone – gain character through the weathering process.

During the 1960s and 1970s in Ticino, Switzerland, an architectural movement sprang up to protest against functionalism despite being partly associated with the forms and ideas of the early Modern Movement.

Relatively liberal building legislation and – in comparison to more northerly regions – less stringent building science stipulations favoured the construction of unusual buildings which were often deliberately juxtaposed to their immediate surroundings. Fair-face concrete and exposed concrete masonry were the dominant building materials of architects like Luigi Snozzi, Dolf Schnebli and Aurelio Galfetti.

Mario Botta, one of the most famous members of the Ticino School, was profoundly influenced by his teacher Carlo Scarpa and his collaboration with Le Corbusier and Louis Kahn. It was from them that he obtained his predisposition for concrete and again and again he cites their typical forms in his works. His school in Mobio Inferiore was built between 1972 and 1977. It is an elongated block of two and three storeys and, made up of a row of identical parts, represents a man-made marshalling factor within the landscape. Inside, this building of fair-face concrete offers unsuspected spatial variety as well as differing views and relationships.

On an international scale, Botta gained most acclaim for his many detached houses, the first of which was built in 1966 as a reinterpretation of Le Corbusier's villas in as struck fair-face concrete. However, he preferred exposed concrete masonry for his later houses. These stand out sharply from the – usually – uninspiring country house architecture of its neighbours and over the years were increasingly characterised by a sort of formalism. These were intended to relate to the traditional Ticino granite rubble stone masonry. His rectangular house in Ligornetto (1976) adopts the reddish-brown-and-white-striped facades originally common in Ticino, and makes use of coloured facade bricks to achieve this effect.

Luigi Snozzi also belonged to the Ticino School. His sports hall in Monte Carasso dates from 1984. The fair-face concrete of the building integrates sensitively within the mature urban landscape. Writing about this building he said: "The fair-face concrete ... takes on a dialectic role with respect to the existing building stock and is at the same time a bonding element reaching beyond the enclosing walls of the complex. The new material leads thereby to a dialogue with the old stone walls and the faded rendering, but without reproducing the original forms and materials by way of nostalgic interpretations." [2]

Representing a complete contrast with the Modern Movement are the concrete structures of the Barcelona-based architectural practice of Ricardo Bofill – "Taller de Arquitectura". They represent an unusual form of publicly assisted housing and are built mainly for clients in Spain and France. Historical references and gigantic facade configurations are intended to prove that publicly assisted housing does not have to mean drab accommodation blocks. However, the structures have a strong link with formalism and are difficult to reconcile with modern interior layouts. These Postmodern projects of Bofill, particularly the housing estates in France, include grand axes and disproportionate structures in which the occupants lose themselves and are just

1.86

1.87

1.88

1.84 José Luis Sert, Miró Museum, Barcelona, 1972-75
1.85 Carlo Scarpa, Brion family cemetery, San Vito d'Altivole, 1970-72
1.86 Mario Botta, school, Mobio Inferiore, 1972-77
1.87 Mario Botta, detached house, Ligornetto, 1975-76
1.88 Ricardo Bofill, "Les Arcades du Lac" housing development, Saint-Quentin-en-Yvelines, 1975-82

# Reinforced concrete in the Modern Movement

1.89

onlookers emphasising the gargantuan nature of these structures. Just a few kilometres from Versailles in the Paris suburb of Saint-Quentin-en-Yvelines, there is a residential estate with about 400 apartments, which imitates the palace itself. Built between 1975 and 1982, the "poor man's Versailles" has no shops and no public amenities. According to the architect, the structure of the space is intended to create symbols and give rise to a new architectural language. The pinkish yellow columns, mouldings and architraves are "natural stone" manufactured in precast concrete.

## Dutch structuralism

In 1966 Herman Hertzberger began work on the design of the offices for the Centraal Beheer insurance company in Apeldoorn, which were built between 1970 and 1973. He was influenced by Aldo van Eyck's orphanage in Amsterdam. The complex is made up of square blocks, which can be combined and extended in both the horizontal and vertical directions. This completely non-hierarchical arrangement is consistent with the underlying democratic concept. The complex represents a town within itself, with open squares, paths and roads, where every employee – contrary to the usual situation in open-plan offices – is given a private island which, to a large extent, he can organise to suit his own requirements. Furthermore, there is constant visual and acoustic contact with other members of staff. A load-bearing reinforced concrete frame provides the flexibility. Concrete masonry forms the infill panels, which remains exposed both internally and externally. The small format of the rough masonry units helped Hertzberger to continue the ideas of structuralism down to the detail level. At the same time, he created a surface which is both dynamic and neutral. Raw concrete masonry is Herman Hertzberger's distinguishing architectural legacy. He used it for his housing, schools and kindergartens, and likewise exposed in the foyer and concert hall of the Vredenburg Music Centre in Utrecht (1978).

1.90

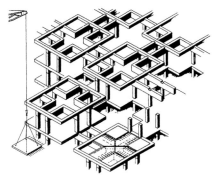
1.91

38

## The use of precast concrete

Mass-produced precast concrete elements have been used increasingly since the 1950s. They are particularly widespread in the socialist countries of the former Eastern bloc, where whole housing estates were built entirely of precast concrete panels using fast-track methods. But the degree of prefabrication grew in Western Europe, too. For example, while many precast concrete elements were being used in the Netherlands, for housing in particular, in West Germany, as it was then, this type of construction became established mainly for industrial and office buildings.

In 1967 the Deutsche Bundespost in Munich built a single-storey shed covering 20,000 m² entirely of precast components. At that time it was the largest building of its kind in the world. Measuring 124 m long and 27.3 m high, the structure consists of two three-part end arches

1.94

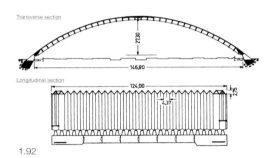
1.92

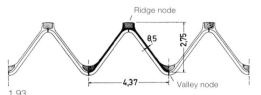
1.93

1.95

1.89 to 1.91  Herman Hertzberger, offices for Centraal Beheer insurance company, Apeldoorn, 1970-72
1.89  General view
1.90  Interior
1.91  Sketch showing overall scheme
1.92  Helmut Bomhard, parcels sorting centre, Munich, 1967, transverse and longitudinal sections through complete arch structure
1.93  Helmut Bomhard, parcels sorting centre, Munich, 1967, section through standard arch
1.94  Otto Steidle, Genter Strasse housing development, Munich, 1977
1.95  Enric Miralles, Carme Pinós, Olympic Games Archery Centre, Barcelona, 1992
1.96  Vittorio Gregotti, school, Palermo, 1989

1.96

1.97

1.98

and 24 standard ones. The latter are each built using two 85 mm thick units sloping towards each other. Each weighs 3.8 t. All 1584 precast parts of the standard arches have the same dimensions. The articulate and large-scale concept of the building, both inside and outside, determine its overall effect.

The housing of Otto Steidle on Munich's Genter Strasse, erected between 1972 and 1974, employs a frame of modular, prefabricated reinforced concrete components. The appended support corbels remain visible everywhere even though they are actually used in only a few places; they thus take on an almost exclusively decorative character. The maximum flexibility in fitting out the interior, which allows individual wishes to be taken into account, contrasts with the regimented primary structure. The work of the Spaniard Ricardo Bofill represents a blatant contrast to this. His neoclassical precast panel structures dictate a large format and do not allow the occupants to exercise any influence.

Gottfried Böhm conceived the office building for Züblin AG in Stuttgart (see p. 258) with a great deal of structural and architectural fantasy. Completed in 1986, the building houses 700 staff and is influenced by the Postmodern Movement. Its red, sculpted concrete elements demonstrate the design possibilities that can be elicited from the use of prefabricated reinforced concrete components.

Eckhard Gerber's Harenberg Tower in Dortmund (see p. 264) demonstrates a completely different form of expression in precast concrete. Employing the sentiments of the Modern Movement, his facade comprises smooth, light-grey precast concrete panels and remind the observer initially of natural stone.

1.99

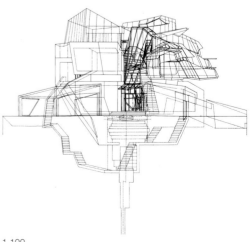
1.100

## Contemporary architecture

The "New York Five", a loosely knit group of New York architects led by Peter Eisenman, has played an important role in recent American architecture. The other members of this group were John Hejduk, Michael Graves, Charles Gwathmey, and Richard Meier, who, of them all, probably remained truest to his purist roots. His architecture is characterised by the clarity of the geometrical shaping, the generous openings and the combination of rounded and right-angled cubes with smooth white surfaces.

On the other hand, a few years later Peter Eisenman, together with – among others – Bernard Tschumi and his design for the Parc de la Villette in Paris, smoothed the way for the Deconstructivist Movement, which emerged during the 1980s. This – the antithesis to Constructivism – was based on the writings of the philosopher Jacques Derrida and strove to transcend accepted logical structures.

Reminiscences of the expressive language of this movement can also be seen in Frank O. Gehry's work in the shape of his irregular, boldly sculpted structures, like the Vitra Design Museum in Weil am Rhein. Even James Stirling exhibits Deconstructivist tendencies. Together with Michael Wilford and Walter Nägeli he conceived the production plant for B. Braun AG in Melsungen (1991) as a small town; angular forms opposing round ones, straight lines and curves intersecting, large enclosed areas contrasting with open zones.

Seldom, however, has there been such a successful symbiosis of expression and utilisation as in the Jewish Museum in Berlin, by Daniel Libeskind. The fragmented architectural language of the structure can also be interpreted allegorically as a symbol of destruction and loss. The highly theoretical conceptual approach of Libeskind is primarily influenced by his understanding of musical ideas. In this case, it was Schönberg's opera "Moses and Aaron" that inspired the expressive "zigzag" of the layout. The sculptural use of fair-face concrete helps give the exhibition rooms their special character. For example, concrete wedges penetrate the rooms, and openings slice, so to speak, through the surfaces.

1.101

1.102

1.97　Frank O. Gehry, Chair Museum, Weil am Rhein, 1989
1.98　James Stirling, office building for Braun AG, Melsungen, 1992
1.99　Richard Meier, city centre development, Ulm, 1993
1.100　Günter Domenig, "Das Steinhaus", Steindorf, 1986, elevation
1.101　Ben van Berkel and Caroline Bos, Möbius House, Het Gooi, 1997
1.102　Daniel Libeskind, Jewish Museum, Berlin, 1996

1.103

1.104

A movement called "textual" architecture (after Peter Eisenman) evolved parallel to Deconstructivism. In terms of expression this was similar but tended to be based on folds and flowing forms rather than fragmentation and fracture. Virtual computer worlds are the preferred medium of these multi-dimensional architectural landscapes. Besides Rem Koolhaas, MVRDV and Zaha Hadid, Ben van Berkel and Caroline Bos have also implemented some of these visions.

The properties of in situ concrete are absolutely predestined for the complicated spatial structures arising from this. In Möbius House in Het Gooi the primary idea is a spatial continuum. The design is based on the mathematical concept of the Möbius strip. This illustrates the abstract idea of infinity as the ends of the strip are turned through 180° and joined together so that there is only one continuous surface and one edge. This performs the structural basis for the building in terms of both style and function. It symbolises our life-cycle and daily routine as a continuous circuit. Consequently, workrooms are also integrated alongside living quarters, with a seamless transition between the two. The homogeneity of the fair-face concrete serves as a bonding element, interrupted only by the openings of shimmering green glass.

Álvaro Siza Vieira from Portugal should be placed somewhere between these movements and the purist influences of the Iberian peninsula. Dense, flowing room sequences accompanied by specially incorporated overhead lighting as well as interweaving volumes of white plastered concrete characterise his buildings, e.g. the house "Villa Nova de Famalicão".

There are, simultaneously, opposing tendencies to these still very complex structures. Here again, concrete plays a major role. This is because in the attempt to achieve maximum possible clarity of expression with a minimalistic use of materials, the details and hence the materiality and surfaces become increasingly significant. Consequently, thanks to its homogeneity and flexibility, concrete has the right credentials.

In this sense, it is probably Tadao Ando from Japan whom we should thank for the current global renaissance in fair-face concrete. His subdued buildings exhibit a perfection in surface finishes that is realised only rarely. The house built in his home town of Osaka in 1990 has a completely plain wall separating it rigorously from the street, but from the inside appears to open out. This impression is partly due to the stairs and courtyards leading us uncompromisingly, and the uniform use of fair-face concrete surfaces. This frugal materiality in conjunction with the precise formwork joints and formwork tie cones has already become Ando's trademark. Another Japanese architect, Toyo Ito, has created an outstanding example of subtle minimalism with his house in Yutenji. The building consists of three very wide reinforced concrete frames which together form a simple "brick". The interior is dominated by the diffuse transparency of the opaque glass panes which are used both for the openings in the facade and for the sliding doors inside the house. The rough concrete surfaces of the floors and ceilings contrast here in an exciting way with the gypsum-clad smooth walls.

The Liechtenstein Art Gallery, by the Swiss architects Morger and Degelo, is an excellent example of specially finished surfaces. The jointless facade of ground and polished fair-face concrete lends the building an almost monolithic effect. Its dark colour and fine texture is due to the combination of basalt, river gravel and a black pigment. Inside the building, the unembellished exhibition rooms are arranged like the sails of a windmill around two central staircases.

Bregenz Art Gallery by the architect Peter Zumthor expresses a similar architectural stance. Hidden behind a homogeneous skin of translucent, acid-embossed glass panels is a genuine concrete construction. However, the building is characterised not only by its coherent conception and precise workmanship, but also by a differentiated interior climate concept. This exploits the storage capacity of the solid concrete slabs. Hot or cold water is pumped through the diaphragm walls below the building and this adjusts the temperature within the building as required. This means that the air-conditioning system can be run at a much lower, less energy-guzzling level. Circulation zones and services are located behind these slabs in order keep to the exhibition rooms in the middle of the building free from any disturbing elements. The fair-face concrete surfaces, together with the terrazzo floors in the same colour, lend the rooms a clear, calm, almost meditative atmosphere.

1.105

1.106

1.107

1.103 Álvaro Siza Vieira, house "Villa Nova de Famalicão", 1984
1.104 Tadao Ando, house, Osaka, 1990
1.105 Toyo Ito, house, Yutenji, 1999
1.106 Morger and Degelo, Liechtenstein Art Gallery, 2000
1.107 Peter Zumthor, Bregenz Art Gallery, 1997

## References

**Books and articles**

[1] Ackermann, Kurt: Industriebau, Deutsche Verlagsanstalt, Stuttgart 1984
[2] Bauen in Beton, 1/86, 1990/91, 199/93
[3] Baukunde des Architekten, vol. 1, 2nd ed., pub. by Deutsche Bauzeitung and Deutscher Baukalender, Verlag Ernst Toeche, Berlin 1893
[4] Behnisch & Partner: Bauten 1952–1992, Galerie der Stadt Stuttgart 1992
[5] Berlage, Hendrik Petrus: Über Architektur und Stil, pub. by Bernhard Kohlenbach, Birkhäuser, Basel 1991
[6] Billington, David P.: Robert Maillart und die Kunst des Stahlbetonbaus (Robert Maillart and the Art of Reinforced Concrete), Verlag für Architektur Artemis, Zurich/Munich 1990
[7] Bolmhard, Helmut: Konstruktion und Bau der Paketumschlaghalle in München, pub. by Deutscher Beton-Verein, Wiesbaden 1967
[8] Busse, H.-B. v., Waubke, N.V., Grimme, R., Mertins, J.: Atlas Flache Dächer, Institut für internationale Architektur-Dokumentation, Munich 1992
[9] Die andere Tradition – Architektur in München von 1800 bis heute, Callwey Verlag, Munich 1981
[10] Doesburg, Theo van: Gesammelte Aufsätze aus Het Bouwbedrijf 1924-1931, Birkhäuser, Basel 1990
[11] Domenig, Günther: Werkbuch, Residenz Verlag, Salzburg 1991
[12] Drechsel, Walther; Neufert, Ernst: Stahlbeton – Bauwerke und Bautechnik, pub. by Deutscher Beton-Verein, Wiesbaden, and Fachverband Zement, Cologne; Bauverlag, Wiesbaden/Berlin, and Beton-Verlag, Düsseldorf 1964
[13] Feuerstein, Günter: New Directions in German Architecture, Studio Vista, London 1968
[14] Frampton, Kenneth: Grundlagen der Architektur – Studien zur Kultur des Tektonischen (Studies in Tectonic Culture), Oktagon Verlag, Munich/Stuttgart 1993
[15] Garnier, Tony: Die ideale Industriestadt, Verlag Ernst Wasmuth, Tübingen 1989
[16] Wright, Frank Lloyd: The meaning of materials – concrete; in: In the causes of architecture – essays by Frank Lloyd Wright for architectural record 1908-1952. Wright, Frank Lloyd; Gutheim, Frederick (eds.), New York
[17] Gössel, Peter; Leuthäuser Gabriele: Architektur des 20. Jahrhunderts, Benedikt Taschen Verlag, Cologne 1990
[18] Hackelsberger, Christoph: Beton – Stein der Weisen? – Nachdenken über einen Baustoff, Vieweg Verlag, Braunschweig/Wiesbaden 1988
[19] Hand, Rudolf: Österreichisch-ungarischer Bauratgeber, Verlag Moritz Perles, Vienna 1894
[20] Haegermann, Gustav; Huberti, Günter; Möll, Hans: Vom Caementum zum Spannbeton, vol 1, Bauverlag GmbH, Wiesbaden/Berlin 1964
[21] Harris, Sir Alan: Eugène Freysinnet – Concrete Pioneer, Concrete Quarterly 1992
[22] Heidrich, Klaus-Dieter; Hofmeister, Herold; Ricken, Herbert: Früher Stahlbeton in Deutschland, Bauzeitung 8/1991
[23] Hersel, Otmar: Die Jahrhunderthalle zu Breslau, Beton- und Stahlbetonbau 12/1987
[24] Huber, Franz: Geschichte des Portlandzements – Zement als Baustoff; in: 100 Jahre Vereinigung der Österreichischen Zementindustrie 1894–1994, Zement + Beton Handels- und Werbegesellschaft, Vienna 1994
[25] Ilkosz, Jerzy: Expressionist inspiration, The Architectural Review 1/1994
[26] Joedicke, Jürgen: Architekturgeschichte des 20. Jahrhunderts, Edition Krämer, Stuttgart/Zurich 1990
[27] Joedicke, Jürgen: Dokumente der modernen Architektur – Schalenbau, Krämer Verlag, Stuttgart 1962
[28] Jones, Bernard E.; Lakeman, Albert: Cassell's Reinforced Concrete, The Waverley Book Company, London 1920
[29] Kasig, W.; Weiskorn, B.: Zur Geschichte der deutschen Kalkindustrie und ihrer Organisationen, pub. by Bundesverband der Deutschen Kalkindustrie, Cologne; Beton-Verlag, Düsseldorf 1992
[30] Kind-Barkauskas, Friedbert: Architektur im Wandel – Beispiele und Meinungen, Beton-Verlag, Düsseldorf 1990
[31] Kind-Barkauskas, Friedbert: Betonbauten in Köln Beiträge zur modernen Architektur, Fertigbau + Industrialisiertes Bauen 4/1985
[32] Klotz, Heinrich (ed.): Vision der Moderne – Das Prinzip Konstruktion, Prestel Verlag, Munich 1986
[33] Knauers Lexikon der modernen Architektur, Droemersche Verlagsanstalt, Munich/Zurich 1963
[34] Koberg, Günter: Robert Maillart – Pionierleistungen in Stahlbeton, Zement & Beton 2/1991
[35] Kultermann, Udo: Die Architektur im 20. Jahrhundert, DuMont Buchverlag, Cologne 1977
[36] Lamprecht, Heinz-Otto: Opus Caementitium – Bautechnik der Römer, 5th ed., Verlag Bau + Technik, Düsseldorf 2001
[37] Lampugnani, Vittorio Magnago und Schneider Romana (ed.): Moderne Architektur in Deutschland 1900 und 1950 – Expressionismus und Neue Sachlichkeit, Verlag Gerd Hatje, Stuttgart 1994
[38] Le Corbusier: Feststellungen zu Architektur und Städtebau, Ullstein Verlag, Frankfurt/Berlin 1964
[39] Le Corbusier, Studio Paperback, Verlag für Architektur Artemis, Zurich 1972
[40] Le Béton en représentation, Editions Hazan, Paris 1993
[41] Marx, Erwin: Die Hochbau-Constructionen, vol. 2, Raumbegrenzende Construktionen, issue 1, Wände und Wand-Oeffnungen, Verlag Arnold Bergsträsser, Darmstadt 1891
[42] von Niebelschütz, Wolf: Züblin-Bau, Stuttgart 1958
[43] Pauser, Alfred: Eisenbeton 1850–1950, Manz Verlag, Vienna 1994
[44] Pehnt, Wolfgang: Die Architektur des Expressionismus, Verlag Gerd Hatje, Stuttgart 1981
[45] Perking, George: Concrete in Architecture, pub. by Cement and Concrete Association, London; Ditchling Press Limited, Sussex 1968
[46] Perret: L'Architecture d'Aujourd'hui, October 1932
[47] Pevsner, Nikolaus: An Outline of European Architecture, Penguin Books 1963
[48] Riley, Terence: Frank Lloyd Wright, Architect, The Museum of Modern Art, New York 1994
[49] Sack, Manfred: Gedanken zur Architektur des 20. Jahrhunderts; in: Betonbau im Wandel der Zeit, Beton-Verlag, Düsseldorf 1986
[50] Sarnitz, August: R.M. Schindler Architekt, Edition Christian Brandstätter, Vienna 1986
[51] Siegel, C.: Strukturformen der Modernen Architektur, Callwey Verlag 1960
[52] Süddeutsche Bauzeitung 12/1908
[53] Thole, J.P.: Auguste Perret, een Pionier in de Franse Beton-Architectur, Cement 4/1987
[54] Trauer: Die Jahrhunderthalle in Breslau, Deutsche Bauzeitung 14/1913
[55] Webb, Peter: François Hennebique – Concrete Pioneer, Concrete Quarterly 1991
[56] Wieschemann, Gerhard; Gatz, Konrad: Betonkonstruktionen im Hochbau, Verlag Architektur + Baudetail and Georg D.W. Callwey, Munich 1968

# Part 2 · Fundamentals

Friedbert Kind-Barkauskas · Jörg Brandt

## Concrete – the material

The composition of concrete
Types of concrete
Properties of concrete
Exposure classes
Reinforcement
Concrete masonry
Application
References

## The concrete surface

Design concepts
The constituents of the concrete mix
The effects of formwork
Finishing the concrete surface
The use of coatings
The effects of the weather
References

## Building science

General
Basic requirements
Interior climate
Energy economy, thermal insulation
The effects of noise, sound insulation
Behaviour in fire, fire protection
An overview of building science requirements
The properties of components
References

# Concrete – the material

Friedbert Kind-Barkauskas

## The composition of concrete

Concrete is produced by mixing together cement, water, aggregates and, if required, concrete admixtures in a mixer. The mixing time is prescribed and is usually longer than one minute. Certain concrete properties can be achieved through different mix ratios. In doing so, it is important to take into account the special qualities of the raw materials and the processing instructions. The most important standard for the production and use of concrete and reinforced concrete, in Germany DIN 1045, was replaced in 2000 at European level by EN 206 part 1 "Concrete – performance, production, placing and compliance criteria", with specific directives for different countries. In Germany this is complemented and defined more precisely for national applications in DIN 1045 part 2. To complement this new standard covering properties, production and quality control of concrete, the rules on the design and use of concrete and reinforced concrete were also revised. Design is dealt with in DIN 1045 part 1, construction in DIN 1045 part 3, and DIN 1045 part 4 is relevant for the production of precast concrete components. This restructuring of the standards was a sensible update of the validity of the old DIN 1045 and a number of subsequent standards and has incorporated them in the widening collection of European standards.

### Cements

Cement is a hydraulic binder. When mixed with water it sets both in the air and underwater to form a water-resistant hydrated cement. According to the European cement standard DIN EN 197 part 1 "Composition, specifications and conformity criteria of common cements" introduced in 2001 we distinguish between, for example, the following main types of cement:

- Ordinary Portland cement (CEM I)
- Portland blastfurnace cement (CEM II/A-S, CEM II/B-S)
- Portland pozzolanic cement (CEM II/A-P, CEM II/B-P)
- Portland pulverised fuel ash cement (CEM II/A-V)
- Portland burnt shale cement (CEM II/A-T, CEM II/B-T)
- Portland limestone cement (CEM II/A-L)
- Portland composite cement (CEM II/B-SV)
- Blastfurnace cement (CEM III/A, CEM III/B)

DIN EN 197 part 1 has considerably increased the number of standardised cements. However, it does not include cements with special properties that, above all, are important at a national level. In Germany these continue to be covered by DIN 1164.

Cements are supplied in various strength classes, as given in table 2.1.1. Besides the printing on the sacks, these can also be identified by sack colour or, if using loose cement in silos, by the colour of the delivery slip. Cement strength classes 32.5, 42.5 and 52.5 are further subdivided according to initial strength into cements with standard initial strength (code letter N) and those with higher initial strength (code letter R).

Table 2.1.2 shows the relationship between cement strength class and type of cement as well as their properties – as a guide to their use.

However, the standard strengths and setting characteristics of the cements do not permit an accurate assessment of their influence on the concrete strength. This depends essentially on other factors such as water/cement ratio, compaction and curing.

Some types of cement covered in German standard DIN 1164 exhibit additional, special properties:

- Low-heat cements (code NW) are particularly suitable for mass concrete
- Sulphate-resisting cements (code HS) must be used, according to DIN 1045, when the sulphate content of the water relevant to the structure exceeds 600 mg $SO_4^{2-}$ per litre or that of the soil exceeds 3000 mg $SO_4^{2-}$ per kilogram
- Cements with a low effective alkali content (code NA) represent an effective preventive measure in the case of alkali-sensitive aggregates, which can be the case in some regions in northern and eastern Germany

White cement is not classed as a special cement to DIN 1164 but instead as a Portland cement with a low iron oxide content (CEM I 42.5 R).

## Concrete – the material

### 2.1.1 Cement strength classes to DIN EN 197 part 1

| Strength class | Compressive strength [N/mm²] | | | |
|---|---|---|---|---|
| | Initial strength | | Standard strength | |
| | 2 days | 7 days | 28 days | |
| 32.5 N | – | ≥ 16 | ≥ 32.5 | ≤ 52.5 |
| 32.5 R | ≥ 10 | – | | |
| 42.5 N | ≥ 10 | – | ≥ 42.5 | ≤ 62.5 |
| 42.5 R | ≥ 20 | – | | |
| 52.5 N | ≥ 20 | – | 52.5 | – |
| 52.5 R | ≥ 30 | – | | |

### 2.1.2 Notes on the use of cements [2]

| Strength class | Type of cement | Properties | | |
|---|---|---|---|---|
| | | Early strength | Heat development | Maturing[1] |
| 32.5 N | Predominantly blastfurnace cement | low | slow | good |
| 32.5 R | Predominantly Portland limestone and Portland blastfurnace cement | normal | normal | normal |
| 42.5 N | Predominantly blastfurnace cement | normal | normal | good |
| 42.5 R | Predominantly ordinary Portland and Portland blastfurnace cement | high | fast | normal |
| 52.5 N | Ordinary Portland cement | high | fast | low |
| 52.5 R | Ordinary Portland cement | very high | very fast | very low |

[1] beyond 28 days

### 2.1.3 Aggregates for concrete [10]

| Type | Oven-dry density [kg/m³] | Examples |
|---|---|---|
| Lightweight aggregates | max. 2200 | Pumice, foamed slag, expanded clay, expanded shale |
| Normal aggregates | from 2300 to 3000 | Natural broken or unbroken dense stone (e.g. sand, gravel, grit); factory-made broken or unbroken dense material (e.g. blastfurnace slag) |
| Heavy aggregates | more than 3100 | Barytes, iron ore, steel shot |

### 2.1.4 Additional designations for aggregates [2]

| Aggregates with | | Additional designation for | |
|---|---|---|---|
| lower grading limit | upper grading limit | unbroken aggregates | broken aggregates |
| – | 4 | sand | crushed rock fine aggregate |
| 4 | 32 | gravel | grit |
| 32 | – | coarse gravel | ballast |

Examples of the designations of various cements:

- Ordinary Portland cement DIN EN 197 part 1 – CEM I 42.5 R: ordinary Portland cement (CEM I) of strength class 42.5 with high initial strength (R)
- Portland blastfurnace cement DIN EN 197 part 1 – CEM II/A-S 32.5 with 6-20% cinder sand (CEM II/A-S) of strength classes 32.5 N and 32.5 R with standard initial strength
- Blastfurnace cement DIN 1164 – CEM III/B 32.5 N – NW/HS: blastfurnace cement with 66-80% cinder sand (CEM III/B) of strength class 32.5 with standard initial strength, low heat of hydration (NW) and high sulphate resistance (HS)

The European cement standard DIN EN 197 part 1 and the German national cement standard DIN 1164 include regulations covering the composition, requirements and conformity criteria of cements. DIN EN 196 specifies the test methods employed to verify the standardised properties.

**Aggregates**

Aggregates must comply with DIN 4226. We distinguish between lightweight, normal and heavy aggregates according to their density (table 2.1.3).

Recycled materials represent a special type of aggregate for concrete. In this case, old concrete from the demolition of concrete structures and components has been prepared for reuse in normal-weight concrete. The production of these aggregates and their use in concrete was initially covered by a directive from the German Reinforced Concrete Committee (DAfStb) but in future will be included in DIN 4226 part 100.

Aggregates are graded according to grain size. In doing so, we specify the smallest and the largest grain sizes in each case, e.g. grade 0/4, 4/8, 8/16 etc. The methods of conveying and working the concrete determine the maximum size of the aggregate in the concrete. The contractor usually determines this according to the method of placing the concrete. The nominal aggregate size should not exceed 1/3, better still 1/5, times the smallest dimension of a component. According to DIN 1045 part 3 the maximum grain size of the aggregate should take into account the pitch of the reinforcing bars designed according to DIN 1045 part 1.

Table 2.1.4 specifies supplementary designations according to the type of aggregate.

# Types of concrete

Aggregates are designated according to the following system:

- geographic origin
- type
- petrographic type
- grade
- requirements categories that deviate from the standard requirements
- if necessary, additional identification data

Aggregates for producing concrete according to the standards must comply with the requirements of DIN 4226. This covers the requirements for grading the aggregate plus physical and chemical properties. Physical properties are, for example, resistance to freezing or frost and de-icing salts. The chemical requirements limit, for example, the content of harmful constituents. If the use and ambient conditions to which the concrete will be subjected mean that the aggregates will have to fulfill further conditions, then this must be agreed. Extra requirements that require the use of a different requirements category according to DIN 4226 can be requested, in particular with regard to

- resistance to freezing
- resistance to frost and de-icing salts
- the proportion of expanding constituents with an organic origin
- the content of water-soluble chloride
- the shape of the grains

Aggregates that do not satisfy the standard requirements regarding certain properties can nevertheless be used for certain concrete applications. The designer establishes the requirements to be met by such aggregates and verifies these by testing trial mixes. Besides the aspects already mentioned in connection with more a demanding specification, lower requirement categories can also affect the proportion of fine aggregate and the amount of lightweight organic impurities.

2.1.5 Concrete admixture action groups and their designations

| Action group | Abbreviation | Colour code |
|---|---|---|
| Plasticiser | BV | yellow |
| Superplasticiser | FM | grey |
| Air entrainer | LP | blue |
| Waterproofer | DM | brown |
| Retarder | VZ | red |
| Accelerator | BE | green |
| Grouting aid | EH | white |
| Stabiliser | ST | violet |

2.1.6 Types of concrete and their applications [10]

| Type of concrete | Aggregates (examples) | Usual density range [kg/m$^3$] | Thermal conductivity [W/mK] | Applications |
|---|---|---|---|---|
| A Concretes with close-grained microstructure | | | | |
| Heavy concrete | Barytes, iron ore, steel shot | 3200-4000 | 2.30 | Radiation shielding |
| Normal-weight concrete | Gravel, grit, slag | 2300-2500 | 2.10 | Reinforced and prestressed concrete constructions |
| Normal-weight concrete masonry units | Gravel, grit, slag | 1600-1800 | 0.92-1.30 | Basement walls, partitions |
| Lightweight concrete 1. LB - Q (with quartz sand) | Expanded clay, expanded shale, foamed slag | 1200-1800 | 0.74-1.56 | Reinforced and prestressed concrete constructions with lower self-weight |
| 2. LB - L (with light-weight sand) | Expanded clay, expanded shale, foamed slag | 1000-1600 | 0.49-1.00[1)] | As above but providing thermal insulation |
| B Concretes with no-fines microstructure | | | | |
| Lightweight concrete blocks 1. Standard masonry mortar | Pumice, expanded clay, expanded shale, foamed slag | 500-1000 | 0.29-0.59 | External walls |
| 2. Lightweight masonry mortar | Pumice, expanded clay, light-weight sand | 500-700 | 0.12-0.25 | External walls with good thermal insulation |
| C Concretes with cellular structure | | | | |
| Gauged bricks of autoclaved aerated concrete | Natural sand | 400-700 | 0.12-0.23 | External walls with good thermal insulation, floor slabs |
| Expanded polystyrene bead concrete bead concrete | Natural sand | 600-800 | 0.22-0.31 | Wall panels, load-bearing layers of industrial floors |

[1)] These values apply only for a moisture content < 5% by vol.

2.1.7 Consistency classes for fresh concrete [3]

| Consistency designation | Class | Mound size in consistency test [mm] | Compacting factor [–] |
|---|---|---|---|
| very stiff | C 0 | – | ≥ 1.46 |
| stiff | C 1 | – | 1.45 ... 1.26 |
|  | F 1 | ≤ 340 | – |
| plastic | C 2 | – | 1.25 ...1.11 |
|  | F 2 | 350 ... 410 | 1.10 ...1.04 |
| soft | C 3 | – | – |
|  | F 3 | 420 ... 480 | – |
| very soft | F 4 | 490 ... 550 | – |
| fluid | F 5 | 560 ... 620 | – |
| very fluid | F 6 | ≥ 630 | – |

Standard consistency:
for in situ concrete – C 3 and F 3
for high-strength concrete – F 3 and softer
The addition of superplasticiser is prescribed for C 3, F 4 and softer
The DAfStb directive for "self-compacting concrete" should be followed for mound sizes > 700 mm

2.1.8 Compressive strength grades for normal-weight and heavy concrete [3]

| Grade | $f_{ck,cyl}$ [1] [N/mm$^2$] | $f_{ck,cube}$ [2] [N/mm$^2$] | Type of concrete |
|---|---|---|---|
| C8/10 | 8 | 10 | Normal-weight concrete |
| C12/15 | 12 | 15 |  |
| C16/20 | 16 | 20 |  |
| C20/25 | 20 | 25 |  |
| C25/30 | 25 | 30 |  |
| C30/37 | 30 | 37 |  |
| C35/45 | 35 | 45 |  |
| C40/50 | 40 | 50 |  |
| C45/55 | 45 | 55 |  |
| C50/60 | 50 | 60 |  |
| C55/67 | 55 | 67 | High-strength concrete |
| C60/75 | 60 | 75 |  |
| C70/85 | 70 | 85 |  |
| C80/95 | 80 | 95 |  |
| C90/105 [3] | 90 | 105 |  |
| C100/115 [3] | 100 | 115 |  |

2.1.9 Compressive strength grades for lightweight concrete [3]

| Grade | $f_{ck,cyl}$ [1] [N/mm$^2$] | $f_{ck,cube}$ [2] [N/mm$^2$] | Type of concrete |
|---|---|---|---|
| LC8/9 | 8 | 9 | Lightweight concrete |
| LC12/13 | 12 | 13 |  |
| LC16/18 | 16 | 18 |  |
| LC20/22 | 20 | 22 |  |
| LC25/28 | 25 | 28 |  |
| LC30/33 | 30 | 33 |  |
| LC35/38 | 35 | 38 |  |
| LC40/44 | 40 | 44 |  |
| LC45/50 | 45 | 50 |  |
| LC50/55 | 50 | 55 |  |
| LC55/60 | 55 | 60 | High-strength lightweight concrete |
| LC60/66 | 60 | 66 |  |
| LC70/77 [3] | 70 | 77 |  |
| LC80/88 [3] | 80 | 88 |  |

[1] $f_{ck,cyl}$ = characteristic strength of cylinders, 150 mm dia., 300 mm long, 28 days old
[2] $f_{ck,cube}$ = characteristic strength of 150 mm cubes, 28 days old
[3] Requires general building authority approval or approval for each case

Examples of aggregate designations:

- DIN 4226-8/16-F1 aggregate
  Aggregate to DIN 4226 with a dense microstructure of nominal size/grade 8/16 that satisfies increased freezing resistance requirements exceeding the standard requirements
- DIN 426-0/2-f10 aggregate
  Aggregate to DIN 4226 with a dense microstructure of nominal size/grade 0/2 that does not satisfy the standard requirements regarding the proportion of fine constituents to DIN 4226

**Concrete additives/admixtures**

Additives are added to the cement during production, and admixtures are added to the concrete during mixing.

Additives are very fine materials such as trass, stone dust, silica dust or pulverised fuel ash (PFA) that influence certain concrete properties. For example, the use of silica dust helps concrete achieve a very high compressive strength exceeding 100 N/mm$^2$, and also improves the impermeability of the concrete microstructure. As concrete additives are generally added in larger amounts, they must be included when calculating the volume.

Admixtures (table 2.1.5) change concrete properties, e.g. its workability, setting, or resistance of the hardened concrete to frost and de-icing salts, by way of chemical and/or physical effects. They are insignificant in terms of volume.

All concrete admixtures and additives not covered by their own standards must be tested and approved if they are to be used in concrete. Such admixtures/additives may only be used with a valid approval symbol and according to the conditions given in the approval documents.

## Types of concrete

Concrete can be adapted to meet specific, different functions through the manner of its production. For example, high loadbearing capacity and good sound insulation require dense concrete; the selection of suitable aggregates achieves the desired grade of concrete with a correspondingly high weight. On the other hand, good thermal insulation properties require the use of porous aggregates, often achieved using natural aggregates (e.g. pumice) or natural materials specially prepared for use in concrete (e.g. expanded clay or expanded shale). If coarse grains of aggregate are not fully cemented, voids amounting to 25-30% by vol. remain. The already low density of such a no-fines concrete microstructure can be lowered even further by introducing cells or slots like those in masonry units. The hydrated cement acting as a binder can also be made lighter and given better thermal insulation properties by including pores (aerated concrete, foamed concrete), or "packaged" air (foamed polystyrene beads). Cement-bound wood-wool lightweight building boards represent another form of thermal insulation.

The use of heavy aggregates such as steel shot result in a heavy concrete useful for providing protection against radiation. Table 2.1.6 provides a summary of the most important types of concrete.

### Normal-weight concrete
We designate concrete with an oven-dry density between 2000 and 2600 kg/m$^3$ as normal-weight concrete. The majority of applications are based on this type of concrete. Generally, concrete is divided into site-batched, ready-mixed and in situ concrete according to its place of production or use, and into seven consistency classes according to its workability (table 2.1.7).

### Lightweight concrete
The distinguishing property of lightweight concrete compared to normal-weight concrete is its lower weight (density ≤ 2000 kg/m$^3$). The density and other properties of lightweight concrete are determined by the properties of the lightweight aggregates used (pumice, expanded clay, expanded shale), the nature of the concrete microstructure (no-fines, close-grained) or the volume of pores (aerated concrete, light foam concrete). It is placed in lifts about 1 m deep and should not be allowed to segregate during compacting. Aerated concrete, light foam concrete and no-fines lightweight concrete are mainly used where thermal insulation is important. Although they have a lower strength than that of normal-weight concrete, this is usually adequate for normal building work. Close-grained lightweight concrete essentially meets the same requirements as normal-weight concrete. Tables 2.1.8 to 2.1.12 contain the most important design information on different concretes.

The design weight for structural analysis (which is lower than that of normal-weight concrete) is established by dividing close-grained structural lightweight concrete into density classes. No-fines, thermally insulating lightweight concrete is defined in terms of its thermal conductivity according to DIN 4108 part 4 and in terms of its airborne sound insulation according to DIN 4109 part 3. The dimensions given in table 2.1.12 are minimum values to which an allowance of 10 mm (5 mm) should be added on the drawings.

### Heavy concrete

Concrete with an oven-dry density exceeding 2600 kg/m³ is designated as heavy concrete. The high oven-dry density is achieved by using heavy aggregates with a density generally exceeding 3000 kg/m³, e.g. baryte, magnetite, haematite or steel shot. Applications for heavy concrete include shielding against radiation, e.g. hospitals, nuclear reactors, or ground slabs where a high self-weight is desirable as protection against flotation.

## Properties of concrete

### Fresh concrete

Fresh concrete is called concrete as long as it can still be worked. It is transported in skips, on conveyor belts or in pipelines (pumped concrete) and placed in the formwork. The measure of its workability (stiffness) is consistency. We distinguish between seven consistency ranges, and each one of these demands its own form of compaction (table 2.1.7). Concrete of consistency F 5 (fluid) generally needs only limited compaction. Consistency range F 6 (very fluid) denotes self-compacting concrete in which gravity alone is sufficient to deaerate the concrete and allow it to flow in a dense layer until it finds its own level.

The workability of fresh concrete must be adapted to suit the circumstances on site. It should not be allowed to segregate upon introducing it into the formwork. Thin sections or heavily reinforced components generally require soft concrete in the middle of consistency range F 3 (mound size in consistency test: 450 ± 30 mm). Even for other components, it is advantageous to use concrete with this or a softer consistency, formerly known as standard consistency.

Numerous concrete properties depend on the amount of cement and water in the concrete, expressed as the water/cement ratio (w/c ratio). The standard therefore specifies upper limits for the water/cement ratio of fresh concrete. These limits are governed by the ambient conditions and their corrosive influence on the concrete or, if included, the reinforcement. For example, in reinforced concrete the water/cement ratio may not exceed 0.75. Only for plain concrete not subjected to any corrosive or aggressive influences is no maximum water/cement ratio specified. For further information see table 2.1.13 A/B.

### Development of strength

The main factors influencing the development of strength in concrete are the properties of the cement, the mix and the age of the concrete, as well as the ambient and curing conditions (temperature, moisture). One factor that has a considerable affect on the early strength of concrete is the water/cement ratio. Strength increases with the age of the concrete. The influence of the type of cement at lower and higher temperatures is illustrated in tables 2.1.14 and 2.1.15.

### Hardened concrete

The most important property of concrete is its compressive strength. This is normally determined by compression tests on test specimens (cubes, cylinders) produced for this purpose or, in special cases, on test cores cut from the structure. The testing of compressive strength is covered in detail in standards, which also specify storage conditions for the test specimens and the testing apparatus. According to the standard, testing is generally carried out after 28 days on a 150 mm cube. The concrete is assigned a grade (strength class) depending on the magnitude of the compressive strength determined in this test. The classification of a component, structure or type of concrete according to a certain grade is carried out using statistical methods and is not normally based on individual values. As an alternative, the strength of the concrete in a structure can also be determined using a rebound hammer. However, this test may only be used under certain conditions.
With the exception of a few standard concrete mixes already specified in the standard, concrete mixes are usually specified according to the standard and the experience of a concrete technologist to suit the intended properties. Tests on trial mixes are used to verify that the desired properties have been achieved.

## Exposure classes

The design and the ambient conditions of the concrete govern more or less all the properties required of hardened concrete in the finished structure. For reinforced concrete, the most common form of concrete construction, ambient conditions can be divided into two corrosive influences:

- those that attack the concrete, and
- those that can damage the reinforcing bars.

These two types of attack on this composite material are divided into exposure classes according to type. Exposure classes are designated with a capital X followed by a zero or a letter, generally standing for the English term for the respective corrosive attack. Exposure class X0 is for concrete that is not subjected to any corrosive influences and does not suffer any corrosion of the steel. Such ambient conditions as defined in the standard apply to concrete permanently in dry interior conditions containing no embedded metal. The exposure classes for corrosion of the rein-

2.1.10  Density classes for lightweight concrete, design values [3]

| Density class | Density range [kg/m³] | Design value for density [kg/m³] plain | reinforced |
|---|---|---|---|
| D 1.0 | 800-1000 | 1050 | 1150 |
| D 1.2 | 1010-1200 | 1250 | 1350 |
| D 1.4 | 1210-1400 | 1050 | 1550 |
| D 1.6 | 1410-1600 | 1650 | 1750 |
| D 1.8 | 1610-1800 | 1850 | 1950 |
| D 2.0 | 1810-2000 | 2150 | 2150 |

2.1.11  Thermal conductivity, characteristic values to DIN 4108

| Density class | Density range [kg/m³] | Characteristic value for thermal conductivity $\lambda_R$[1] [W/mK] |
|---|---|---|
| D 1.0 | up to 900<br>up to 1000 | 0.44<br>0.49 |
| D 1.2 | up to 1100<br>up to 1200 | 0.55<br>0.62 |
| D 1.4 | up to 1300<br>up to 1400 | 0.70<br>0.79 |
| D 1.6 | up to 1500<br>up to 1600 | 0.89<br>1.0 |
| D 1.8<br>D 2.0 | up to 1800<br>up to 2000 | 1.3<br>1.6 |

[1] Values apply only to aggregates with porous microstructure without the addition of quartz sand

2.1.12  Concrete cover to reinforcement depending on exposure class[1] [3]

| Exposure class | Bar dia $d_s$[2] [mm] | min. dim. [mm] | nom. dim. [mm] |
|---|---|---|---|
| XC1 | up to 10<br>12, 14<br>16, 20<br>25<br>28 | 10<br>15<br>20<br>25<br>30 | 20<br>25<br>30<br>35<br>40 |
| XC2, XC3 | up to 20<br>25<br>28 | 20<br>25<br>30 | 35<br>35<br>45 |
| XC4 | up to 25<br>28 | 25<br>30 | 40<br>45 |
| XD1, XD2, XD3[3] | up to 28 | 40 | 55 |
| XS1, XS2, XS3 | up to 28 | 40 | 55 |

[1] If several exposure classes apply to one component, the exposure class with the highest requirements apply in each case
[2] The equivalent diameter $d_{sv}$ applies for dowel bars
[3] Special corrosion protection measures for the reinforcement may be necessary in individual cases for exposure class XD 3

2.1.13 A  Limiting values for the composition and properties of concrete [3]

| Exposure class | Max. w/c or w/c (eq) | Min. $f_{ck}$[1] [N/mm²] | Min. $z$[2] [kg/m³] | Min. $z$[3] (including additives) [kg/m³] | Min. p (min. air content) [% by vol.] | Other requirements |
|---|---|---|---|---|---|---|
| No risk of corrosion or attack ||||||||
| X0 | – | C8/10 | – | – | – | – |
| Reinforcement corroded by carbonation ||||||||
| XC1 | 0.75 | C16/20 | 240 | 240 | – | – |
| XC2 | 0.75 | C16/20 | 240 | 240 | – | – |
| XC3 | 0.65 | C20/25 | 260 | 240 | – | – |
| XC4 | 0.60 | C25/30 | 280 | 270 | – | – |
| Reinforcement corroded by chlorides ||||||||
| XD1 | 0.55 | C30/37[5] | 300 | 270 | – | – |
| XD2 | 0.50 | C35/45[5] | 320[6] | 270 | – | – |
| XD3 | 0.45 | C35/45[5] | 320[6] | 270 | – | – |
| Reinforcement corroded by chlorides from seawater ||||||||
| XS1 | 0.55 | C30/37[5] | 300 | 270 | – | – |
| XS2 | 0.50 | C35/45[5] | 320[6] | 270 | – | – |
| XS3 | 0.45 | C35/45[5] | 320[6] | 270 | – | – |
| Frost attack with and without de-icing salts ||||||||
| XF1 | 0.60 | C25/30 | 280 | 270 | – | $F_4$[9] |
| XF2 | 0.55[7] | C25/30 | 300 | [7] | [4] | $MS_{25}$[9] |
|  | 0.50[7] | C35/45 | 320 | [7] | – |  |
| XF3 | 0.55 | C25/30 | 300 | 270 | [4] | $F_2$[9] |
|  | 0.50 | C35/45 | 320 | 270 | – |  |
| XF4 | 0.50[7] | C30/37 | 320 | [7] | [4] [8] | $MS_{18}$[9] |

Footnotes to table 2.1.13 A:
[1] Min. compressive strength grade (min. $f_{ck}$) does not apply for lightweight concrete.
[2] Cement content (min. z) may be reduced by 30 kg/m³ in the case of 63 mm max. aggregate size; in this case footnote [6] does not apply.
[3] The conditions according to DIN EN 206 part 1 or DIN 1045 part 2 section 5.2.5 must be observed when including additives.
[4] Average air content of fresh concrete directly prior to placing:
max. aggregate size 8 mm ≥ 5.5% by vol.; max. aggregate size 16 mm ≥ 4.5% by vol.;
max. aggregate size 32 mm ≥ 4.0% by vol.; max. aggregate size 63 mm ≥ 3.5% by vol.;
single values may not drop below these values by more than 0.5% by vol.
[5] Air-entrained concrete is classed one grade lower owing to the simultaneous requirement from exposure class XF.
[6] Min. z = 300 kg/m³ for mass concrete components (smallest dimension 800 mm).
[7] The addition of type II additives is permitted; adding to cement content or w/c ratio is not permitted.
[8] Low-slump concrete with w/c ≤ 0.40 may be produced without air pores.
[9] Aggregates with standard requirements and additional resistance to frost or frost and de-icing salts (see DIN 4226 part 1).

Footnotes to table 2.1.13 B:
[1] Min. compressive strength grade (min. fck) does not apply for lightweight concrete.
[2] Cement content (min. z) may be reduced by 30 kg/m³ in the case of 63 mm max. aggregate size; in this case footnote [6] does not apply.
[3] The conditions according to DIN EN 206 part 1 or DIN 1045 part 2 section 5.2.5 must be observed when including additives.
[4] Aggregates: moderately rough surface, stocky shape; ≤ 4 mm predominantly quartz or equivalent hardness; > 4 mm with high abrasion resistance; coarse grading if possible.
[5] Air-entrained concrete is classed one grade lower owing to the simultaneous requirement from exposure class XF.
[6] Max. cement content z = 360 kg/m³, but not for high-strength concrete.
[7] Concrete must be protected, if necessary with a special assessment for a special solution.

forcing bars alone are class XC for carbonation and XS and XD for attack by chlorides in seawater or de-icing salts (table 2.1.16 A/B).

Carbonation is always a potential problem when concrete is exposed to moisture and carbon dioxide ($CO_2$) in the atmosphere, as is the case for any component in the open air. The carbon dioxide in the atmosphere penetrates the hardened concrete and reacts with the moisture available there and the alkaline constituents in the hydrated cement. The alkalinity of the concrete, which protects the steel against corrosion, is lowered in the course of this reaction. Once this reaction reaches the steel, it begins to corrode. The formation of rust causes an increase in volume, which leads to spalling and other characteristic damage, and weakens the structural reinforcement. Chlorides too, e.g. from seawater or de-icing salts, promote corrosion of the reinforcing bars when they penetrate the concrete cover and come into contact with the steel. The corrosive effect of chlorides in conjunction with moisture is considerably more aggressive than the corrosion mechanism of carbonation.

Exposure classes relating to concrete attack, which can also damage plain concrete components, are class XA for a chemical attack, XF for attack by frost or combined frost and de-icing salts, and XM for abrasion. The corrosion mechanism of frost is brought about by water in the pores near the surface expanding as it freezes in very cold weather. The bursting effect of the ice can initially weaken the concrete microstructure and, if the process is allowed to continue, eventually destroy the microstructure. This destruction gradually migrates from the surface into the interior of the material and leads to the familiar pattern of damage. The use of de-icing salts on frozen concrete surfaces can aggravate this mechanism quite considerably. The reason for this is that the energy relationships can also lead to the formation of ice in deeper-lying regions of the microstructure and hence damage. Exposure class XM primarily covers concrete surfaces that are intended to be subjected to extreme wear during use, e.g. pavements for tracked vehicles.

An experienced designer can recognise the corrosive influences acting on the intended structure and allocate each component to an exposure class according to the types of attack and loads to be expected. The exposure classes are further subdivided into three or four subclasses (e.g. XC 1, XC 2, XC 3 and XC 4) according to severity and intensity of attack in order to avoid overdesign. Decades of experience with reinforced concrete construction, extensive research and investigations of damage have led to the exposure classes being coupled with requirements for

2.1.13 B  Limiting values for the composition and properties of concrete [3]

| Exposure class | Max. w/c or w/c (eq) | Min. $f_{ck}$[1] [N/mm²] | Min. $z$[2] [kg/m³] | Min. $z$[3] (including additives) [kg/m³] | Min. p (min. air content) | Other requirements [% by vol.] |
|---|---|---|---|---|---|---|
| Concrete subject to abrasion[4] | | | | | | |
| XM1 | 0.55 | C30/37[5] | 300[6] | 270 | – | – |
| XM2 | 0.55 | C30/37[5] | 300[6] | 270 | – | surface of concrete treated |
| XM3 | 0.45 / 0.45 | C35/45[5] / C35/45[5] | 320[6] / 320[6] | 270 / 270 | – | hard material to DIN 1100 |
| Concrete exposed to aggressive chemical environment | | | | | | |
| XA1 | 0.60 | C25/30 | 280 | 270 | – | – |
| XA2 | 0.50 | C35/45[5] | 320 | 270 | – | – |
| XA3[7] | 0.45 | C35/45[5] | 320 | 270 | – | – |

2.1.14  Standard figures for strength development of concrete at constant +20 °C (after Wischers and Dahms) [10]

| Cement strength class | Strength in % of 28-day compressive strength after | | | | |
|---|---|---|---|---|---|
| | 3 days | 7 days | 28 days | 90 days | 180 days |
| 52.5 N, 52.5 R, 42.5 R | 70-80 | 80-90 | 100 | 100-105 | 105-110 |
| 42.5 N, 32.5 R | 50-60 | 65-80 | 100 | 105-115 | 110-120 |
| 32.5 N | 30-40 | 50-65 | 100 | 110-125 | 115-130 |

2.1.15  Standard figures for strength development of concrete at constant +5 °C (after Wischers and Dahms) [10]

| Cement strength class | 5 °C strength in % of compressive strength at 20 °C after | | |
|---|---|---|---|
| | 3 days | 7 days | 28 days |
| 52.5 N, 52.5 R, 42.5 R | 60-75 | 75-90 | 90-105 |
| 42.5 N, 32.5 R | 45-60 | 60-75 | 75-90 |
| 32.5 N | 30-45 | 45-60 | 60-75 |

## Concrete – the material

### 2.1.16 A  Ambient conditions classes for reinforcement corrosion, to DIN 1045 part 2

| Class | Description of ambient conditions | Examples for allocating exposure classes (for information only) | Min. compressive strength grade |
|---|---|---|---|

**No risk of corrosion or attack**
Exposure class X0 can be allocated to components without reinforcement or embedded metal in ambient conditions not corrosive to concrete:

| Class | Description of ambient conditions | Examples for allocating exposure classes | Min. compressive strength grade |
|---|---|---|---|
| X0 | For concrete without reinforcement or embedded metal; all exposure classes except frost attack with or without de-icing salts, abrasion or chemical attack | Foundations without reinforcement and not subject to frost  Internal components without reinforcement | C8/10  LC12/13 |
| | For concrete without reinforcement or embedded metal; very dry | No applications possible in Germany | |

**Corrosion of reinforcement triggered by carbonation**
When concrete containing reinforcement or other embedded metal is exposed to the air and moisture, the exposure classes must be allocated as follows:

| Class | Description of ambient conditions | Examples for allocating exposure classes | Min. compressive strength grade |
|---|---|---|---|
| XC 1 | Dry or permanently wet | Components in interiors with normal humidity (including kitchens, bathrooms and utility rooms in residential buildings)  Concrete permanently immersed in water | C16/20  LC16/18 |
| XC 2 | Wet, seldom dry | Parts of water tanks; components in foundations | |
| XC 3 | Moderate moisture | Components to which the outside air can gain access frequently or permanently, e.g. open sheds; interiors with high humidity, e.g. commercial kitchens and bathrooms, laundries, wet areas of indoor pools, cattle stalls | C16/20  LC20/22 |
| XC 4 | Alternately wet and dry | External components subjected to direct rainfall | C25/30  LC25/28 |

Note: The moisture conditions relate to the condition within the concrete cover to the reinforcement or other embedded metal; in many cases it can however be assumed, that the conditions in the concrete cover correspond to those of the surroundings.
In these cases, allocating the class according to the ambient conditions may be assumed to be equivalent. This need not be the case if there is a barrier layer between the concrete and its surroundings.

**Corrosion of reinforcement caused by chlorides, except for those in seawater**
When concrete containing reinforcement or other embedded metal is exposed to water containing chloride, including de-icing salts, but excluding seawater, the exposure classes must be allocated as follows

| Class | Description of ambient conditions | Examples for allocating exposure classes | Min. compressive strength grade |
|---|---|---|---|
| XD 1 | Moderate moisture | Components in the proximity of spray water from traffic pavements  Detached garages | C30/37  LC30/33 |
| XD 2 | Wet, seldom dry | Brine baths  Components exposed to industrial waste water containing chloride | C35/45  LC35/38 |
| XD 3 | Alternately wet and dry | Parts of bridges subjected to frequent splashing water  Traffic pavements, parking decks | C35/45  LC35/38 |

**Corrosion of reinforcement caused by chlorides in seawater**
When concrete containing reinforcement or other embedded metal is exposed to seawater containing chlorides or salt-laden sea air, the exposure classes must be allocated as follows:

| Class | Description of ambient conditions | Examples for allocating exposure classes | Min. compressive strength grade |
|---|---|---|---|
| XS 1 | Salt-laden air but no direct contact with seawater | External components in coastal regions | C30/37  LC30/33 |
| XS 2 | Underwater | Components in port facilities constantly under water | C35/45  LC35/38 |
| XS 3 | Within tidal range, splashing and spray water | Quay walls in port facilities | C35/45  LC35/38 |

## Exposure classes

2.1.16 B  Ambient conditions classes for concrete attack, to DIN 1045 part 2

| Class | Description of ambient conditions | Examples for allocating exposure classes (for information only) | Min. compressive strength grade |
|---|---|---|---|

Frost attack with or without de-icing salts
When saturated concrete is exposed to many frost-thaw cycles, the exposure classes must be allocated as follows:

| Class | Description | Examples | Min. compressive strength grade |
|---|---|---|---|
| XF 1 | Moderate water saturation, without de-icing salts | External components | C25/30<br>LC25/28 |
| XF 2 | Moderate water saturation, with de-icing salts | Components subjected to spray or splashing water from traffic pavements treated with de-icing salts, unless XF 4 components subjected to spray from seawater | C35/45<br>LC25/28 |
| XF 3 | High water saturation, without de-icing salts | Open water tanks; components within the splash zone of freshwater | C35/45<br>LC25/28 |
| XF 4 | High water saturation, with de-icing salts | Traffic pavements treated with de-icing salts; predominantly horizontal components subjected to splashing water from traffic pavements treated with de-icing salts; rotating scraper tracks in sewage treatment works; seawater components within the splash zone | C30/37 (air-entrained)<br>LC30/33 |

Concrete attacked by chemicals
When concrete is exposed to attack by chemicals in natural soils, groundwater, seawater, to DIN 1045 part 2 table 2, and waste water, the exposure classes must be allocated as follows:

| Class | Description | Examples | Min. compressive strength grade |
|---|---|---|---|
| XA 1 | Weak chemical attack to DIN 1045 part 2 table 2 | Tanks in sewage treatment works; fertiliser silos | C25/30<br>LC25/28 |
| XA 2 | Moderate chemical attack to DIN 1045 part 2 table 2, and marine structures | Components which can come into contact with seawater; components in soils aggressive to concrete | C35/45<br>LC35/38 |
| XA 3 | Severe chemical attack to DIN 1045 part 2 table 2 | Industrial waste-water plants with chemically aggressive waste water; silage fodder silos and feeding tables in agriculture; cooling towers with exhaust gas ducting | C35/45<br>LC35/38 |

Note: In the case of XA 3 and ambient conditions outside the limits of DIN 1045 part 2 table 2, the presence of other aggressive chemicals, chemically contaminated soil or water, high water flow velocities and the effects of chemicals to DIN 1045 part 2 table 2, requirements for concrete or protective measures are prescribed in these application rules, DIN 1045 part 2 section 5.3.2.

Concrete subjected to abrasion
When concrete is exposed to considerable mechanical loading, the exposure classes must be allocated as follows:

| Class | Description | Examples | Min. compressive strength grade |
|---|---|---|---|
| XM 1 | Moderate abrasion | Loadbearing or stabilising industrial floors carrying the loads of pneumatically tyred vehicles | C30/37<br>LC30/33 |
| XM 2 | Severe abrasion | Loadbearing or stabilising industrial floors carrying the loads of forklift trucks with pneumatic or solid rubber tyres | C30/37<br>LC30/33 |
| XM 3 | Very severe abrasion | Loadbearing or stabilising industrial floors carrying the loads of forklift trucks with elastomeric or steel rollers; surfaces often subjected to the loads of tracked vehicles; hydraulic structures in contact-loaded waters, e.g. stilling basins | C35/45<br>hard material to DIN 1100<br>LC35/38 |

2.1.17 Curing of concrete [2]

| Type | Measures | Outside temperature [°C] | | | | |
|---|---|---|---|---|---|---|
| | | below -3 | -3 to +5 | 5 to 10 | 10 to 25 | above 25 |
| Sheeting / curing film | Cover or spray on curing film plus wetting; wet timber formwork; protect steel formwork against direct sunlight | | | | | ☐ |
| | Cover or spray on curing film | | | ☐ | ☐ | |
| | Cover or spray on curing film, plus thermal insulation; use of thermally insulating formwork, e.g. timber, is advisable | | ☐[1] | | | |
| | Cover and insulate; surround working area (tent) or heat (e.g. radiant heater); plus: maintain concrete at +10 °C for min. 3 days | ☐[1] | | | | |
| Water | Keep moist by way of continuous wetting | | | | ☐ | |

[1] Prolong curing and striking times by the number of days of frost; protect concrete against precipitation for at least 7 days.

2.1.18 Minimum duration of curing in days[1] without more accurate verification of strength in the zone near the surface, to DIN 1045 part 3.

| Surface temperature[3] [4] $\vartheta$ [°C] | Strength development of concrete[5] $r = f_{cm2}*/f_{cm28}{}^{3)}$ | | | |
|---|---|---|---|---|
| | $r \geq 0.50$ | $r \geq 0.30$ | $r \geq 0.15$ | $r < 0.15$ |
| $\geq 25$ | 1 | 2 | 2 | 3 |
| $25 > \vartheta \geq 15$ | 1 | 2 | 4 | 5 |
| $15 > \vartheta \geq 10$ | 2 | 4 | 7 | 10 |
| $10 > \vartheta \geq 5^{6)}$ | 3 | 6 | 10 | 15 |

[1] Prolong curing time appropriately for workability time > 5 h.
[2] Double values for exposure class XM.
[3] Intermediate values may be interpolated.
[4] The air temperature may be used instead of the surface temperature of the concrete.
[5] Determined according to DIN 1048 part 5 from average values of compressive strength after 2 and 28 days, either by means of a suitability test or from the known ratios of concretes of comparable composition (same cement, same water/cement ratio).
[6] For temperatures < 5 °C, prolong curing time by time for temperatures < 5 °C.

All exposure classes[2] except X0 and XC 1, for which at least 1/2 day must be allocated.

the concrete mixes and the properties of the hardened concrete that must be achieved. State-of-the-art concrete must exhibit such properties to be able to withstand the corresponding degree of attack. We talk of whole-life design when applying this principle. This means besides designing for load-carrying capacity and serviceability, the third essential design criterion is to guarantee that the service life of a planned concrete structure is as long as possible. So right at the start of classifying a structure or components according to exposure classes, the designer is given a specification for the concrete mix and minimum composite strength which must be maintained during the construction of the structure.

The corresponding exposure classes prescribe protective measures (coatings, inclusion of hard materials) for exceptional cases, where concrete is subjected to particularly severe chemical attack or mechanical damage (acids, very severe abrasion) which it cannot withstand without damage, even after maximising its properties.

Concrete for retaining aqueous liquids is an exception to this system. This type of concrete is always required when watertight components or structures are being constructed. Special provisions in the standards cover the mixes and strength properties of such concretes. Exposure classes are not applicable in such instances.

## Establishing the concrete properties

A concrete is defined either by specifying its overall properties or its precise composition. The standard therefore distinguishes between designed and prescribed concrete mixes. A reinforced concrete structure is generally built in the following sequence:

- Design phase
- Appointment of a contractor to carry out the work
- Ordering the concrete from a ready-mixed concrete supplier
- Delivery of the concrete as required
- Placing, compaction and curing of the ready-mixed concrete

The persons involved in this process usually belong to the following three groups:

- Developer/designer
- Contractor
- Concrete producer/supplier (ready-mixed concrete works)

The ready-mixed concrete supplier must always be informed of either the required properties (designed mix) or the exact composition (prescribed mix) of the concrete when it is ordered in order that the "right" concrete is supplied. Designed mixes are the norm when ordering ready-mixed concrete.

The sum of the specified properties of a concrete is made up of the properties that ensue in the course of the design process. These properties may be determined by the exposure class or the strength called for in the structural analysis. And they are exclusively properties of the hardened concrete in the completed structure. Other properties of the concrete are determined by the contractor. These concern, above all, the properties of the fresh (wet) concrete, which affect the workability of the concrete and depend on site operations, e.g. methods of conveying and placing, type of formwork, degree of reinforcement. This results in a specification for the consistency and the maximum aggregate size of the concrete to be used.

Another important concrete property for site operations is its strength development. To minimise costs, the contractor generally prefers a rapid strength development so that the formwork can be struck as early as possible. Swift reuse of the formwork means being able to continue the construction of adjoining components without delay.

2.1.19 Concrete cover to reinforcing bars depending on exposure classes[1)4)] to DIN 1045 part 1

| Exposure class | Bar dia. $d_s$[2)] [mm] | Min. dim. $c_{min}$ [mm] | Nom. dim. $c_{nom}$ [mm] |
|---|---|---|---|
| XC1 | up to 10<br>12, 14<br>16, 20<br>25<br>28 | 10<br>15<br>20<br>25<br>30 | 20<br>25<br>30<br>35<br>40 |
| XC2, XC3 | up to 20<br>25<br>28 | 20<br>25<br>30 | 35<br>40<br>45 |
| XC4 | up to 25<br>28 | 25<br>30 | 40<br>45 |
| XD1, XD2, XD3[3)] | up to 28 | 40 | 55 |
| XS1, XS2, XS3 | up to 28 | 40 | 55 |

[1)] If several exposure classes apply to one component, the exposure class with the highest requirements applies in each case.
[2)] The equivalent diameter $d_{sv}$ applies for dowel bars.
[3)] Special corrosion protection measures for the reinforcement may be necessary in individual cases for exposure class XD3.
[4)] The requirements of [3] table 5 apply to prestressing tendons.

The concrete cover must be increased in the case of:
- Components of lightweight concrete
  Additional requirement: $c_{min}$ must be at least 5mm larger than the diameter of the largest porous lightweight aggregate, except for exposure class XC 1.
- Concrete subjected to abrasion
  As an alternative to the additional requirements to be met by the aggregate, there is the option of increasing the minimum concrete cover to the reinforcement $c_{min}$ (sacrificial concrete).
  Standard values for sacrificial concrete: for XM 1: $\Delta c_{sac} = +5$ mm
  for XM 2: $\Delta c_{sac} = +10$ mm
  for XM 3: $\Delta c_{sac} = +15$ mm
- Concrete cast against uneven surfaces
  increase cover allowance:
  - generally by the height of the unevenness
  - min. increase $\Delta c_{uneven} \geq +20$ mm
  - casting directly against the subsoil: $\Delta c_{uneven} \geq +50$ mm

The concrete cover may be reduced in the case of:
- Components with a high compressive strength $f_{ck}$
  when $f_{ck}$ is 2 grades higher than that required: by 5 mm.
  Exception: no reduction permitted for XC 1.
- Components with a mechanical interlock between precast and in situ concrete:
  $c_{min} \geq 5$ mm in precast component; $c_{min} \geq 10$ mm in in situ component,
  but when using the reinforcement in the as-built condition, the values for cmin in the table apply.
- Appropriate quality control:
  Reductions of, as a rule, 5 mm, corresponding to DBV memo "Concrete cover and reinforcement" are permitted during planning, design, production and construction.

## Concrete – the material

2.1.20 Dimensions, formats and quantities required for masonry of lightweight concrete blocks to DIN 18151 and 18152

|   | 1 | 2 | 3 | 4 | 5 | 6 | 7 | 8 | 9 |
|---|---|---|---|---|---|---|---|---|---|
|   | Dimensions | | | Format | | Quantity/m² [3] | | Quantity/m³ [3] | |
|   | Length[1] [mm] | Width[2] [mm] | Height [mm] | Code | DF | Units [No. off] | Mortar [l] | Units [No. off] | Mortar [l] |
| 1 | 245 | 175 | 238 | 17.5 k | 6 | 16.0 | 17 | 92 | 99 |
| 2 | 370 | | 175 | 17.5 mx | 6¾ | 14.2 | 18 | 81 | 103 |
| 3 | | | 238 | 17.5 m | 9 | 10.7 | 15 | 61 | 85 |
| 4 | 495 | | 175 | 17.5 x | 9 | 10.7 | 17 | 61 | 95 |
| 5 | | | 238 | 17.5 | 12 | 8.0 | 14 | 46 | 77 |
| 6 | 245 | 240 | 175 | 24 kx | 6 | 21.3 | 28 | 89 | 117 |
| 7 | | | 238 | 24 k | 8 | 16.0 | 24 | 67 | 99 |
| 8 | 370 | | 175 | 24 mx | 9 | 14.2 | 25 | 59 | 102 |
| 9 | | | 238 | 24 m | 12 | 10.7 | 21 | 45 | 85 |
| 10 | 495 | | 175 | 24 x | 12 | 10.7 | 23 | 45 | 95 |
| 11 | | | 238 | 24 | 16 | 8.0 | 18 | 33 | 77 |
| 12 | 245 | 300 | 175 | 30 kx | 7½ | 21.3 | 35 | 71 | 117 |
| 13 | | | 238 | 30 k | 10 | 16.0 | 30 | 53 | 99 |
| 14 | 370 | | 175 | 30 mx | 11¼ | 14.2 | 31 | 47 | 102 |
| 15 | | | 238 | 30 m | 15 | 10.7 | 26 | 36 | 85 |
| 16 | 495 | | 175 | 30 x | 15 | 10.7 | 29 | 36 | 95 |
| 17 | | | 238 | 30 | 20 | 8.0 | 23 | 27 | 77 |
| 18 | 245 | 365 | 175 | 36.5 kx | 9 | 21.3 | 43 | 58 | 117 |
| 19 | | | 238 | 36.5 k | 12 | 16.0 | 36 | 44 | 99 |
| 20 | 370 | | 175 | 36.5 mx | 13½ | 14.2 | 37 | 39 | 102 |
| 21 | | | 238 | 36.5 m | 18 | 10.7 | 31 | 29 | 85 |
| 22 | 495 | | 175 | 36.5 x | 18 | 10.7 | 35 | 29 | 95 |
| 23 | | | 238 | 36.5 | 24 | 8.0 | 28 | 22 | 77 |

[1] The lengths given here apply to masonry units laid "brick to brick" (i.e. without mortar to the perpends). For units laid with perpends, lengths decreased by 5 mm (240/365/490 mm) are in accordance with the standard.
[2] Unit width is generally equal to wall thickness
[3] The quantities required are for wall thickness = unit width. The numbers of units are theoretical values without allowance for waste during laying. The figures given for mortar requirements are rough estimates; they include an allowance of 15% for waste during laying.

2.1.21 Types of cement-bound units and prefabricated panels [8]

| 1 | 2 | 3 | 4 | 5 | 6 | 7 | 8 |
|---|---|---|---|---|---|---|---|
| Type of unit | DIN | Code | Grade [N/mm²] | | | | Density class [kg/m³] |
| Hollow wall elements of lightweight concrete | 18148 | Hpl | 2 | – | – | – | 600-1400 |
| Perforated units of lightweight concrete | 18149 | Llb | – | 4 | 6 | 12 | 600-1600 |
| Hollow blocks of lightweight concrete | 18151 | Hbl | 2 | 4 | 6 | – | 500-1400 |
| Solid bricks and blocks of lightweight concrete | 18152 | V Vbl | 2 | 4 | 6 | 12 | 500-2000 |
| Hollow concrete blocks | 18153 | Hbn | – | 4 | 6 | 12 | 1200-1800 |
| Wall elements of lightweight concrete | 18162 | Wpl | Tensile bending strength $\beta_{BZ} \geq 0.8$ | | | | 800-1400 |
| Granulated slag aggregate units, solid and perforated units | 398 | HSV/HSL | – | – | 6 | 12[1] | 1000-2000 |
| Autoclaved aerated concrete blocks and gauged bricks | 4165 | PB | 2 | 4 | 6 | 8 | 400-800 |
| Autoclaved aerated concrete panels and gauged panels | 4166 | PBpl, PPpl | Tensile bending strength $\beta_{BZ} \geq 0.4$ | | | | 500-800 |

[1] Granulated slag aggregate units are also available in grades 20 and 28.

So the properties of a concrete are first specified by the designer when deciding on load-carrying capacity, serviceability and durability. And in a second stage, by the contractor when planning site operations in terms of rational conveying, placing and working of the fresh concrete.

**Proof of conformity**
The construction of a reinforced concrete building almost always results in a relationship of responsibility between the concrete supplier and the contractor, who is ordering, buying and working with the concrete. Quality control employing the principle of verifying conformity and compliance takes this relationship into account.

According to this quality control principle the producer of the concrete verifies the conformity of the production with the specified properties. This is achieved by taking and testing samples at regular intervals and evaluating the results of the tests. Compressive strength and, if applicable, resistance to water penetration are the most important properties requiring verification, as well as other properties. Lightweight concrete must also be checked to ensure that it falls within the specified density class. Every concrete must reliably attain the hardened concrete properties, but at the same time must also satisfy the requirements regarding its fresh concrete properties, primarily its workability, within the permissible deviations of the mix. The producer checks these important properties regularly at the works, evaluates the results and records them. The standard specifies the permissible deviations of individual properties.

Concrete that fulfils the conformity requirements within the specified tolerances is deemed to comply with the specified properties. In the case of unacceptable deviations or if certain properties lie outside the tolerances, the standard specifies measures for modifying the production and thereby guaranteeing the conformity. Concrete produced in a ready-mixed concrete plant is also tested regularly by outside centres in order to guarantee that measures to verify conformity are being carried out properly.

A contractor who orders and buys a designed mix from a ready-mixed concrete supplier can generally assume that the product received on site complies with the specification. The standard therefore only provides for the compressive strength of the concrete, the most important property of hardened concrete, to be checked on site.

The test for compliance is carried out by the contractor, who orders and buys the concrete, by counterchecking the concrete received at the building site. The aim is to confirm that the product supplied agrees with the specified and warranted properties. DIN 1045 part 3 annex A.2 deviates from DIN EN 206 part 1 in that it specifies a testing regime in which besides sampling and frequency of testing it also specifies compliance (identification) criteria for the compressive strength.

**Curing concrete**
Thorough curing for an adequate period of time is indispensable for ensuring that concrete, near the surface as well, attains the intended properties reliably. Curing is crucial to the durability of components and structures. The aim of curing is to ensure that the concrete has enough water during the strength development phase for forming the crystals essential for its strength. In addition, the heat of hydration that ensues during setting should be maintained until the tensile strength required for accommodating internal stresses due to temperature differences has been reached.

DIN 1045 part 3 specifies in detail minimum curing times for concrete in days depending on exposure class and the surface temperature of the component during the strength development phase (table 2.1.17).

The strength development of concrete is expressed as a figure which describes the relationship between the 2-day and the 28-day strengths of samples stored under standardised conditions.

Exceptions to this are concretes of exposure classes X0, XC 1 and XM, whose surface strength has to satisfy lower or higher requirements due to the loads expected. For instance, concretes of exposure class X0 and XC 1, contrary to the provisions in the standard, need only be cured for at least 1/2 day when the temperature of the surface does not drop below +5 °C. Concrete surfaces subjected to abrasion corresponding to exposure class XM must be cured until the strength of the concrete near the surface has reached 70% of the target strength.

Methods of curing are:

- Leaving the concrete in the formwork
- Covering with vapour-tight sheeting
- Applying water-retaining coverings
- Maintaining a visible film of water on the surface of the concrete
- Applying liquid curing agents

These methods may also be combined, e.g. covering with sheeting and simultaneous thermal insulation measures (table 2.1.17). Experience has shown that spraying with water and concurrent excessive temperature differentials can, in certain circumstances, damage the

2.1.22  Dead loads of concrete components, to DIN 1055 part 1 [kN/m³]

| Concrete | Density class [kg/m³] | | | | | | | | | | | Stone density up to 2.7 g/cm³ |
|---|---|---|---|---|---|---|---|---|---|---|---|---|
| | 400 | 500 | 600 | 700 | 800 | 1000 | 1200 | 1400 | 1600 | 1800 | 2000 | |
| Normal-weight concrete with close-grained microstructure to DIN 1045<br>B 5 and B 10<br>B 15 and B 55 | | | | | | | | | | | | 23<br>24 |
| Reinforced concrete of normal-weight concrete with close-grained microstructure to DIN 1045<br>B 15 and B 55 | | | | | | | | | | | | 25 |
| Lightweight concrete with close-grained microstructure to DIN 4219 | | | | | | 10.5 | 12.5 | 14.5 | 16.5 | 18.5 | 20.5 | |
| Lightweight reinforced concrete to DIN 4219 | | | | | | 11.5 | 13.5 | 15.5 | 17.5 | 19.5 | 21.5 | |
| Lightweight concrete with wood chip aggregate | 5 | 6 | 7 | 8 | | | | | | | | |
| Lightweight concrete with no-fines structure to DIN 4232 | | | | | | 10 | 12 | 15 | 16 | 18 | 20 | |
| Reinforced aerated concrete to DIN 4223 | 6.2 | 7.2 | 8.4 | 9.5 | | | | | | | | |

2.1.23  Dead loads for mortar and finishes, to DIN 1055

| Material | Characteristic value [kN/m³] |
|---|---|
| Masonry mortars, plaster and rendering | |
|   cement mortar | 21 |
|   cement-trass mortar | 21 |
|   mortar with plaster binder and masonry cement | 21 |
|   lime-cement mortar | 20 |
|   lime-trass mortar | 20 |
|   lime mortar, lime-gypsum mortar, anhydrite mortar | 18 |
|   gypsum mortar, without sand | 12 |
|   lightweight mortar with lightweight aggregates mortar density ≤ 1 g/cm³ | 10 |
| Floor coverings and wall finishes | |
|   reconstituted stone panels | 24 |
|   ceramic wall tiles (including mortar bed) | 19 |
|   ceramic floor tiles (including mortar bed) | 22 |
|   natural stone slabs (including mortar bed) | 30 |
|   terrazzo | 24 |
|   cement screed | 22 |

2.1.24  Dead loads for masonry of factory-made units, to DIN 1055

| Density [g/cm³] | 0.5 | 0.6 | 0.7 | 0.8 | 0.9 | 1.0 | 1.2 | 1.4 | 1.6 | 1.8 | 2.0 | 2.1 | 2.2 | 2.5 |
|---|---|---|---|---|---|---|---|---|---|---|---|---|---|---|
| Characteristic value [kN/m³] | | | | | | | | | | | | | | |
| normal-weight masonry mortar | 7 | 8 | 9 | 10 | 11 | 12 | 14 | 15 | 17 | 18 | 20 | 21 | 22 | 25 |
| lightweight masonry mortar | 6 | 7 | 8 | 9 | 10 | 11 | 13 | 14 | 16 | 17 | 19 | 20 | 21 | 24 |

# Concrete – the material

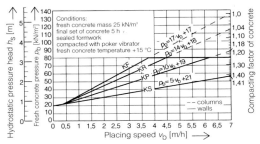

A fresh concrete mass of 25 kN/m³ corresponds to a fresh concrete density of 2500 kg/m³.

2.1.25 Fresh concrete pressure on vertical formwork, to DIN 18218

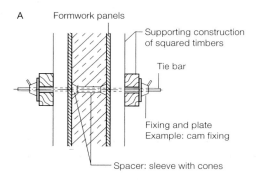

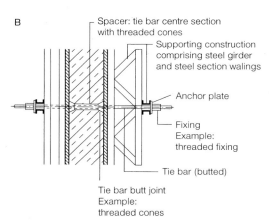

2.1.26 Examples of formwork ties, to DIN 18216

A  Formwork tie with cam fixing
B  Formwork tie with threaded fixing

newly placed concrete because that leads to sudden, partial cooling of the concrete, which in turn gives rise to restraint stresses and hence to the formation of cracks. Other methods of curing are therefore generally preferred.

## Reinforcement

### Concrete cover

The minimum dimensions for concrete cover to reinforcement mainly depend on the ambient conditions to which the component is exposed and the diameter of the reinforcement (table 2.1.19). The dimensions given in the table must be maintained; the nominal dimension is critical. DIN 1045 part 1 specifies minimum values for concrete cover $c_{min}$ [mm] related to the exposure class for corrosion of the reinforcement. In order to ensure that this minimum concrete cover is reliably attained on site, an allowance $\Delta c$ [mm] is allocated to each minimum value. The nominal concrete cover $c_{nom}$ is based on the minimum cover $c_{min}$ [mm] and the allowance $\Delta c$ [mm].

The drawings showing the layout of the reinforcement must include details of the cover $c_v$ to bar nearest concrete surface for the concrete cover to every bar and every layer of reinforcement. This dimension is based on the amount of room in the formwork and the designer should specify a sensible dimension. It may not be less than the nominal concrete cover $c_{nom}$.

When using reinforcing bars of a very large diameter, it should be ensured that the minimum concrete cover $c_{min}$ is not less than the respective bar diameter. Care must be taken in components of lightweight concrete to ensure that the minimum concrete cover is 5 mm larger than the nominal diameter of the largest porous grain of a lightweight aggregate, except for exposure class XC 1. The concrete cover of concrete that is intended to be subject to wear and abrasion can diminish over time. In this case the minimum concrete cover $c_{min}$ for exposure class XM 1 should be increased by 5 mm, for XM 2 by 10 mm and for XM 3 by 15 mm. If a reinforced component is cast against an uneven surface, the allowance must be increased by the height of the unevenness, but not less than 20 mm. The allowance should be increased by 50 mm in the case of reinforced concrete cast directly against a graded soil surface.

The concrete cover should also take into account architectural surface features, e.g. textures, coarse exposed aggregate finishes, and the allowance should be increased accordingly.

The rules on concrete cover must also be adhered to on those reinforced parts that are not relevant to load-carrying capacity and serviceability.

The minimum concrete cover may be reduced by 5 mm if the concrete strength used for a component is at least two grades higher than the minimum compressive strength required for the relevant exposure class.

### Spacing

The clear space between parallel reinforcing bars must be at least 20 mm. It must also be at least equal to the largest bar diameter. If the largest grain diameter of the aggregate $d_g$ is more than 16 mm, the clear spacing between parallel individual bars must be at least $d_g + 5$ mm. During the design it is often forgotten that the actual overall diameter of ribbed bars, i.e. to the outside edges of the ribs, is greater than the nominal diameter. The height of a longitudinal rib for a III S bar (steel grade BSt 420 S) is, for example, 10% of the nominal diameter.

### Vibrator openings

Vibration is the usual method of compacting concrete. The shape and dimensions of the component determine the type of vibration used: vibration tampers, internal/immersion vibrators (poker vibrators, in more confined spaces slimline poker vibrators), external/clamp vibrators (thin components with formwork to both sides), table vibrators for standard components in concrete works, spun concrete facilities for centrosymmetric cross-sections. Openings must be left for inserting an internal vibrator; these should be positioned according to DIN 1045 part 3 when fixing the reinforcement. Bars arranged in separate layers should be placed one above the other with sufficient room to insert a vibrator. The inclusion of vibrator openings may require a different arrangement of the reinforcement and so can influence the design. For example, closely spaced top reinforcement in beams, particularly at intersections over columns, can lead to difficulties in practice.

### Concreting openings

Placing concrete in column and wall formwork at free-fall heights which could lead to segregation of the concrete demands the use of a tremie pipe (DIN 1045 part 3 section 8.5). When inserting tremie pipes or concrete pump lines between reinforcing bars, a space 200 mm in diameter is required. This may alter the arrangement of the reinforcement and that can have an effect on the design as well as the cross-section of the component; such details must be taken into account during design.

## Concrete masonry

Many different types of cement-bound masonry units are available (table 2.1.20). We distinguish between hollow blocks, solid bricks, solid blocks, hollow wall elements and wall elements. The units are mostly made of lightweight concrete with lightweight aggregates, but can also be produced from normal-weight concrete and autoclaved aerated concrete. Depending on the type of unit, hollow blocks have cells and solid blocks have slots, arranged in rows. Grip aids and openings may be provided to facilitate easier handling during laying.

Dimensions, density and strength represent further classification features for masonry units (table 2.1.21). The thermal insulation properties of the units vary greatly and depend mainly on the aggregates, the type of sand used, the voids (cells, slots), the proportion of joints and the type of mortar used. The reader is referred to the specialist literature for details of the thermal insulation, sound insulation and fire resistance of masonry made from lightweight concrete units. [11]

## Application

### Loading assumptions

The values for determining the dead loads of concrete components can be found in DIN 1055 part 1 "Design loads for buildings: stored materials, building materials and structural members, dead load and angle of friction". These are used in analysing the structure for the ultimate and serviceability limit states.

### Concrete

The oven-dry density of concrete essentially depends on the type of aggregate used (table 2.1.3). For lightweight concrete this is max. 2000 kg/m³, for normal-weight concrete between 2000 and 2600 kg/m³, and for heavy concrete more than 2600 kg/m³. Table 2.1.22 provides a summary of characteristic values for the dead loads of components made from in situ and precast concrete. Generally, the figures should be increased by 1 kN/m³ for fresh concrete.

### Mortar

Table 2.1.23 contains characteristic values for masonry mortar, plaster and rendering, as well as floor coverings and wall finishes.

### Masonry

The characteristic values to DIN 1055 part 1 given in table 2.1.24 apply to masonry of factory-made masonry units (solid bricks and blocks, hollow blocks etc.).

2.1.27 Equivalent loads for vertical imposed loads when concreting, to DIN 1055

| Means of conveying concrete | Equivalent load [kN/m²] |
|---|---|
| Conveyor belt, hoist, barrow | ≥ 1.5 |
| Crane skip or other similar plant | |
| 250 l | ≥ 2.5 |
| 500 l | ≥ 5.0 |
| 1000 l | ≥ 10.0 |
| 1500 l | ≥ 15.0 |
| Concrete pump | ≥ 5.0 |

Other values may be used when an accurate analysis is carried out.

2.1.28 Permissible loads on formwork ties with wedge or cam fixing or combined wedge/cam fixing, to DIN 18216

| Tie dia. [mm] | Area A [cm²] | Permissible load F [kN] |
|---|---|---|
| 6 | 0.28 | 3 |
| 8 | 0.50 | 6 |
| 10 | 0.79 | 10 |
| 12 | 1.13 | 14 |
| 14 | 1.54 | 20 |
| 16 | 2.01 | 25 |

Non-circular tie cross-sections are to be allocated to the next lower line in the table according to their smallest loadbearing cross-sectional area.

2.1.29 Permissible loads on formwork ties with non-standardised nuts and tie bars of grade St 37-2 or St 37-3 steel, to DIN 18216

| Thread diameter[1] [mm] | Permissible load F [kN] |
|---|---|
| 10 | 6 |
| 12 | 8 |
| 16 | 16 |
| 20 | 25 |
| 24 | 35 |
| 27 | 45 |

[1] For standard threads to DIN ISO 898 part 1 or with a corresponding core cross-section.

2.1.30 Modulus of elasticity for hydrated cement, aggregate and concrete (after Bonzel) [10]

| Modulus of elasticity | [N/mm²] |
|---|---|
| $E_z$ of hydrated cement | 5000–20 000 |
| $E_g$ of concrete aggregate | 20 000–100 000 |

## Concrete – the material

2.1.31 Strength and deformation coefficients (moduli of elasticity) of normal-weight concrete, to DIN 1045 part 1

| $f_{ck,cube}$ | 15 | 20 | 25 | 30 | 37 | 45 | 50 | 55 | 60 | 67 | 75 | 85 | 95 | 105 | 115 | N/mm² |
|---|---|---|---|---|---|---|---|---|---|---|---|---|---|---|---|---|
| $E_{cm}$ | 25.8 | 27.4 | 28.8 | 30.5 | 31.9 | 33.3 | 34.5 | 35.7 | 36.8 | 37.8 | 38.8 | 40.6 | 42.3 | 43.8 | 45.2 | kN/mm² |
| | $E_{cm} = 9.5(f_{ck} + 8)^{1/3}$ ||||||||||||||||

2.1.32 Strength and deformation coefficients (moduli of elasticity) of lightweight concrete, to DIN 1045 part 1

| $f_{ck,cube}$ | 13 | 18 | 22 | 28 | 33 | 38 | 44 | 50 | 55 | 60 | 66 | N/mm² |
|---|---|---|---|---|---|---|---|---|---|---|---|---|
| $E_{cm}$ | $E_{cm} = 9.5(f_{ck} + 8)^{1/3} \times (\varphi/2200)$, with $\varphi$ = dry gross density of concrete |||||||||||| |

2.1.33 Coefficients of thermal expansion for concrete [10]

| Material | Coefficient of thermal expansion $\alpha_T$ | |
|---|---|---|
| | in mm/mK | in mm/mmK |
| Normal-weight concrete, reinforced concrete, masonry of lightweight concrete units | 0.010 | $1.0 \times 10^{-5}$ |
| Lightweight concrete, lightweight reinforced concrete, masonry of aerated concrete units | 0.008 | $0.8 \times 10^{-5}$ |

2.1.34 Final creep coefficients and final shrinkage strains for concrete components, 1) and 2) to DIN 4227

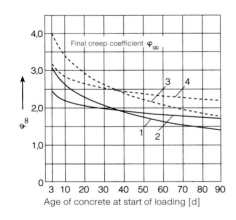 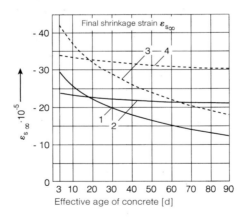

| Line | Location of component | Thickness of concrete $d_m = 2A/U$ [3] |
|---|---|---|
| 1 | damp, in the open air | small (≤ 100 mm) |
| 2 | (relative humidity ≈ 70%) | large (≥ 800 mm) |
| 3 | dry, in interior conditions | small (≤ 100 mm) |
| 4 | (relative humidity ≈ 50%) | large (≥ 800 mm) |

[1] The table and associated diagrams apply to consistency range KP. The figures should be decreased by 25% for consistency range KS and increased by 25% for consistency range KR. The initial consistency may be used when employing superplasticisers.

[2] The table and associated diagrams apply to concrete that hardens at normal temperature and that uses cement of strength class 32.5 R and 42.5 R. The effect of creep on cement with a slow set (32.5 and 42.5) can be taken into account by taking the standard values for half the age of the concrete at start of loading, or for cement with a very rapid set (52.5 and 52.5 R) by taking 1.5 times the value.

[3] A = area of concrete cross-section; u = perimeter of component exposed to the atmosphere.

### Formwork

Horizontal formwork panels and their supports must be designed for the load of the fresh concrete to be placed plus the embedded reinforcement. The dead load of fresh or hardened reinforced concrete is taken as 25 kN/m³. An equivalent load is used to cover all other loads that occur during concreting operations. The magnitude of this equivalent load should be based on the method of conveying the concrete and the nominal content of such plant, e.g. a crane skip (table 2.1.27).

Vertical formwork should be designed for the horizontal pressure of the fresh concrete. This depends on the speed of placing the concrete and its consistency. Figure 2.1.25 is a diagram for determining the pressure of fresh concrete according to DIN 18218. When using this standard published in 1980, it should be borne in mind that some terms have been changed in the new concrete standard, and that concretes with fluid and very fluid consistencies are now available. A mound size exceeding 600 mm in the consistency test exceeds the range of applicability of the standard. In this case the formwork should be designed to accept the full hydrostatic concrete pressure, unless experience dictates otherwise.

This also applies in principle to the permissible loads on formwork ties, which are summarised in tables 2.1.28 and 2.1.29 according to type (figure 2.1.26).

### Permissible stresses

The safety plan of DIN 1045 part 1 compares the actions exerted on a component against the resistance of that component. The prescribed methods of analysis for the serviceability and ultimate limit states simulate design situations in which the actions exerted on the component and its resistance are increased and decreased respectively by means of partial safety factors. These design models are used to verify load-carrying capacity and serviceability. The maximum permissible stresses in concrete and steel are limited when verifying the serviceability limit state. DIN 1045 part 1 limits the compressive stresses in concrete components for exposure classes XD, XF and XS, when no constructional measures are provided to increase the permissible stresses, to max. $0.6 f_{ck}$. However, the maximum concrete compressive stress for the corresponding loading case is limited to $0.45 f_{ck}$ for components whose serviceability, load-carrying capacity or durability is substantially influenced by irreversible deformations under load (creep). The stress in the reinforcement is to be limited to $0.8 f_{yk}$ when subjected to the direct effects of external actions. If the stress is xclusively due to indirect effects (restraint), a maximum value of $1.0 f_{yk}$ is permissible.

## Deformations

Loads, external actions or internal processes cause deformations in fresh and hardened concrete which may be reversible or irreversible. Besides the nature, magnitude and duration of the action, the degree of actual strength-development and the moisture content of the concrete, these deformations are affected to a great extent by the properties of the aggregate, which accounts for about 70% of the volume of concrete.

Elastic deformations
The modulus of elasticity of concrete – the measure of the elastic behaviour of concrete under uniaxial stresses – results from the modulus of elasticity of the hydrated cement and the modulus of elasticity of the aggregate. It can be estimated from the volume proportions (table 2.1.30). Corresponding characteristic values for the moduli of elasticity of normal-weight and lightweight concrete are given in the rules governing reinforced and prestressed concrete (tables 2.1.31 and 2.1.32). The same modulus of elasticity may be used for compression and tension in all analyses to verify serviceability.

Thermal expansion and curvature
A rise in temperature increases and a drop in temperature reduces the volume of a building material. This results in an elongation or shortening respectively, measured by the change in length $\Delta l$:

$$\Delta l = \pm \alpha_T \cdot \Delta T \cdot l$$

where:
$\Delta l$ = change in length as elongation (+) or shortening (-) of component [mm]
$\alpha_T$ = coefficient of thermal expansion based on the ratio of the temperature strain $\varepsilon_T$ to the associated temperature change $\Delta T$ [mm/mmK]
$\Delta T$ = rise in temperature (+) or drop in temperature (-) [K]
$l$ = original length of component [mm]

Coefficients of thermal expansion for normal-weight concrete and lightweight concrete according to DIN 1045 part 1 are given in table 2.1.33. Temperature curvature happens with uneven temperature loads. The deformation can be expressed as an index f, which is calculated from the thickness d and the length l of a component, as well as the temperature difference $\Delta T$ between the top and bottom faces:

$$f = \alpha_T \cdot \Delta T / d \cdot l^2 / 8$$

Creep and shrinkage
Creep is the term for the increase in deformation over time of a component subjected to a permanent stress. Shrinkage denotes the shortening of a component while it is drying out.

Verification of creep and shrinkage is not usually required for masonry and reinforced concrete constructions. By contrast, the effects of creep and shrinkage must be taken into account in prestressed concrete work as the critical section sizes or stresses may otherwise be altered very unfavourably.

Creep and shrinkage of concrete depend primarily on the following influences:

- the humidity of the surrounding air
- the dimensions of the component
- the concrete mix

In addition, creep is also affected by the following factors:

- the degree of actual strength-development of the concrete at the onset of the action (degree of maturity)
- the duration and magnitude of the action

As the effects of creep and shrinkage only need to be taken into account for the final condition (time $t = \infty$), even for prestressed concrete construction, the final creep coefficients and final shrinkage strains given in table 2.1.34 may be used for simplicity in the analyses.

Total deformation
It is often necessary to estimate the magnitude of the total anticipated deformation of a component in order to establish the necessary constructional measures. For stresses due to axial forces resulting from temperature loads, creep and shrinkage, total deformation is:

$$\Delta l = l \cdot [\pm \alpha_T \cdot \Delta T \pm (1+\varphi_\infty) \cdot \sigma_0/E_b \pm \varepsilon_{S\infty}]$$

where:
$\Delta l$ = change in length [mm] (shortening -, elongation +)
$l$ = length of component [mm]
$\alpha_T$ = coefficient of thermal expansion of concrete to table 2.1.33 [mm/mmK]
$\Delta T$ = temperature difference [K] (rise +, drop -)
$\sigma_0$ = constant concrete stress [N/mm$^2$] (compression -, tension +)
$E_b$ = modulus of elasticity of concrete [N/mm$^2$]
$\varphi_\infty$ = final creep coefficient of concrete from figure 2.1.34
$\varepsilon_{S\infty}$ = final shrinkage strain of concrete from figure 2.1.34 (water absorption +, drying out -)

The stress-dependent deformations (load +, creep -) generally amount to about 3/4 of the total deformation. Those not related to the stress (shrinkage +, temperature -) on the other hand, only 1/4 of the total deformation.

## References

### Books and articles

[1] Bauteilkatalog – Planungshilfe für dauerhafte Betonbauteile nach der neuen Normengeneration, Bauberatung Zement (ed.), pub. by Bundesverband der Deutschen Zementindustrie, Cologne; Verlag Bau+Technik GmbH, Düsseldorf 2001
[2] Bayer, Edwin; Kampen, Rolf: Beton-Praxis: Ein Leitfaden für die Baustelle, pub. by Bundesverband der Deutschen Zementindustrie, Cologne; Verlag Bau+Technik GmbH, Düsseldorf 1999, 8th, revised edition
[3] Beton-Herstellung nach Norm – Die neue Normengeneration, Bauberatung Zement (ed.), pub. by Bundesverband der Deutschen Zementindustrie, Cologne; Verlag Bau+Technik GmbH, Düsseldorf 2001, 13th, revised edition
[4] Beton-Kalender, editions 1-43, Verlag Wilhelm Ernst & Sohn, Berlin 1958-2001
[5] Beton-Prüfung nach Norm, Bauberatung Zement (ed.), pub. by Bundesverband der Deutschen Zementindustrie, Cologne; Beton-Verlag GmbH, Düsseldorf 1996, 10th, revised edition
[6] Betontechnische Berichte, Forschungsinstitut der Zementindustrie (ed.), Düsseldorf; pub. by Gerd Thielen, Verlag Bau+Technik GmbH, Düsseldorf 1960-2001
[7] Beton+Fertigteil-Jahrbuch, editions 1-49, Bauverlag GmbH, Wiesbaden/Berlin 1952-2001
[8] Brandt, Jörg; Moritz, Helmut: Bauphysik nach Maß – Planungshilfen für Hochbauten aus Beton, pub. by Bundesverband der Deutschen Zementindustrie, Cologne; Beton-Verlag GmbH, Düsseldorf 1995
[9] Iken, Hans; Lackner, Roman; Zimmer, Uwe: Handbuch der Betonprüfung, Beton-Verlag GmbH, Düsseldorf 1994, 4th, revised edition
[10] Kind-Barkauskas, Friedbert: Beton Atlas; with: Jörg Brandt, Frieder Kluckhohn, Rudolf Krieger, Gottfried Lohmeyer, Helmut Moritz; pub. by Bundesverband der Deutschen Zementindustrie, Cologne; Beton-Verlag GmbH, Düsseldorf 1980/vol. 1, 1984/vol. 2
[11] Mauerwerk-Kalender – Taschenbuch für Mauerwerk, Wandbaustoffe, Schall-, Wärme- und Feuchtschutz, Peter Funk (ed.), Verlag Wilhelm Ernst & Sohn, Berlin 1975-2001
[12] Pauser, Alfred: Beton im Hochbau – Handbuch für den konstruktiven Vorentwurf, Verlag Bau+Technik GmbH, Düsseldorf 1998
[13] Weber, Robert; Tegelaar, Rudolf: Guter Beton – Ratschläge für die richtige Betonherstellung, pub. by Bundesverband der Deutschen Zementindustrie, Cologne; Verlag Bau+Technik GmbH, Düsseldorf 2001, 20th, revised edition
[14] Weigler, Helmut; Iken, Hans-W.; Lackner, Roman R.: Europäische Regel für Beton, Beton-Verlag GmbH, Düsseldorf 1993
[15] Wendehorst, R.: Baustoffkunde, pub. by D. Vollenschaar, Curt R. Vincentz Verlag, Hanover 1998, 25th, revised edition
[16] Wesche, Karlhans: Baustoffkunde für tragende Bauteile – Bd. 2 Beton, Bauverlag GmbH, Wiesbaden/Berlin 1993, 3rd, revised edition
[17] Zement-Merkblätter. Bearb.: Bauberatung Zement, pub. by Bundesverband der Deutschen Zementindustrie, Cologne; Verlag Bau + Technik GmbH, Düsseldorf 1954-2001
[18] Zement-Taschenbuch, editions 1-49, pub. by Verein Deutscher Zementwerke, Düsseldorf; editions 1-48, Bauverlag GmbH, Wiesbaden/Berlin 1911-1950 (Zementkalender) 1952-1984; edition 49, Verlag Bau+Technik GmbH, Düsseldorf 2001

### Standards and directives

DIN 398 Granulated slag aggregate concrete blocks; solid, perforated, hollow blocks, edition 1976
DIN 1045 Concrete, reinforced and prestressed concrete structures
  part 1: Design, edition 2001
  part 2: Concrete; specification, properties, production and conformity, edition 2001
  part 3: Execution of structures, edition 2001
  part 4: Additional rules for the production and conformity control of prefabricated elements, edition 2001
DIN 1048 Testing concrete
  part 1: Testing of fresh concrete, edition 1991
  part 2: Testing of hardened concrete (specimens taken in situ), edition 1991
  part 4: Determination of the compressive strength of hardened concrete in structures and components, edition 1991
  part 5: Testing of hardened concrete (specimens taken in mould), edition 1991
DIN 1053 Masonry
  part 1: Design and construction, edition 1996
  part 2: Masonry strength classes on the basis of suitability tests, edition 1996
DIN 1055 Design loads for buildings
  part 1: Stored materials, building materials and structural members, dead load and angle of friction, edition 1978
  part 2: Soil characteristics, specific weight, angle of friction, cohesion, angle of wall friction, edition 1976
  part 3: Live loads, edition 1971
  part 4: Imposed loads, wind loads on buildings not susceptible to vibration, edition 1986
  part 5: Live loads, snow load and ice load, edition 1975
DIN 1164 Cement with special properties – composition, requirements, proof of conformity, edition 2000
DIN 4030 Assessment of water, soil and gases for their aggressiveness to concrete
  part 1: Principles and limiting values, edition 1991
  part 2: Collection and examination of water and soil samples, edition 1991
DIN 4103 Internal non-loadbearing partitions
  part 1: Requirements, testing, edition 1984
DIN 4108 Thermal protection and energy economy in buildings
  part 1: Thermal insulation in buildings; quantities and units, edition 1981
  part 2: Minimum requirements for thermal insulation, edition 2001
  part 3: Protection against moisture subject to climate conditions; requirements and directions for design and construction, edition 2001
DIN 4165 Autoclaved aerated concrete blocks and flat elements, edition 1996
DIN 4166 Autoclaved aerated concrete slabs and panels, edition 1997
DIN 4172 Modular coordination in building construction, edition 1955
DIN 4211 Masonry cement – specifications, control, edition 1995
DIN 4226 Aggregates for concrete and mortar
  part 1: Aggregates of normal and dense structure (heavy aggregates), edition 2001
  part 2: Aggregates of porous structure (lightweight aggregates), draft 2000
  part 3: Testing of heavy and lightweight aggregates, edition 1983
  part 4: Inspection, edition 1983
  part 100: Recycled aggregates, draft 2000
DIN 4227 Prestressed concrete
  part 1: Structural members made of normal-weight concrete, with limited concrete tensile stresses or without concrete tensile stresses, draft 1999
DIN 4232 No-fines lightweight concrete walls; design and construction, edition 1987
DIN 4235 Compacting of concrete by vibrating
  part 1: Vibrators and vibration mechanics, edition 1987
  part 2: Compacting by internal vibrators, edition 1987
  part 3: Compacting by external vibrators during the manufacture of precast components, edition 1978
  part 4: Compacting of in situ concrete by formwork vibrators, edition 1978
  part 5: Compacting by surface vibrators, edition 1978
DIN 18000 Modular coordination in building, edition 1984
DIN 18148 Lightweight concrete hollow boards, ed. 2000
DIN 18151 Lightweight concrete hollow blocks, ed. 1987
DIN 18152 Lightweight concrete solid bricks and blocks, edition 1987
DIN 18153 Normal-weight concrete masonry units, edition 1989
DIN 18162 Lightweight concrete wall-boards, unreinforced, edition 2000
DIN 18202 Dimensional tolerances in building construction – buildings, edition 1997
DIN 18216 Formwork ties; requirements, testing, use, edition 1986
DIN 18217 Concrete surfaces and formwork surfaces, edition 1981
DIN 18218 Pressure of fresh concrete on vertical formwork, edition 1980
DIN 18330 Contract procedures for building works
  part C: General technical specifications for building works; masonry works, edition 2000
DIN 18331 Contract procedures for building works
  part C: General technical specifications for building works; concrete and reinforced concrete works, edition 2000
DIN EN 196 Methods of testing cement
  part 1: Determination of strength, edition 1995
DIN EN 197 Cement
  part 1: Composition, specifications and conformity criteria of common cements, edition 2001
  part 2: Conformity evaluation, edition 2000
DIN EN 206 Concrete
  part 1: Specification, performance, production and conformity, edition 2001
DIN V ENV 1992 Eurocode 2; Design of concrete structures
  part 1-1: General rules and rules for buildings, edition 1992

DAfStb-Richtlinien (German Reinforced Concrete Committee directives):
Alkalireaktion; vorbeugende Maßnahmen gegen schädliche Alkalireaktion im Beton (alkaline reactions; preventive measures against damaging alkaline reactions in concrete), edition 1997
Beton mit rezykliertem Zuschlag (concrete with recycled aggregates)
  part 1: Betontechnik (concrete technology), ed. 1998
  part 2: Betonzuschlag aus Betonsplitt und Betonbrechsand (concrete aggregates of concrete chippings and crushed concrete), edition 1998
Beton mit verlängerter Verarbeitungszeit (Verzögerter Beton); Eignungsprüfung, Herstellung, Verarbeitung und Nachbehandlung (concrete with prolonged working time [retarded concrete]; suitability test, production, handling and curing), edition 1995
Fließbeton; Herstellung, Verarbeitung und Prüfung (flowable concrete; production, handling and testing), edition 1995
Hochfester Beton (high-strength concrete), edition 1995
Anwendung Europäischer Normen im Betonbau; Richtlinien zur Anwendung von Eurocode 2; Planung von Stahlbeton- und Spannbetonbauwerken (application of European standards in concrete construction; guidelines for the application of Eurocode 2; design of reinforced and prestressed concrete structures), edition 1995
Schutz und Instandsetzung von Betonbauteilen (protection and maintenance of concrete components)
  part 1: Allgemeine Regelungen und Planungsgrundsätze (general rules and design principles), draft 2000
  part 2: Bauprodukte und Anwendung (building products and applications), draft 2000
  part 3: Anforderungen an die Betriebe und Überwachung der Ausführung (requirements for contractors and monitoring of workmanship), draft 2000
Nachbehandlung von Beton (curing of concrete), ed. 1984

# The concrete surface

**Friedbert Kind-Barkauskas**

## Design concepts

A structure is created by the combination and interaction of different materials, each of which possesses its own particular architectural language. The appearance and the characteristics of the processing and mouldability of materials are crucial for the overall impression of individual components and the whole structure. It is the task of the architect to incorporate each material according to its characteristics, to recognise its special merits and to allow these to become evident in use. Combinations of materials can considerably enlarge the design repertoire and enable particularly interesting aesthetic effects.

The formwork moulds the concrete, which is plastic and takes on any desired shape until it has set. Design using this material demands a great deal of expertise and imagination from architects and engineers; but also self-restraint – for the possibilities that this versatile material offers them are almost infinite.

To use a material correctly means not only exploiting its structural possibilities but also its architectural options. Surface textures, shiny – matt, smooth – rough, coarse – fine, to name just a few alternatives, are specific to the material. They can express different materials or certain aspects of a single material. Textures can reinforce or dilute or conceal. They determine the overall visual impression, which can be varied further by using colour. Like texture, colour can also exploit the light-dark contrast. It is only the effects of light and shade that can bring a wall to life and create an interesting overall visual impression.

It is also possible to achieve surface textures in relation to the whole structure by giving certain details of the construction, e.g. joints, a special architectural emphasis. However, the selected individual texture of the design element must always be harmonised with the appearance of the surrounding structure, and very often also with its urban effect.

## The constituents of the concrete mix

Fresh concrete is concrete that can still be worked as a plastic mixture. It contains various materials in different proportions, which can have consequences for the appearance of the hardened concrete, too. Therefore, in trying to achieve a certain concrete surface, special care is necessary when choosing suitable raw materials and additives for a concrete mix.

### The effects of different cements

The natural colour of the hydrated cement is primarily determined by the colour of the cement and is also responsible for the appearance of the finished concrete. This is, however, not an indication of quality but rather the outcome of the raw materials used, the type of cement, the grinding fineness and the method of production. Certain colour fluctuations are therefore also possible among cements of the same strength class from the same supplier. The concrete mix and method of working have an even more pronounced influence on the colour of the hardened concrete. For example, local, limited fluctuations in the water/cement ratio, variations in the grading of the aggregate, formwork with different absorption rates, the release agents used and the varying degree to which the fresh concrete is compacted.

The dark grey shade of ordinary Portland cement (CEM I) is mainly due to the relatively high iron oxide content. This cement is produced by grinding Portland cement clinker to a fine consistency and adding gypsum and/or anhydrite plus, if required, inorganic mineral substances. The addition of finely ground cinder sand results in the somewhat lighter-coloured Portland blastfurnace (CEM II/A-S and B-S) and blastfurnace (CEM III/A and B) cements. Besides cement clinker, these generally contain 6-80% by wt cinder sand. Portland composite cement (CEM II/B-SV) contains up to 20% by wt cinder sand plus 10-20% by wt pulverised fuel ash (PFA). The production of white Portland cement requires the use of raw materials low in iron oxide (limestone and china clay) and also a decrease in the amount of colouring constituents, mainly calcium alumino ferrite, in the production process. Portland burnt shale cement (CEM II/A-T and B-T), covered by DIN 1164 and DIN EN 196, also has a reddish colour. This cement contains 6-35% by

# The concrete surface

(a) taller

(b) lower

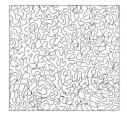
(c) two-dimensional

(d) material texture

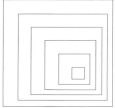
(e) depth

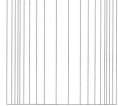
(f) curvature

2.2.1  Graphic effects produced by formwork textures

wt oil shale combustion residue as well as cement clinker. It can be used like the other standardised cements without restrictions.

**The use of different aggregates**
Most of the aggregates used in concrete are obtained naturally as rounded material from rivers or gravel pits. Quarries supply materials broken down into irregular shapes. Both types can occur as sand, gravel or grit. The effects of the natural colours of the fine particles and the coarser grains depend on the surface finish of the concrete. Only the finest grains affect the appearance of the concrete in the case of "as-struck" surfaces (i.e. without any further treatment). The reason for this is that the actual concrete microstructure is concealed behind a thin covering layer composed of cement and the ultra-fine grains of aggregate. On treated concrete surfaces the appearance is affected by the natural colours of both the coarse and fine grains of aggregate. In Germany virtually all types of aggregate in the most diverse colours can be obtained from indigenous natural sources. The interesting colours of limestone, quartz, granite and porphyry make these aggregates popular for the production of fair-face concrete.

Aggregates for normal-weight concrete must comply with DIN 4226 part 1. Their oven-dry density is generally between 2600 and 2900 kg/m$^3$. We distinguish between natural aggregates with rounded and broken grains, and factory-made, mineral aggregates. The strength of the aggregate must allow the production of the customary grades of concrete. Adequate resistance to frost and de-icing salts is the most important property for exposed concrete surfaces.

Aggregates for lightweight concrete must comply with DIN 4226 part 2. Their oven-dry density may not exceed 2200 kg/m$^3$. The raw materials are pumice, foamed slag, expanded clay and expanded shale. The surfaces of components made of lightweight concrete are not usually worked after hardening. The texture of the formwork alone has an effect here – which demands very thorough advance planning in the case of fair-face concrete surfaces.
A glazed or opaque colour coating may be applied subsequently for architectural purposes.

**The use of coloured pigments**
It is very easy to give concrete a coloured finish with the help of coloured pigments. Iron oxide pigments are preferred primarily for red, yellow, brown and black shades, chromic oxide and hydrated chromium sequioxide pigments for green colours, and pigments based on mixed crystals, e.g. cobalt-aluminium-chromium oxide, for blue. Colouring the concrete in this way is permanent and weather-resistant. The use of grey cement results in subdued, darker colours, whereas white cement produces brighter, purer colours. A light surface texture intensifies the colour (figure 2.2.2).

## The effects of formwork

After hardening, concrete shares the properties and appearance of stone. An impression of the formwork remains visible on the surface. This can be a rough-sawn or specially designed board texture, a smooth or patterned surface. The variations in the visible appearance that the formwork can lend the concrete are virtually infinite.

**Formwork textures**
A concrete surface is moulded each time in a very particular way by the formwork employed. Different types of formwork and different materials lead to different patterns (table 2.2.11). Besides the actual surfaces, it is mainly the joints between the individual formwork panels which are conspicuous. These can be arranged as a continuous line or offset depending on how the formwork and planning grid is coordinated. Large steel, timber and plastic panels naturally create completely different grid effects to those of formwork made up of small elements or individual boards. Modern proprietary formwork systems employ framed standard panels, mostly of multi-ply boards, measuring, for example, 2.5-3.3 m high x 1.2/1.25-2.4/2.5 m wide. Casings with film-coated plywood panels measuring 2.5 x 5.2 m are available in the customary standard sizes.

The positions of formwork ties, which remain visible in the finished concrete surface, are predefined in the case of framed formwork panels owing to their steel construction. Using casings they can be planned in accordance with both structural and architectural requirements.

With timber formwork the method of working, e.g. rough-sawn or planed, plays a major role in the appearance of a fair-face concrete surface (figures 2.2.3-2.2.7).

The visual effect of formwork textures is generally based on exploiting particular graphic effects. Figure 2.2.1 illustrates a number of basic patterns. Closely spaced vertical lines (a) make a surface appear taller and narrower; widely spaced horizontal lines (b) reduce the height and make a surface appear longer; irregular textures (c) produce a two-dimensional, wallpaper effect; and textures reminiscent of materials (d) lead to associations with those materials. Certain arrangements of lines can give depth to a flat surface (e), while other arrangements convey the impression of curvature (f).

## The effects of formwork

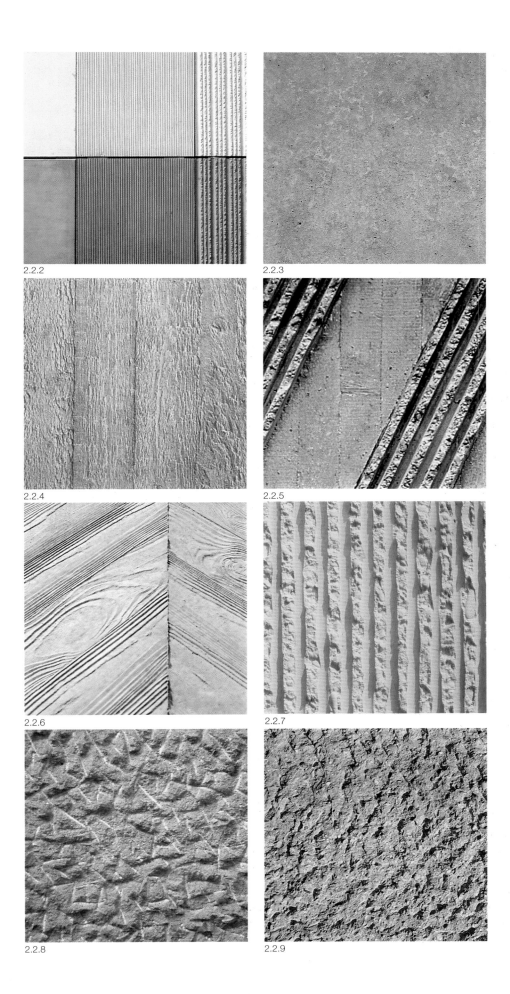

2.2.2   Coloured concrete with 2% iron oxide yellow, with white and grey cement

2.2.3   Flat concrete surface, Betoplan formwork, grey cement

2.2.4   Rough-sawn board finish, non-planed boards, grey cement

2.2.5   Proprietary formwork, Reckli textured formwork No. 1/21 Malta, rough-sawn board finish, Portland burnt shale cement

2.2.6   Proprietary formwork, timber texture, Reckli textured formwork No. 2/23 Alster, grey cement

2.2.7   Proprietary formwork, stone texture, Reckli textured formwork No. 2/30 Havel, white cement

2.2.8   Embossed concrete finish, limestone aggregate, grey cement

2.2.9   Point-tooled concrete finish, Rhine gravel aggregate, grey cement

# The concrete surface

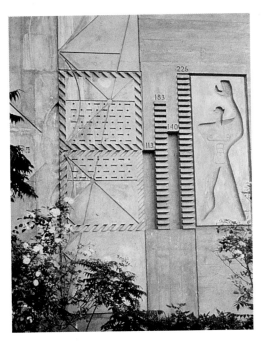

2.2.10  Joints, ornamentation and relief used to decorate a surface

Besides surface textures dependent on the material, architectural ornamentation can also be used to achieve special effects. Such ornamentation is able to "dissolve" surfaces and seemingly "erase" those surfaces from the overall picture. Ornamentation mediates between the simple geometric forms of the building components and the imagination of the observer.

Artistic ornamentation usually entails the use of depth, the third dimension, to create reliefs. These can break up and animate large areas and so lend the respective component an unmistakable appearance. The creation of ornamentation and reliefs in concrete surfaces generally requires particularly elaborate formwork. It is just as important here to provide the necessary concrete cover to the reinforcement as it is to take care when striking such complicated shapes (figure 2.2.10).

### Requirements to be met by formwork

According to DIN 18217 a concrete surface which is to remain visible must satisfy certain requirements regarding its appearance. Fair-face concrete therefore demands very careful preparation of the formwork as well as the concrete itself. In particular, work on the building site plays a vital role. The formwork specification calls for:

- plumbed, uniform, essentially sealed surfaces
- a more or less even colouring within each discrete area
- dimensional accuracy within the specified tolerances
- inconspicuous and properly formed construction joints
- proper positioning of formwork ties

As the setting of the concrete is a chemical process, the material of formwork should neither influence this process nor be influenced by it. Timber in the various forms in which it is used as well as steel and plastic fulfil these requirements. The cleanliness of the inside of the forms is particularly important when high demands are being placed on fair-face concrete surfaces. A tightly jointed form prevents the cement paste from leaking out during placing and compacting the concrete. This in turn avoids discoloration, dusting and streaking. The method of joining the boards should therefore be laid down in the tender documents (table 2.2.12). Timber formwork must be prewetted prior to use so that the boards swell up and thereby seal the joints. However, this effect must be taken into account when assembling the formwork because otherwise distortion can occur.

For economic reasons, conventional formwork made of timber boarding is supplemented by framed formwork panels or casings. Critical here is the dimensional accuracy and an accurate fit between the standard elements used.

### The effects of release agents

Release agents are employed mainly to ensure that the formwork can be detached from the surface of the concrete without damage and to protect the material of the formwork. They can help to avoid most of the damage that can occur upon striking the forms, and improve the appearance of exposed surfaces (table 2.2.13). Blemishes and varying shades of grey on fair-face concrete surfaces are often the result of inadequate application of the release agent. For instance, applying waxes and pastes – which subsequently harden – by hand can lead to marks caused during the application remaining visible on the finished concrete surface. It is therefore better to use mechanical plant, e.g. floor polishers, buffing wheels, to apply such release agents.

Spraying apparatus is used for liquid release agents in order to achieve a more even coating.

### Finishing the concrete surface

An intrinsic characteristic of concrete components is their ability to be treated so that they resemble natural stonework. Treatment is carried out on site on the finished structure but above all during the fabrication of precast concrete components according to the definitions in DIN 18500. This standard applies to precast concrete components with an imitation stone or other special finish. The methods of treatment also include working the fresh concrete surface with, for example, brushes or rollers.

The term "reconstituted stone" generally applies to any components made from plain or reinforced concrete which have been worked to give the impression of natural stone.

The various methods for working the surface of the concrete involve exposing the aggregate to a greater or lesser degree. This results in great variations in colour. Different techniques lead to differentiated refraction of the light on the surface of the grains and thus cause a surface to appear either lighter or darker. The aggregate accounts for more than 80% of the colour of an exposed-aggregate surface. The remaining areas of hydrated cement owe their colour to the ultra-fine particles of the cement used or, additionally, coloured pigments.

The requirements of DIN 1045 and DIN EN 206 part 1 are valid for the concrete cover to reinforcement beneath special concrete surfaces. The dimension for the concrete cover is the smallest distance between the surface of the reinforcement and the finished surface of the component, i.e. the deepest point of the treatment for an exposed aggregate finish. Combinations of various treatments are possible by taking into account the necessary cover allowance, e.g. grinding plus sandblasting (4 mm + 2 mm = 6 mm). [2] [23]

**Manual treatments**
Manual techniques for working the surface like a stonemason remove the top layer of cement from the concrete.

This produces a rough, brighter surface on which the grains of aggregate, some broken, are visible. The use of white cement, coloured aggregates or coloured pigments with this type of surface finish leads to special effects which are then complemented by the effects of light and shade. [2] [23]

Embossing
According to DIN 18500 an embossed surface is one which has been worked with a stonemason's hammer or bolster chisel to a depth of about 5-6 mm. The stonemason's hammer is generally used for rough working of a stone surface and can be used only with relatively soft concrete, e.g. with limestone aggregate of low strength. This is a type of bolster chisel for working the edges of the component (figure 2.2.8).

Point tooling
According to DIN 18500 a point-tooled surface is one which has been worked with a pointed chisel to a depth of about 5-10 mm. This technique involves working the entire surface, blow by blow, with a stonemason's or club hammer and a pointed chisel. The edges should be chamfered or given some other appropriate treatment in order to achieve a proper edge. The ensuing roughness of the surface makes the concrete appear lighter (figure 2.2.9).

Bush hammering
According to DIN 18500 a bush-hammered surface is one which has been worked with a bush hammer to a depth of about 6 mm. The surface is hammered evenly either by hand or by machine to expose the aggregate. This results in a considerably lighter, variegated concrete surface (figures 2.2.14 and 2.2.15).

Comb chiselling
According to DIN 18500 a comb-chiselled surface is one which has been worked with a bolster chisel to a depth of about 4-5 mm. The initially smooth surface is chiselled either by hand or by machine with even, parallel strokes to expose the aggregate and achieve a lighter surface. This method of working is not possible with components containing hard aggregates (figure 2.2.16).

2.2.11 Applications for various types of formwork to "ZM Schalung für Beton" (formwork for concrete)

| Type of formwork | Formwork material | Applications | Typical number of reuses with suitable treatment |
|---|---|---|---|
| Palling boards | Fir or spruce with bark edges and resin pockets | Fair-face concrete | 2-3 |
| Rough boards | Fir or spruce with rough-sawn surface | Concrete without special requirements for exposed faces | 4-5 |
| Smooth boards | Fir or spruce with planed surface | Concrete with special requirements for exposed faces | up to 10 |
| Boards, profiled one side | Fir or spruce with sandblasted or flame-cleaned surface | Fair-face concrete with timber texture | up to 10 |
| Panel of boards | Fir or spruce, impregnated, standard size 1500 x 500 mm | Concrete without special requirements for exposed faces | up to 50 |
| Resincoated plywood | Blockboard or laminboard made from softwood, coated with resin | Concrete without special requirements for exposed faces | up to 30 |
| Filmcoated plywood | Blockboard or laminboard made from softwood, with sodium or kraft paper | | |
| Polyestercoated plywood | Laminboard made from softwood, with polyester coating | Smooth concrete | up to 100 |
| Laminated plastic panels | Melamine or phenol coating on blockboard or laminboard | | 80-100 |
| Polysulphide formwork | Polysulphide | Textured fair-face concrete | 30-50 |
| Rubber formwork | Polypropylene silicone rubber | Textured fair-face concrete (rubber dies), pipe production (inflatable formwork) | up to 50 |
| Polystyrene formwork | Rigid expanded polystyrene foam | Textured fair-face concrete, as void former for system flooring and cut-outs | 1-5 |
| Steel formwork | Steel | Concrete without special requirements for exposed faces | up to 500 |
| Rolled steel pipes | Steel strip with helical folded seams | Fair-face concrete | 1 |

2.2.12 Different board joints for formwork to "ZM Schalung für Beton" (formwork for concrete)

| Type of joint | Effect |
|---|---|
| Overlap | Formation of fins possible |
| Tongue and groove | Grout-tight formwork, difficult to reuse (tongues break off easily) |
| Triangular tongue and groove | Formation of fins possible |
| Scarfed V-tongue and groove | Grout-tight formwork, easy to reuse |
| Butt | Formation of fins possible |

2.2.13 Release agents for various types of formwork to "ZM Schalung für Beton" (formwork for concrete)

| Type of formwork | timber | | | metal | plastic | |
|---|---|---|---|---|---|---|
| Surface of formwork | rough sawn | planed smooth | coated with synthetic resin | very smooth | very smooth | textured |
| Absorption behaviour of formwork | absorbent | | not absorbent | | | |
| Mineral oil lacquer, emulsion | very suitable | | not suitable | | | |
| Mineral oil with separating agents (physical-chemical) moderately | suitable | | very suitable | | | |
| Liquid wax moderately | suitable | | very suitable | moderately suitable | very suitable | moderately suitable |
| Paste | not suitable | moderately suitable | very suitable | moderately to very suitable | | not suitable |

### Mechanical treatments

Other methods for finishing the concrete surface are mainly used in the production of precast concrete components. On the one hand there are textures (sawing and splitting) that depend on the method of production, and on the other pure surface treatments (grinding, fine grinding and polishing) in which, again, the top layer of cement is removed in order to allow the aggregate to be featured as an architectural element. The reconstituted stone produced in this way is suitable for a variety of applications, e.g. floors, spandrel panels, cornices, window sills, claddings to columns, walls and abutments. [2] [23]

### Sawing

According to DIN 18500 a sawn surface is one which has been roughly sawn but given no further treatment. In this method stone saws or stone rift saws are employed on plain concrete blocks to produce either parallel or circular marks. The resulting concrete surface is dense and the colour reflects the overall picture of hydrated cement and aggregate (figure 2.2.17). The surface may be worked further by means of grinding, fine grinding or polishing.

### Splitting

According to DIN 18500 a split surface resembles a quarry-faced surface. Splitting is usually employed without further treatment on plain concrete elements, e.g. masonry units or facing bricks, which are then classed as "split stones" or "ashlar stones". The prefabricated components are cut in splitting machines, which produces quarry-faced surfaces with light, animated textures (figure 2.2.18).

### Grinding

According to DIN 18500 a ground surface is one which has been ground once – without filling – to a depth of about 4 mm. Score marks and pores may still be visible with this method. The depth of working depends on the diameter of grain to be exposed because grinding is usually continued until the largest grain diameter has been exposed. Grinding removes the top layer of the concrete. The overall colour is a mixture of the hydrated cement and aggregate (figure 2.2.19).

### Fine grinding

According to DIN 18500 a fine-ground surface is one which has been worked to a depth of about 5 mm by grinding followed by filling and fine grinding if necessary. This is therefore an continuation of the coarsely ground concrete surface described above and results in a noticeably more intense colouring (figure 2.2.20).

### Polishing

According to DIN 18500 a polished surface is one which has been ground and fine-ground to a depth of about 5 mm using the finest abrasive grit and until a shiny finish has been achieved.

Fine grinding does not itself remove any additional material. Contrasting with this "natural polishing" is the gloss effect that can be produced by applying additional means of treatment (paraffins or resins). This is generally known as "wax polishing" even when other substances are used, and is not a surface treatment but rather a surface finish. Polishing results in a concrete surface with a very much more intense colouring (figure 2.2.21).

Finishing the concrete surface

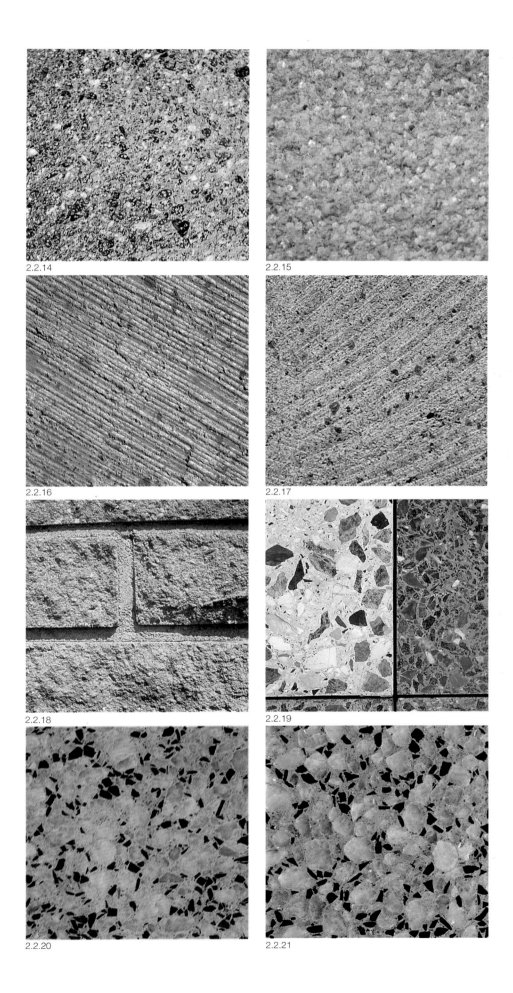

2.2.14  Bush-hammered concrete surface, coloured aggregate, grey cement

2.2.15  Bush-hammered concrete surface, light-coloured aggregate, Portland burnt shale cement

2.2.16  Comb-chiselled concrete surface, coloured aggregate, grey cement

2.2.17  Sawn concrete surface, light-coloured aggregate, coloured, white cement

2.2.18  Split concrete surface, light-coloured aggregate, white cement

2.2.19  Ground concrete surfaces, light and dark aggregates, grey and white cements

2.2.20  Fine-ground concrete surface, light and dark aggregate, white cement

2.2.21  Polished concrete surface, light and dark aggregate, white cement

71

**Technical treatments**
In contrast to the mechanical methods of treating the surface, the technical methods of treatment, e.g. blasting with solid particles and flame cleaning, require complicated plant and apparatus. Again, the aim of the work is to achieve a particularly fine or coarse surface texture for various applications. [2] [23]

Blasting
According to DIN 18500 a blasted surface is one from which the topmost, fine layer of cement laitence has been removed to a depth of about 1-2 mm by means of blasting, e.g. with sand, steel shot, corundum or a water-sand mixture. This method of working is frequently known as "sandblasting" even when other materials are used. It is popular for precast as well as for in situ concrete. The particles are ejected onto the surface of the hardened concrete under high pressure and thus remove the outer layers of material. The aggregate so exposed is slightly roughened and so appears lighter. However, this treatment does not make the sandstone-like appearance of the concrete any more vulnerable to soiling (figure 2.2.22).

Flame cleaning
According to DIN 18500 a flame-cleaned surface is one from which the original top layer has been removed to a depth of about 4-8 mm by means of intense flame treatment. After the concrete has hardened, its surface is subjected to a flame at about 3200 °C. The outer layer of cement laitence and, if applicable, limestone aggregates melt at this high temperature. In the case of quartzitic aggregates the uppermost caps of the grains spall off owing to the temperature stresses to produce an intensely rugged, colourful surface (figure 2.2.23).

Exposed aggregate finishes
Probably the most common method of working the surface is to brush and wash away the uppermost layer of cement. The diverse aggregates employed allow a variety of architectural effects to be achieved – whether angular or rounded, light or dark, uniform colour or variegated, exposed aggregate finishes always appear different. The use of a retarder applied to the formwork means that the surface of the concrete can only be exposed to a depth of about 1 mm. This produces a very interesting appearance resembling sandstone, which can be enhanced by using appropriate aggregates and coloured pigments. [2] [23]

**Brushing and washing**
According to DIN 18500 a brushed and washed surface is one from which the uppermost fine layer of cement laitence has been removed by brushing and washing to a depth exceeding 2 mm, usually about 4-6 mm. A concrete surface is brushed and washed either while it is still fresh or, with the help of a surface retarder, after the concrete has hardened. In the "negative method" the surface to be brushed and washed is always in contact with the formwork, i.e. side or underside, in the "positive method" not in contact with the formwork, i.e. top face. Fair-face concrete behind formwork always requires a surface retarder to be applied to the formwork, but the direct brushing and washing method is possible on top faces not in contact with any formwork (figures 2.2.24 and 2.2.25).

When using aggregates with a size exceeding 50 mm, brushing and washing is carried out using the sand bed technique. This involves laying the coarse – usually round – grains in a sand bed within the formwork prior to concreting in order that, for brushing and washing, they are not completely surrounded by cement paste. Generally, no more than half the diameter of the largest aggregate may be exposed by brushing and washing. This method of working allows the natural colour of the respective aggregate used to be exploited to best effect.

Light brushing and washing
According to DIN 18500 a lightly brushed and washed surface is one from which the uppermost layer of fine surface laitence has been removed by brushing and washing to a depth of no more than 2 mm. Both the positive and negative methods can be used with this technique. The difference lies only in the visual effect because in this case only the tips of the grains of aggregate are exposed. The size of aggregate used when light brushing and washing is to be employed is generally 16 mm, or occasionally 8 mm if a particularly fine-grained texture is to be achieved (figure 2.2.26).

Acid etching

This is a special way of exposing the aggregate, in which the surface of the concrete is washed with diluted acid followed by water. According to DIN 18500 an acid-etched surface is one from which the uppermost layer of cement laitence has been removed by chemical means (acids) to a depth of about 0.5 mm. This method is employed when the concrete surface merely requires a slight roughness. Here again, the final texture, with individual aggregates exposed and cleaned, is similar to that of sandstone. Acid etching is carried out only in concrete works and almost exclusively on relatively small areas because only in this way can the environmental problems associated with the use of diluted acids be handled reliably (figure 2.2.27).

## The use of coatings

People's attitudes to the design of structures change with the technical and economic possibilities and with fashion. Just 25 years ago, large, daring structures, at the limits of what was feasible in engineering terms but coupled with very low production costs and fast-track methods, were considered to be the standards to aim for in construction. The fashions of the day preferred engineering geometries and unclad building materials. Today, a change in values can be seen. These include the move towards quality, economic assessments that take into account durability and a different attitude to the appearance of structures in general. The preference for large, imposing structures that dominate the urban or rural landscape is giving way to the concept of optimum integration within the surroundings. The trend is towards a stronger structural form and colour provided by combinations of materials, claddings or coatings. [9] [18]

### Conditions for coating

The performance of uncoated concrete with respect to durability, resistance and appearance must be taken into account during the design. This depends on planning, construction, concrete mix, the care taken during production, placing and curing, as well as the intensity of the actions to which the concrete is subjected. Concrete components designed and constructed with a reasonable amount of care in accordance with the current state of knowledge are highly durable and resistant. Therefore, coatings are necessary only in exceptional cases. Concrete does not require a coating even to withstand "acid rain". Coatings should always be rejected if they are only being called for to compensate for anticipated inadequacies in workmanship. However, they can be useful for reducing the vulnerability of components to unavoidable shortcomings in workmanship and appearance. Even with careful planning and workmanship, it is difficult to prevent efflorescence and, especially, soiling on regularly saturated fair-face concrete components. In these cases planned impregnation and coatings can contribute to a better appearance.

However, colour cannot be a substitute for inadequate architecture. It should therefore be used only to reinforce and complement. The nature of the colour scheme can be related to the material or autonomous. The desired effect can be either to highlight or harmonise within a larger context. Every use of colour should form part of the design and stem from an intrinsic need.

### Types of coating

Concrete components must be protected according to DIN 1045 or DIN EN 206 part 1 if they are permanently exposed to very severe chemical attack according to DIN 4030, e.g. acids. In this case, however, the thickness of the coating must extend to several millimetres because every flaw leads to an undermining of the coating. Epoxy resins, coal tar epoxy resins and polyurethanes, acid-resistant plastic foils or ceramic finishes, are employed.

Coatings represent an advantage for concrete components in buildings or civil engineering projects when they aim to prevent

- the dark discoloration caused by rain (difference between wet and dry patches)
- the accumulation of dust and dirt
- the occurrence of efflorescence
- unsightly blemishes denoting the prevailing drainage routes for rainwater

However, the minimum specification for a coating material in these and other cases is that it must remain weather-resistant for at least 20 years without noticeable changes and thereafter be easy to overhaul. This also applies to glazed, thin coatings (sealants), which can be applied in such a way that they are invisible to the layman. They also offer the option of preserving the appearance of, for example, surfaces coloured and/or post-treated with pigments because the coating has the same colour as the concrete.

Coloured paints can be used on concrete surfaces to emphasise, in particular, the effect of larger areas, a number of identical surfaces, or special components on the structure as a whole. The paints to be used must always be compatible with the concrete substrate. Coloured paints have to satisfy a large number of requirements based on the stipulations of DIN 55945. Even those paints applied to a concrete surface for artistic reasons generally have to satisfy the following requirements:

- Resistance to alkaline reactions originating from the concrete
- Good bonding within the paint system
- Painting over with the same paint
- Resistance to the effects of the weather
- In some circumstances resistance to industrial atmospheres and/or substances dissolved in water
- Resistance to UV light
- A low tendency to attract dirt
- Sufficient water vapour permeability
- In some circumstances resistance to flowing water
- In some circumstances resistance to washing or scrubbing

The painting or coating materials that can be used are primarily those mineral, silicate, dispersed synthetic or polymer resin paints, which harden in the air or hydraulically. Hydrophobic impregnations have the effect of reducing the absorption of water by capillary action only for a limited time. However, the water vapour permeability of the concrete surface remains unchanged because an impervious film does not develop. The water-repellent effect observed at the beginning disappears over time without the level of protection decreasing. Sealants partially fill the capillary pores near the surface of the concrete substrate and at the same time form a thin, congruous surface film. They also reduce water absorption and diffusion of carbon dioxide. Coatings are able to cover a very broad and diverse spectrum of properties, but one of the main features is always reducing or preventing the penetration of water (figures 2.2.28 and 2.2.29).

## The effects of the weather

Facades age. This is partly due to the materials used. The question is: how will the appearance of a building change over the course of time? If the appearance of a structure is to remain acceptable over a long period, it is essential to consider the soiling that will inevitably occur over the years and the controlled drainage of rainwater over the facade. Exposure to or shelter from the wind are among the critical factors affecting the amount of rainwater reaching certain areas and determining where deposits of dirt can build up. It is particularly important to observe the flow pattern of rainwater as this can reveal where dirt is washed away and, in certain circumstances, show if it is deposited again at other positions (figures 2.2.30-2.2.34).

The gradient or inclination of a concrete surface is also an important factor. A vertical surface attracts relatively little water and is easily rinsed clean. If a surface is inclined backwards, it attracts considerably more water. But the cleansing effect is somewhat reduced; dirt is often deposited again along the bottom edges. A surface inclined forwards remains the cleanest and generally remains dry. However, the upper part must be designed in such a way that no water drains over the surface because otherwise the flow pattern very quickly becomes apparent here and thus spoils the overall appearance of the facade (figure 2.2.35).

Observing these planning rules is an important foundation for proper designing and building with concrete. The correct construction details can prevent the weather or the environment causing uncontrolled soiling of a concrete facade in the long term. This does not call for sophisticated solutions but rather a full understanding of the principles involved.

2.2.22 Blasted concrete surface, Singhofener quartzite aggregate, 0-16 mm, white cement, 0.2% iron oxide yellow

2.2.23 Flame-cleaned concrete surface, Rhine gravel aggregate, 0-16 mm, white cement, 3% titanium oxide

2.2.24 Brushed and washed concrete surface, coloured aggregate with rounded grains, grey cement

2.2.25 Brushed and washed concrete surface, coloured aggregate with angular grains, Portland burnt shale cement

2.2.26 Lightly brushed and washed concrete surface, Rhine sand and porphyry aggregate, 0-16 mm, white cement, 1% iron oxide red

2.2.27 Acid-etched concrete surface, light-coloured aggregate, white cement, 0.2% iron oxide yellow

2.2.28 Concrete surface with opaque coating, acrylic resin paint

2.2.29 Concrete surface with transparent glaze, mineral paint

## The concrete surface

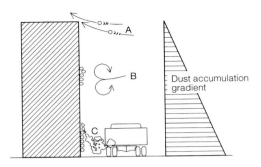

2.2.30  Accumulation of dust on facade surfaces – A) high wind velocity, low dust accumulation, B) increased dust accumulation due to turbulence, and C) traffic

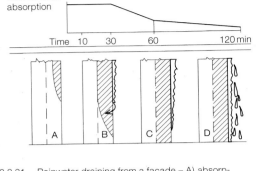

2.2.31  Rainwater draining from a facade – A) absorption, B) onset of draining plus absorption, C) substrate saturated plus draining, D) draining plus dripping

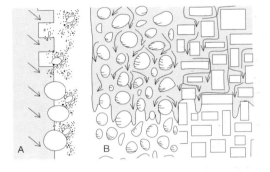

2.2.32  Deep surface textures form dust traps (A) but also regulate the uniform draining of water

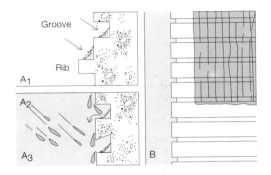

2.2.33  Horizontal grooves and ribs form dust traps (A1, A2, A3) but also regulate the uniform draining of water

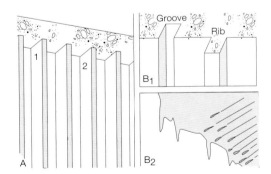

2.2.34  Vertical grooves and ribs (A1, A2, B1) prevent the uneven incidence of rainwater caused by side winds (B2)

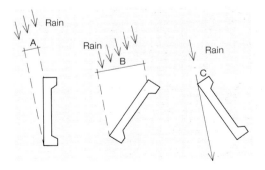

2.2.35  Different amounts of rainwater on surfaces inclined at different angles (A, B, C)

## References

**Books and articles**

[1] Belz, Walter: Materialgerecht, werkgerecht – eine skizzenhafte Betrachtung, Deutsche Bauzeitung 9/1987
[2] Betonfassaden; Facades Study Group, Fachvereinigung Deutscher Betonfertigteilbau, Bonn, and Dyckerhoff Weiss Marketing & Vertriebsgesellschaft, Wiesbaden 1993
[3] Döring, Wolfgang; Meschke, Hans-Jürgen; Kind-Barkauskas, Friedbert; Schwerm, Dieter: Fassaden – Architektur und Konstruktion mit Betonfertigteilen, Verlag Bau+Technik GmbH, Düsseldorf 2000
[4] Durability of Appearance of Concrete Facades, pub. by Cembureau – The European Cement Association, Brussels 1993
[5] Ebeling, Karsten: Wunsch und Wirklichkeit – Ausschreibung von Sichtbetonarbeiten, Deutsche Bauzeitung 10/2000
[6] Franke, Horst; Schaarschmidt, Birgit: Sichtbeton – eine zugesicherte Eigenschaft?, Beton 4/2000
[7] Gestalten mit Farbe, pub. by Bayer AG, Leverkusen 1993
[8] Grube, Horst: Oberflächenschutz von Stahlbeton, Beton 10/1981
[9] Grube, Horst; Kind-Barkauskas, Friedbert: Beschichtungen auf Beton, Beton 12/1991
[10] Hahn, Ulrich: Farbiger Beton – die Natur hat den Zuschlag, Betonwerk+Fertigteil-Technik 1/1998
[11] Heeß, Stefan: Ausschreibungshinweise für farbigen Sichtbeton, Dyckerhoff Weiss Marketing & Vertriebsgesellschaft, Wiesbaden 2000
[12] Huberty, J. M.: Fassaden in der Witterung, Beton-Verlag GmbH, Düsseldorf 1983
[13] Kind-Barkauskas, Friedbert: Gestalten mit Beton, Beton 12/1986
[14] Kind-Barkauskas, Friedbert: Gestalterische Aspekte der Behaglichkeit, Beton 12/1983
[15] Kind-Barkauskas, Friedbert: Gestaltung von Betonfassaden – Entwicklungen und Tendenzen, Fertigteilbau+Industrialisiertes Bauen 3/1983
[16] Kind-Barkauskas, Friedbert; Richter, Thomas: Weg frei für Beton – Sichtbeton bei Verkehrsbauten, Leonardo-online 2/2000
[17] Klose, Norbert: Alterung von Betonbauteilen, Betonwerk+Fertigteil-Technik 9/1986
[18] Lamprecht, Heinz-Otto; Kind-Barkauskas, Friedbert; Pickel, Ulrich; Otto, Horst; Schmincke, Peter; Schwara, Herbert: Betonoberflächen – Gestaltung und Herstellung, Expert-Verlag GmbH, Grafenau 1984
[19] Mayer, Roland; Bährle, Peter: Sichtbeton – keine Patentrezepte, sondern individuelle Planung, Das Architekten-Magazin 9/2000
[20] Luley, Hanspeter; Kampen, Rolf; Kind-Barkauskas, Friedbert; Klose, Norbert; Tegelaar, Rudolf: Instandsetzen von Stahlbetonoberflächen, Beton-Verlag GmbH, Düsseldorf 1997
[21] Pickel, Ulrich: Architekturbeton – mehr als Beton; in: Beton+Fertigteil-Jahrbuch 1993, Bauverlag, Wiesbaden/Berlin 1993
[22] Pickel, Ulrich: Hinweise für die Ausschreibung bei Betonwerksteinarbeiten; in: Beton + Fertigteil-Jahrbuch 1994, Bauverlag GmbH, Wiesbaden/Berlin 1994
[23] Pickel, Ulrich: Oberflächenbearbeitung; in: Betonfassaden, Facades Study Group, Fachvereinigung Deutscher Betonfertigteilbau, Bonn, and Dyckerhoff Weiss Marketing & Vertriebsgesellschaft, Wiesbaden 1993
[24] Plenker, Heinz-Herbert: Dosierung und Verteilung von Pigmenten in Beton, Concrete Precasting Plant and Technology 9/1991
[25] Schmincke, Peter: Sichtbeton – gewusst wie, Beton 7/1990
[26] Trüb, Ulrich: Die Betonoberfläche, Bauverlag GmbH, Wiesbaden/Berlin 1973
[27] Zimmermann, Thomas: Künstliche Oberflächen, Der Architekt 12/1990

**Standards and directives**

DIN 1045 Concrete, reinforced and prestressed concrete structures
    part 1: Design, edition 2001
DIN 4226 Aggregates for concrete and mortar
    part 1: Aggregates of normal and dense structure (heavy aggregates), edition 2001
    part 2: Aggregates of porous structure (lightweight aggregates), draft 2000
    part 3: Testing of heavy and lightweight aggregates, edition 1983
    part 4: Inspection, edition 1983
DIN 18216 Formwork ties; requirements, testing, use, edition 1986
DIN 18500 Cast stones; terminology, requirements, testing, inspection, edition 1991
DIN EN 206 Concrete
    part 1: Specification, performance, production and conformity, edition 2001

DBV-Merkblätter (German Concrete Society data sheets):
Betondeckung und Bewehrung (concrete cover and reinforcement), edition 1997
Betonschalungen (formwork for concrete), edition 1999
Trennmittel für Beton (release agents for concrete)
    part A: Hinweise zur Auswahl und Anwendung (advice on selection and application), edition 1997
    part B: Prüfungen (testing), edition 1999
    (in conjunction with German Cement Industry Association:) Sichtbeton – Ausschreibung, Herstellung und Abnahme von Beton mit gestalteten Ansichtsflächen (fair-face concrete – tendering, production and inspection of concrete with special exposed surfaces), edition 1997

DAfStb-Richtlinien (German Reinforced Concrete Committee directives):
Schutz und Instandsetzung von Betonbauteilen (protection and maintenance of concrete components)
    part 1: Allgemeine Regelungen und Planungsgrundsätze (general rules and design principles), draft 2000
    part 2: Bauprodukte und Anwendung (building products and applications), draft 2000
    part 3: Anforderungen an die Betriebe und Überwachung der Ausführung (requirements for contractors and monitoring of workmanship), draft 2000
Nachbehandlung von Beton (curing of concrete), edition 1984

Cement data sheets:
Schalung für Beton (formwork for concrete), edition 1999
Sichtbeton – Gestaltung von Betonoberflächen (fair-face concrete – design of concrete surfaces), edition 1999

# Building science

Jörg Brandt

2.3.1 Influences on external components and their functions

2.3.2 Thermal comfort in relation to physiological, intermediary and physical influences

### General

Building science is a separate discipline within building technology. It deals mainly with the interior climate, thermal insulation, moisture control, acoustics and fire protection. The aim is to achieve a good standard within the building, preserve the building and use environmentally compatible, energy-saving technologies.

The quality of the interior climate has a considerable influence on people's health and performance; we spend more than two-thirds of our lives indoors living and working. Health care measures are becoming more and more important in the planning of buildings in which people will live or work. In the past, epidemics like plague, typhoid and cholera prompted health care activities. Today, domestic hygiene is concerned with the identification and elimination of the causes of modern-day illnesses. This includes carrying out research into how the built environment affects our health and specifying interior climates conducive to our performance and general well-being. Essentially, hygiene demands are founded on the temperature and humidity of the interior air, the movement of the air, daylight and artificial lighting, the effects of noise and pollutants, ventilation and heating. The layout and the interior fittings of the building also play a role. When planning a building, this calls for, above all, the need to observe building science requirements and relationships.

Components are subjected to different requirements depending on the type of building and its use, as well as the position of the component within the structure. These requirements must always be given priority. Only when all these requirements are satisfied in economic terms can a final decision being made about the choice of material and design of the components (figure 2.3.1).

However, the design of the building and the building science properties of the components forming the building envelope can only create the framework conditions. Within this framework, important interior climate factors can be controlled within optimum ranges by means of heating and ventilation.

### Basic requirements

The planning and arrangement of the built environment must reflect the needs of people. The building codes of Germany's federal states incorporate the basic requirements for planning and erecting buildings, taking into account the health of the public and that of the building's occupants. Clause 3 of the Model Building Code states:
"Buildings shall be designed, erected, converted and maintained such that public safety or public order, in particular the lives and health of the public, are not endangered."

This general statement is described in more detail in clause 16 as follows:
"Buildings must be constructed in such a way that dangers or unreasonable disadvantages do not arise as a result of water, moisture, saprogenous substances, the effects of the weather, vegetable or animal pests, or other chemical or physical influences."

Requirements for the building, components and building materials can be derived from these basic requirements. Satisfying these requirements creates the framework conditions for a healthy interior climate. The basic requirements are therefore a skeleton, which will be filled in and followed as far as possible in interdisciplinary cooperation.

### Interior climate

The interior climate results from interaction of numerous factors. The most important of these are the temperature, movement, composition and contamination of the interior air, the surface temperatures of the enclosing components, their moisture content and behaviour in moist conditions, and the radioactivity of components and interior air. On the following pages we deal with these interior climate factors, demonstrate threshold conditions and derive requirements or measures relevant for the construction.

## The interior climate

**Thermal comfort**

A person's physical and mental performance is normally at its best when he or she feels neither too hot nor too cold. In this respect, the thermal balance of the body is in equilibrium at a constant internal body temperature of about 37 °C. This equilibrium state of thermal comfort depends on a number of factors, which can be divided into physical, physiological and intermediary conditions. Of the 21 influencing factors listed in figure 2.3.2, six are classed as primary and dominant, eight as supplementary and seven as secondary, presumed factors.

Taking the primary and dominant factors, the following physical influences also have an effect on the planning of the building:
- the temperature of the interior air
- the surface temperatures of the enclosing components
- the relative humidity of the interior air
- the movement of the air in the vicinity of the occupants

So thermal comfort can be described by means of quantifiable variables that vary within certain limits depending on clothing, degree of activity and subjective sensations.

If these variables are placed in relation to the temperature of the interior air, we obtain the comfort zones shown in figures 2.3.3-2.3.7.

**Thermal insulation necessary for hygienic conditions**

We can derive thresholds and ranges from this and according to [33] for comfort factors in winter and summer to cover occupied rooms (table 2.3.10).

The U-values given in the table are based on an internal air temperature of 22 °C, an external air temperature of -10 °C and a thermal surface resistance coefficient of 8 W/m²K on the inside.

Requirements in winter
The lowest U-values in table 2.3.10 occur in lines 3 and 5 (0.75 W/m²K), i.e. for the requirement that the difference in temperature between interior air and surface of component, or between floor and ceiling, should not exceed 3 K. This value is far below the maximum value of 1.39 W/m²K given in DIN 4108 "Thermal insulation in buildings", which specifies values for minimum thermal insulation.

The considerably higher U-value of the window (≤ 1.40 W/m²K) in line 4 is a concession to what is feasible in technical and economic terms; a lower window U-value would be better. It must be accepted here that the heat exchange by way of radiation between a person and the surface of the window is particularly high and that an unpleasant, one-way loss of heat by radiation from the body to the window takes place near the windows.

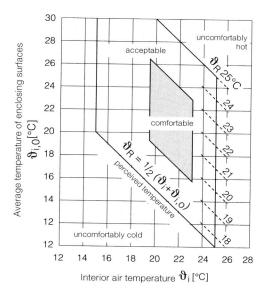

2.3.3 Comfort zone of people in enclosed spaces in relation to the interior air temperature and the average temperature of the enclosing surfaces [19]

Applicability:
relative air humidity $\varphi_i$ = 30-70%
air movement v = 0-0.02 m/s
temperature of all enclosing surfaces essentially equal (after H. Reiher and W. Frank)

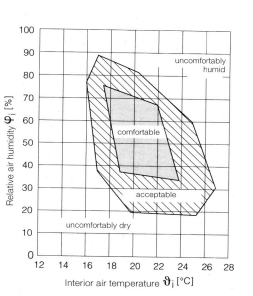

2.3.4 Comfort zone of people in enclosed spaces in relation to the interior air temperature and the relative humidity of the air [19]

Applicability:
average temperature of enclosing surfaces
$\vartheta_{i,0}$ = 19.5-23 °C
air movement v = 0-0.02 m/s
(after F.P. Leusden and H. Freymark)

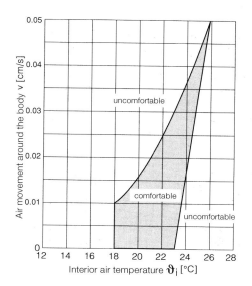

2.3.5 Comfort zone of people in enclosed spaces in relation to the interior air temperature and the movement of the air [19]

Applicability:
average temperature of enclosing surfaces
$\vartheta_{i,0}$ = 19.5-23 °C
relative air humidity $\varphi_i$ = 30-70%
(after Rietschel-Raiss)

# Building science

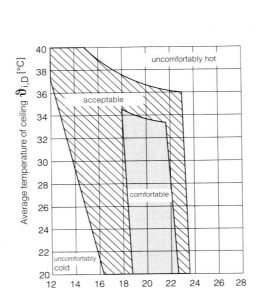

2.3.6 Comfort zone of people in enclosed spaces in relation to the interior air temperature and the floor temperature [19]

Applicability:
relative air humidity $\varphi_i$ = 30-70%
air movement v = 0-0.02 m/s
(after H. Reiher and W. Frank)

2.3.7 Comfort zone of people in enclosed spaces in relation to the interior air temperature and the ceiling temperature [19]

Applicability:
relative air humidity $\varphi_i$ = 30-70%
air movement v = 0-0.02 m/s
(after H.G. Wenzel and A. Müller)

2.3.8 Water content of a concrete for external components, B 25/KR, cement content 300 kg/m³ [25]

| State of concrete | Water content [kg/m³] |
|---|---|
| Fresh concrete<br>  unbonded water | 175 |
| Hardened concrete – 28 days old<br>(at 70% hydration)<br>  bonded water<br>  dried out<br>  unbonded water | <br><br>85<br>25-45<br>65-45 |
| hardened concrete – 3-6 months old<br>(at 90% hydration)<br>  bonded water<br>  dried out<br>  unbonded water | <br><br>105<br>35-50<br>35-20 |
| DIN 4108 – thermal insulation<br>  practical moisture content | 50 |

Figure 2.3.9 shows how the window areas can affect the comfort zones in a room. The diagram clearly shows the boundary of the comfortable zone on the window side. If the room is used as an office, the desk must be positioned within this zone and the back of the chair for the person seated at the desk must face towards the two internal walls. This will counteract the aysmmetric, unhealthy radiation of heat away from the body towards the window.

Requirements in summer
It is desirable for the components of the building to possess an effective heat storage capacity on the inner surfaces in order to act as a heat buffer and prevent the surface temperatures in a room from exceeding the maximum values given in table 2.3.10, as well as to prevent great fluctuations in the interior temperature over the course of a day. The temperature amplitude ratio (TAV) is a parameter that expresses the efficiency of the heat storage capacity of external components. For example, TAV = 0.1 means that only 10% of the temperature amplitude on the outside of a component is transmitted to the inside. A low TAV, especially for roof and floor slabs, has a beneficial effect on the interior temperature in summer. The TAV value of an external wall should therefore not exceed 0.15 and that of a roof 0.10. Components with usable storage masses on the inside fulfil this requirement. These include all masonry walls, multi-layer walls and concrete floor slabs and roofs with layers of thermal insulation in the middle or on the outside. If thermal insulation is attached to the inside face of building materials with a heat storage capacity, the TAV values are generally much higher than 0.15.

**Water and moisture**
To guarantee the conditions for long-term thermal comfort, components must be protected against becoming saturated. The thermal insulation capacity of components decreases as the moisture content increases; therefore, interior surface temperatures also fall. In winter damp external components can be affected by frost. Moisture on internal surfaces leads to mould growth, and spores can have a detrimental effect on the health of occupants.

Moisture due to construction process
The components of new buildings contain water that, for example, stems from mixing water in mortar and concrete, or from rainwater soaked up by unprotected masonry before the finishes are applied to the building. More recent studies of the drying behaviour of normal-weight concrete, external leaves and hollow blocks of no-fines aerated concrete have revealed that the evaporation of water in the first months progresses very rapidly. After four to six months, the drying rate drops to below 5 g/m² per day. Therefore, the release of water vapour from the components does not have a negative effect

# The interior climate

on the interior climate in the long term. Table 2.3.8 shows the typical drying behaviour of concrete. Generally, concrete reaches its practical moisture content while construction work is still continuing.

Moisture in components below ground level
Components adjacent to the ground can be subjected to hydrostatic and non-hydrostatic pressure. Drainage, horizontal and vertical waterproofing or the use of an impermeable concrete and the provision of external tanking are just some of the techniques employed to prevent the components and the walls above from becoming saturated.

Driving rain
Water entering the external wall during periods of driving rain should be able to evaporate to the outside air again during the intervening dry periods. The lower the partial diffusion resistance sd of the surface layer, e.g. rendering, the faster is this evaporation. Such a surface layer should therefore, in terms of protection against rain, be waterproof or water-repellent, but at the same time remain as permeable as possible for water vapour.
The driving rain load on buildings or parts of buildings is divided into groups.
Group I (low driving rain load) includes regions with low wind speeds and < 600 mm precipitation p.a., but also sites particularly sheltered from the wind even though they have a higher annual rainfall.
Group II (moderate driving rain load) includes regions with 600-800 mm precipitation p.a., as well as sheltered sites with a higher annual rainfall. This group also covers high-rise buildings and buildings on exposed sites that would otherwise fall into group I.
Group III (high driving rain load) includes regions with > 800 mm precipitation p.a. and also exposed, windy sites that would otherwise fall into group II.
Protective measures, e.g. appropriate coatings and rendering or ventilated facing leaves, are necessary, appropriate to the driving rain load.

Table 2.3.11 shows that wall constructions with surfaces of dense concrete provide good protection against driving rain. In contrast, masonry and aerated concrete components require special protective measures to suit the driving rain load, e.g. a waterproof or water-repellent rendering, coatings or ventilated facing leaves.

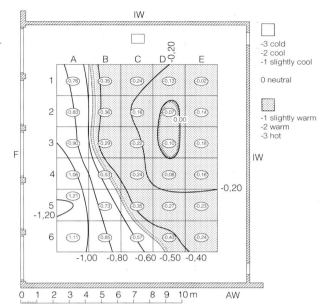

2.3.9 The influence of windows on thermal comfort in a room
F = window, AW = external wall, IW = internal wall
◯ = optimum comfort value subjectively assessed as "neutral" (= "comfortable", subjective value 0.00)
E/E = boundary of comfort zone on window side, determined from the normal, average width of the individual thermal comfort zone [33]

2.3.10 Limiting values and ranges for comfortable conditions in occupied rooms and requirements for external components

|  | Factor influencing comfort |  | Season | Thresholds Ranges |  | U-value [W/m²K] | TAV 1 |
|---|---|---|---|---|---|---|---|
| 1 | Interior air temperature | $\vartheta_i$ | winter summer | 20-22 ≤ 25 | °C °C |  |  |
| 2 | Average temperature of enclosing surfaces | $\vartheta_{i,m}$ | winter summer | ≥ 17 ≤ 25 | °C °C | ≤ 1.25 | ≤ 0.15 |
| 3 | Temperature difference between interior air and surface of component ($\vartheta_i = 21$ °C) | $\Delta\vartheta$ | winter | ≤ 3 | K | ≤ 0.75 |  |
| 4 | Temperature difference of opposing vertical components | $\Delta\vartheta$ | winter (wall) (window) | ≤ 5 | K K | ≤ 1.25 ≤ 1.40 |  |
| 5 | Temperature difference between floor and ceiling | $\Delta\vartheta$ | winter | ≤ 3 | K | ≤ 0.75 |  |
| 6 | Temperature of floor surface | $\vartheta_i$ | winter summer | 17-26 ≤ 26 | °C °C | ≤ 1.25 | ≤ 0.15 |
| 7 | Temperature of ceiling surface | $\vartheta_i$ | winter summer | 17-34 ≤ 34 | °C °C | ≤ 1.25 | ≤ 0.10 |
| 8 | Heat flow | q | winter | ≤ 40 | W/m² | ≤ 1.25 |  |
| 9 | Relatively humidity of interior air | $\varphi_i$ | winter summer | 40-60 40-60 | % % |  |  |
| 10 | Movement of the air around the body | VL | winter summer | ≤ 0.15 ≤ 0.30 | m/s m/s |  |  |

2.3.11 Allocation of wall construction groups and loading groups to DIN 4108

| Column | 1 | 2 | 3 |
|---|---|---|---|
| Line | Group I<br>low driving rain load | Group II<br>moderate driving rain load | Group III<br>high driving rain load |
| 1 | With rendering, without special requirements for driving rain protection, to DIN 18550 pt 1 on:<br>• external walls of masonry, wall panels, concrete or similar<br>• wood-wool lightweight building boards installed according to DIN 1102 (with reinforced joints)<br>• multi-ply lightweight building boards installed according to DIN 1104 pt 2 (fully reinforced) | With water-retardant rendering to DIN 18550 pt 1, or a synthetic resin plaster on:<br>• external walls of masonry, wall panels, concrete or similar<br>• wood-wool lightweight building boards installed according to DIN 1102 (with reinforced joints)<br>or<br>• multi-ply lightweight building boards with wood-wool layers for plastering, ≥ 15 mm thk, installed according to DIN 1104 pt 2 (fully reinforced)<br>• multi-ply lightweight building boards with wood-wool layers for plastering, < 15 mm thk, installed according to DIN 1104 pt 2 (fully reinforced), with premixed mortar to DIN 18557 | With water-repellent rendering to DIN 18550 pt 1, or a synthetic resin plaster on:<br>• external walls of masonry, wall panels, concrete or similar |
| 2 | Single-leaf facing masonry to DIN 1053 pt 1, 310 mm thk[1] | Single-leaf facing masonry to DIN 1053 pt 1, 375 mm thk[1] | Double-leaf facing masonry with cavity to DIN 1053 pt 1[2]; double-leaf facing masonry without cavity to DIN 1053 pt 1 with facing bricks |
| 3 | | External walls with claddings bedded in mortar to DIN 18515 | External walls with built-in claddings and plaster undercoat to DIN 18515 and water-repellent mortar[3]; external walls with claddings bedded in mortar and plaster undercoat to DIN 18515 and water-repellent mortar[3] |
| 4 | | | External walls with dense concrete outer layer to DIN 1045 and DIN 4219 pt 1 & 2 |

[1] If the thermal insulation required is provided solely by an additional, included layer of thermal insulation, the masonry can be allocated to the next higher loading group.
[2] The cavity must comply with DIN 1053 part 1. Cavity insulation may only be installed according to the standards covering such work or requires a special analysis of the serviceability, e.g. furnished by way of a building authority approval.
[3] Water-repellent mortar must exhibit a water absorption coefficient w ≤ 0.5 kg/m²h$^{0.5}$, calculated according to DIN 52617.

Condensation on internal surfaces
To avoid condensation forming on internal surfaces, the surface temperatures of the components should be kept above the dewpoint of the interior air by adequate thermal insulation. The moisture that occurs due to the use of the building should be removed by ventilation in order to keep the relative air humidity in the desirable range of 40-60%. In the central European climate, external concrete components always require adequate thermal insulation. The degree of insulation depends on the function of the building and the component.

Thermal bridges caused by the geometry and the details of the building are particularly at risk of surface condensation, as are poorly ventilated corners behind curtains or cupboards. The increase in condensation damage to building surfaces in recent years is connected with endeavours to save heating energy: interior air temperatures are reduced, the window frames sealed and hence "automatic" air changes diminished. The interior is not adequately ventilated, e.g. by opening the windows regularly (brief, intensive ventilation).

Condensation as a result of water vapour diffusion
The amount of interstitial condensation in a component in winter is not a problem when the thermal insulation effect is not seriously impaired, there is no frost damage and the component can dry out again in summer. The material and the position of the dewpoint in the component is important here and should be checked.
In single-layer components, e.g. walls of lightweight concrete masonry, water vapour diffusion usually takes place without any problems. In multi-layer constructions the principle to be followed is that the diffusion resistance of the individual layers should decrease from inside to outside. Very porous vapour-permeable building materials in multi-layer components often require special detailing to provide ventilation at the rear or vapour barriers on the inside to prevent the components becoming saturated.
DIN 4108 part 3, with standardised assumptions for climate data, is used to verify that the water vapour diffusion is harmless. If controlled water vapour diffusion cannot be proved for a construction, then the sequence of the layers may need to be rearranged or a vapour barrier included as well.

Verification according to DIN 4108 part 3 is not required for the following components:

- Masonry to DIN 1053 part 1 of factory-made units without additional thermal insulation as single- or double-leaf masonry, faced or rendered or covered with cladding attached with mortar or built into the masonry according to DIN 18515 (proportion of joints ≥ 5%), and double-leaf masonry with an air cavity to DIN 1053 part 1 with or without additional thermal insulation
- Masonry to DIN 1053 part 1 of factory-made units with thermal insulation attached to the outside and rendering with mineral binders to DIN 18550 parts 1 and 2, or a synthetic resin plaster – where the diffusion-equivalent air layer thickness $s_d$ of the plaster is ≤ 4.0 m – or with a ventilated cladding
- Masonry to DIN 1053 part 1 of factory-made units with thermal insulation attached to the inside, where – including internal plaster – $s_d ≥ 0.5$ m, and rendering, or with ventilated cladding
- Masonry to DIN 1053 part 1 of factory-made units with wood-wool lightweight building boards to DIN 1101 attached to the inside, plastered or clad, constructed externally as facing masonry (no engineering bricks to DIN 105) or rendered or with ventilated cladding
- Walls of dense lightweight concrete to DIN 4219 parts 1 and 2 without additional thermal insulation
- Walls of reinforced aerated concrete to DIN 4223 without additional thermal insulation, with a synthetic resin plaster ($s_d ≤ 4.0$ m), or with ventilated cladding or with a ventilated facing leaf
- Walls of no-fines lightweight concrete to DIN 4232 plastered both sides, or with ventilated cladding on the outside, without additional thermal insulation
- Walls of normal-weight concrete to DIN 1045 or dense lightweight concrete to DIN 4219 parts 1 and 2, with thermal insulation on the outside and rendering with mineral binders to DIN 18550 parts 1 and 2, or a synthetic resin plaster or a cladding or a facing leaf
- Roofs with a vapour barrier ($s_d ≥ 100$ m) below or in the layer of thermal insulation, where the thermal resistance of the layers of construction below the vapour barrier is max. 20% of the total thermal resistance
- Single-skin roofs of aerated concrete to DIN 4223 without a vapour barrier on the underside

No analysis is required for roofs with a ventilated space above the thermal insulation provided the following conditions are satisfied:

a) Roofs with a pitch ≥ 10°
- The unobstructed ventilation cross-section of each of the openings included on two opposite eaves is ≥ 2‰ of the associated pitched roof area, but ≥ 0.02 m² per metre of eaves
- The ventilation opening at the ridge is ≥ 0.5‰ of the total pitched roof area
- The unobstructed ventilation cross-section within the area of the roof above the thermal insulation in the finished condition is ≥ 0.02 m² per metre perpendicular to the direction of flow with a clear height ≥ 20 mm
- The diffusion-equivalent air layer thickness $s_d$ of the layers of construction below the ventilated space is, depending on the length of the rafter "a":

$a ≤ 10$ m : $s_d ≥ 2$ m
$a ≤ 15$ m : $s_d ≥ 5$ m
$a ≥ 15$ m : $s_d ≥ 10$ m

b) Roofs with a pitch < 10°
- The unobstructed ventilation cross-section of openings included on at least two opposite eaves is ≥ 2‰ of the total plan area of the roof
- The height of the unobstructed ventilation cross-section within the roof above the thermal insulation in the finished condition is ≥ 50 mm
- The diffusion-equivalent air layer thickness $s_d$ of the layers of construction below the ventilated space is ≥ 10 m

c) In roofs with, for example, existing vapour barriers ($s_d ≥ 100$ m), these should be arranged in such a way that the thermal resistance of the layers of the construction below the vapour barrier is max. 2‰ of the total thermal resistance (area between roof members should be used for roofs with adjacent areas of different thermal insulation).

d) In roofs with monolithic slab constructions or layered roof constructions, the layer of thermal insulation should be positioned as the uppermost layer below the ventilated space.
- On the underside of roofs of aerated concrete to DIN 4223 without additional thermal insulation and without a vapour barrier

Water vapour sorption
The sorption behaviour of the components forming the envelope of the building is significant for interiors in which large amounts of water vapour accumulate temporarily, e.g. bathrooms, kitchens, classrooms, places of assembly. In such cases the components should be able to absorb moisture quickly from the air, store this and then release it later in order to compensate for the otherwise severe fluctuations in the relative humidity of the interior air.

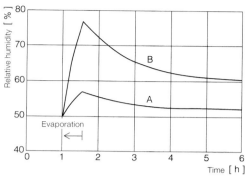

2.3.12  The change in the relative air humidity upon the evaporation of 200 g water within 30 min in two rooms measuring 4 x 2.5 m but with different properties for the inner surfaces of walls and ceilings:
room A = lime plaster,
room B = oil-based paint on plaster [26]

2.3.13 Water vapour absorption coefficients, results of investigations according to [27]

| Material | Water vapour absorption coefficient d [g/m²h⁰·⁵] |
|---|---|
| Concrete B 15 | 11 |
| Concrete B 25 | 9 |
| Concrete B 45 | 8 |
| Lime-cement plaster, sheet metal | 13 |
| Lime-cement plaster, B 15 concrete | 10 |
| Lime-cement plaster, lightweight clay bricks | 10 |
| Lime-gypsum plaster, sheet metal | 13 |
| Lime-gypsum plaster, B 15 concrete | 13 |
| Gypsum-sand plaster, sheet metal | 11 |
| Wallpaper, B 15 concrete | 12 |
| Wallpaper, lightweight clay bricks | 12 |
| Dispersion paint, wallpaper, B 15 concrete | 11 |
| Dispersion paint, wallpaper, lightweight clay bricks | 11 |
| Timber | |
| Spruce, pine, beech (untreated) | 20-25 |
| Oak (untreated) | 12 |
| Oak (wax finish) (floors) | 3 |
| Textiles | |
| Curtains of natural fibres, 0.15-0.30 kg/m² | 5-15 |
| Carpets of natural fibres | 30-36 |
| Carpets of synthetic fibres | 15 |

The following test illustrates the process of water vapour sorption:
200 g of water evaporates in 30 min in each of two rooms measuring 4 x 4 x 2.5 m having 50% r.h. Accordingly, the relative humidity must climb to 79%. In room A with lime plaster on the walls and the ceiling the relative humidity only rises to 58%. In room B an oil-based paint was applied to the plaster so that it could no longer absorb any water vapour. The relative humidity rose almost to the predicted 79% here. Four hours after evaporation the relative humidity in room B was still 10% above that in room A (figure 2.3.12).
Only the layers near the surface, i.e. plaster and wallpaper, participate in these processes. This can be seen clearly in table 2.3.13, which is based on the results of corresponding studies. Which material is used as the substrate for surface layers is completely irrelevant to the water vapour sorption of components.

**Natural radioactivity**
Human beings have been exposed to ionising radiation throughout their existence. Cosmic radiation emanates from thermonuclear fusion in the sun; its magnitude depends on the altitude of the site. Part of the ionising radiation originates from radioactive substances in the soil, including uranium, radium and thorium. This radiation causes a load on the human body, specified as the whole-body dosage in mSv/a. Together, cosmic and terrestrial radiation in Germany measures between 0.3 mSv/a on a ship at sea and 2.0 mSv/a on granite rocks at an altitude of 3000 m (figure 2.3.14).

In addition to this radiation acting on the body from outside there is the contribution from radioactive substances in the air we inhale, indoor and outdoors, or take in with our food. This means that in Germany the average natural radioactive load is 2.4 mSv/a (table 2.3.15).

Radioactivity of concretes
All natural rocks and soils contain small amounts of potassium 40, radium 226 and thorium 232, whose decomposition produces the radioactive noble gas radon. Building materials produced from these, including concretes, therefore emit radiation according to the raw materials and release small amounts of radon into the interior air.
The specific radioactivity of concretes depends on their composition. The influence of the cement on the radioactivity of the concrete is low, corresponding to its proportion by weight. The influence of the aggregate is by far the most critical.
The natural radioactivity of the concretes common in Germany is very low, even in comparison to other building materials.

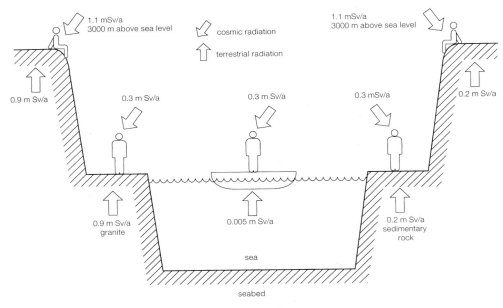

2.3.14 Natural radioactivity in the open air

2.3.15 Average natural equivalent dosage H of the population of Germany in 1989; figures in mSv/a according to [15]

| Origin | H | |
|---|---|---|
| Cosmic radiation | approx. 0.3 | |
| Terrestrial radiation from outside | approx. 0.5 | |
|   in the open air | | approx. 0.43 |
|   in buildings | | approx. 0.57 |
| Inhalation of radon in residential buildings | approx. 1.3 | |
| Natural radioactive substances in the body | approx. 0.3 | |
| Total exposure to natural radiation | approx. 2.4 | |

The release of radon from a component depends on the radium and thorium content of the material as well as on its porosity and the practical moisture content. The radon concentration in the interior air is also very heavily dependent on the ventilation [11]. Keller comes to the following conclusion in [24]:

"Building materials contribute only a small amount to the radiation exposure of the occupant. On the basis of these results, continuing to assess the effects of radiation from building materials in terms of their concentration of naturally radioactive substances is no longer justifiable. Due to their composition and structure (fused "glass beads"), building materials with a radionuclide content slightly higher than average (e.g. pumice or slag) release significantly less radon than, for example, highly porous building materials with a much lower radionuclide content. As the contribution of customary building materials to the total radiation exposure in buildings is relatively small, building materials should be assessed – if at all – with regard to their sealing effect towards radon diffusion in the course of radon refurbishment measures."

Radon in interiors
Investigations in existing buildings have proved that the radon load in interiors comes principally from the soil and not from the building materials used. The shielding effect of concrete ground slabs and basement walls means that such buildings exhibit markedly lower radon concentrations than similar older buildings with compacted clay floors and basement walls of natural stone masonry with a high joint proportion. In Germany it is recommended that the radon concentration in the interior air over a longer period should not exceed 250 Bq/m$^3$. The average value is about 50 Bq/m$^3$.

**Airborne pollutants and air change rate**
Providing the occupants with an adequate supply of fresh air demands a minimum air change rate of 0.5-0.8 per hour under normal living conditions. This means that, for reasons of hygiene, the volume of air in a residential unit must be replaced by fresh air 0.5-0.8 times every hour. At the same time, this minimum air change rate ensures that particles in the air are removed and their concentration kept below certain threshold values, and so do not present any hazards. Sources of pollutants in residential accommodation include:

- human secretions, principally water vapour, carbon dioxide, odours, microbes
- water vapour and odours caused by washing, bathing, cooking, houseplants
- the constituents of cleaning and washing agents
- gases from heating appliances such as ovens, cookers and open fires
- gases from building materials and interior fittings such as timber preservative, formaldehyde, radon

So the minimum air change rate can also be determined by the accumulation of pollutants and their removal. Depending on the use of the building, the density of occupation and activities, the avoidance of condensation on the surfaces of components, and the amount of water vapour can determine the minimum air change rate.

2.3.16 Number of air changes b according to the water vapour scale for different absolute moisture contents of the external air h

| | External air | | | Internal air | | | Difference | No. of air changes |
|---|---|---|---|---|---|---|---|---|
| | $\vartheta_a$ | $\varphi$ | h | $\vartheta_i$ | $\varphi$ | h | $\Delta h$ | $\beta$ |
| | °C | % | g/m$^3$ | °C | % | g/m$^3$ | g/m$^3$ | h$^{-1}$ |
| 1 | −10 | 80 | 1.64 | +22 | 50 | 9.26 | 7.62 | 0.5 |
| 2 | ±0 | 80 | 3.70 | +22 | 50 | 9.26 | 5.56 | 0.7 |
| 3 | +10 | 70 | 6.27 | +22 | 50 | 9.26 | 2.99 | 1.3 |

2.3.17 Quantities of moisture removed from the interior by vapour diffusion and air changes, after [26]

| Temperature of outside air [°C] | Quantity of moisture removed from the interior [g/h] due to vapour diffusion through external wall | due to a single air change |
|---|---|---|
| −20 | 5.5 | 436 |
| −10 | 4.8 | 378 |
| 0 | 3.2 | 242 |
| 10 | 0.4 | 15 |

# Building science

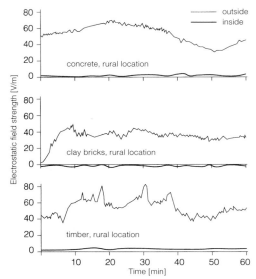

2.3.18 Electrostatic fields in the open air and in unoccupied interiors with components of reinforced concrete, clay bricks and timber [28]

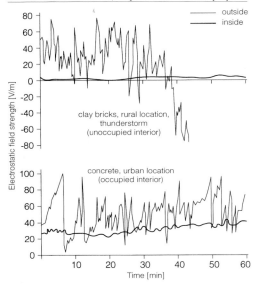

2.3.19 Electrostatic fields in the open air and in interiors [28]

a) upper half of diagram: during a thunderstorm, unoccupied interior, rural location
b) lower half of diagram: during fine weather, occupied interior, urban location

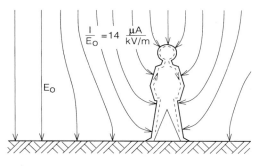

2.3.20 Atmospheric electricity fields do not penetrate the body (after Hauf)

This minimum air change rate based on hygienic and building science considerations is determined by the amount of water vapour and pollutants in the interior or exterior air, i.e. on the difference in the constituents between the fresh air and the exhaust air.

Table 2.3.16 shows that to remove the same amount of water vapour the air change rate must increase as the temperature of the outside air rises. In spring and autumn it reaches (for +10 °C, 70% r.h.) 2.6 times the value of the corresponding external air figure in winter (for -10 °C, 80% r.h.).

The air change rate can be achieved by organising through-ventilation via windows and doors depending on the layout of the building, or it must be guaranteed by means of mechanical ventilation.

"Breathing walls"

The components forming the building envelope do not "breathe" to provide an exchange of air between inside and outside. Such behaviour by the components is indeed undesirable because this would lead to an uncontrolled loss of heat and render the sound insulation inadequate. Nevertheless, again and again we hear the expressions "breathing activity" or "breathing ability" referred to as a particularly advantageous property of components with respect to improving the quality of the interior air. However, if diffusion processes are really meant here, these are negligible in comparison to the exchange of air through ventilation.

The following comparison is considered in [26]:

"The desiccation effect both of vapour diffusion and the result of air changes increases as the temperature of the outside air drops. A comparison of the efficiency of the two effects (vapour diffusion and air changes) was carried out based the following assumptions:

Room (4 x 6 x 2.6 m) with two external walls consisting of 240 mm vertically perforated clay brick masonry (diffusion resistance index $\mu$ = 10), plastered inside, rendered outside

| | |
|---|---|
| window area: | 6 m² |
| air change rate: | single |
| interior air temperature: | 22 °C |
| interior air humidity: | 40% |
| exterior air humidity: | 80% |

The amounts of moisture listed in table 2.3.17 were removed from the room under these conditions according to the temperature of the outside air.

Consequently, the quantity of moisture transported by diffusion at winter outside temperatures is only 1-3% of that removed by air changes. In carrying out this experiment, favourable conditions were used when calculating the diffusion effect, i.e. two external walls and relatively vapour-permeable masonry. In other cases the diffusion proportion could be even lower. The conclusion is that in terms of removing moisture from the interior, the effect of vapour diffusion through external components can be totally disregarded. The interior air humidity is not influenced noticeably by the vapour permeability of the external walls."

### Electric and magnetic fields

It is frequently claimed that electrical and magnetic fields exert particular influences on the health of humans. Up to now this has not been proved scientifically for customary field strengths. Building materials, which in physical terms are only semiconductors, shield the enclosed space against external electrical fields. Therefore, no electrical fields can be measured within buildings, regardless of whether the external components are made from concrete, clay bricks or timber (figures 2.3.18 and 2.3.19). This also applies to the higher field strengths present during a thunderstorm. In contrast, in an occupied room the friction caused by the movement of people alone generates electrical fields equal to those of a thunderstorm and completely independent of the building materials employed.

The moist skin of the human body also has a shielding effect. This works like a Faraday cage and keeps the insides of our bodies free from the effects of a natural electrical field (figure 2.3.20).

Man-made alternating fields arise around alternating and three-phase current cables in interiors and beneath power lines. Studies throughout the world have shown that these non-natural alternating fields only penetrate the body in extremely low proportions (approx. 1:10 million) such that even under 380 000 V power lines only minimal electrical fields build up in the body. These account for about one-millionth of the natural voltages that can excite our nerve and brain cells. They therefore have no effect. [13]

The strength of the electrical fields inside buildings for living and working are only about one-thousandth of those beneath power lines. These fields lie well below strengths that could have any biological effect. Only in the case of strong fields with a very high frequency, e.g. in the microwave range, are there any risks. The effect of these is to cause excessive warming of the biological tissue of the parts of the body affected.

Magnetic fields penetrate all components regardless of the materials used; they also penetrate the human body. The only known natural magnetic field is the geomagnetic constant field that aligns a magnetised needle;

its field strength is 0.05 mT. Although investigations seem to indicate that some species, e.g. fish and birds, can perceive this magnetic field and use it as an aid to orientation, humans have not been shown to possess such abilities. Natural magnetic fields to not have a detrimental effect on our health.

The effect of non-natural magnetic fields generated by alternating currents have also been investigated worldwide. If indeed effects have been observed at all in interiors and in the open air, these fluctuate within the natural physiological ranges of the functions of our organs. The magnetic field strengths measured remain below 0.1 mT; they are therefore hardly any stronger than the natural magnetic constant field. In experiments, such field strengths have been proved to have no effect. Even in the vicinity of electricity substations, no magnetic alternating fields occur which could be damaging to our health. [13]

"Earth radiation theories" are widespread but lack any physical foundation.

It should be noted that neither the presence nor the morbific effect of earth radiation or geopathogenic stimulation zones can be claimed to have been proved. [13]

**Energy economy, thermal insulation**

Saving and handling energy intelligently are important goals in our modern society. Reducing the consumption of energy also always contributes to preserving the environment.

In construction there are two areas in which energy is consumed:
- in the production of building materials, components and buildings
- in the use of buildings

**Energy requirements for producing components**
The energy required to produce materials and components can be determined by analysing the process chain. It is expressed as the primary energy content per tonne and per cubic metre of building material or per square metre of building component. Table 2.3.21 illustrates comparisons based on this concept. It shows that, for example, lightweight concrete components with pumice aggregate exhibit the lowest primary energy content compared to similar components made from other materials. In this case nature has supplied the energy for producing the lightweight aggregate in advance. Even components made from normal-weight concrete have a very low primary energy content. However, this rises with the amount of reinforcement in reinforced concrete owing to the high energy content of steel.

2.3.21 Primary energy content of building materials, examples according to [29]

| Material | Density [kg/m$^3$] | Primary energy content [kWh/t] | Primary energy content [kWh/m$^3$] |
|---|---|---|---|
| Concrete masonry units (pumice aggregate) | 700 | 290 | 203 |
| Calcium silicate masonry units | 1400 | 242 | 339 |
| Concrete masonry units (expanded clay aggregate) | 700 | 678 | 475 |
| Aerated concrete masonry units | 550 | 863 | 475 |
| Lightweight clay bricks (polystyrene void formers) | 800 | 681 | 545 |
| Concrete roof tiles | 2300 | 206 | 474 |
| Clay roof tiles | 2000 | 754 | 1508 |
| Normal-weight concrete, grade B 25 | 2300 | 196 | 451 |

2.3.22 Energy-saving buildings, design advice

| Purpose | Measures | Examples | Purpose | Measures | Examples |
|---|---|---|---|---|---|
| Reducing heat losses in winter | Choice of location | Wind direction, compass direction, topology, subsoil, water table | Compensating for heat losses | Heating | Correct sizing of system |
| | | | | | High degree of efficiency |
| | | | | | Temperature-based control |
| | Consideration of microclimate | Driving rain, incidence of sunlight, shade, wind, cold air dam, fog | | | Temperature lowered at night |
| | Orientation of building | Small areas facing north and prevailing wind direction | | Passive use of solar energy | Orientation of large building areas to face south |
| | | | | | Living rooms facing south |
| | Choice of compact building form | Smaller ratio of external area to volume | | | Large window areas facing south |
| | | | | | Provision of interior components that store heat |
| | Plan layout | Rooms with high temperatures facing south, living rooms | | | Concrete walls behind glass as heat collectors |
| | | Rooms with lower temperatures facing north, formation of buffer zones, storage rooms | | | Concrete walls behind transparent thermal insulation as heat collectors |
| | | Lobbies to external doors | | Active use of solar energy | Concrete components constructed as energy absorbers (massive absorber heating system) |
| | Thermal insulation | Low U-values for external components | | | |
| | | Avoidance of thermal bridges | | | External walls, roofs, garden walls, garages |
| | | Small window areas facing north and prevailing wind | | | Interior components constructed to store energy Walls, ceilings, floors |
| | | Temporary thermal insulation to window areas by way of roller and folding shutters | Protecting against heat in summer | Reducing ingress of solar radiation | Exernal components shaded by planting Provision of roof overhang |
| | Ventilation | Maintain minimum air change rate | | | |
| | | Enable brief, intensive ventilation via windows | | | Shading to windows by way of roller or folding shutters, awnings |
| | | Sealed joints in windows, avoidance of incidental ventilation | | Reducing temperature gradient | Low TAV, especially for roofs |
| | | Inclusion of mechanical ventilation, ducts, shafts, fans | | Heat storage | Provision of interior components that store heat |
| | | Provide heat recovery | | Ventilation | Day: min. air change rate |
| | | | | | Night: higher air change rate |

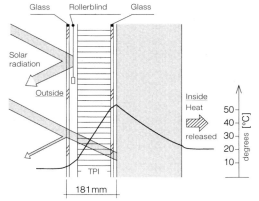

2.3.23 Comparison of transparent and opaque thermal insulation [1]

### Energy-saving buildings

The energy required in buildings for heating and ventilation in winter, and perhaps for air-conditioning in summer, is, taken over the entire useful life of a building, generally much higher than that required to construct the building. The possibility for savings is thus correspondingly greater. Table 2.3.22 provides a summary of advice on the planning and operation of energy-saving buildings. These reveal that good thermal insulation of components is only one factor – albeit an important one – among many that determine the thermal balance of a building.

Energy economy measures
Measures for reducing the consumption of fuel for heating are far more effective when associated statutory provisions are established, for both new buildings and the existing building stock. Space heating accounts for about 40% of the primary energy requirement in Germany. To reduce this energy requirement, the building industry must apply the following and other measures:

*Reduction of thermal losses*
Decreasing the heating requirement of buildings calls for optimised design according to location, orientation and functions, the prudent use of thermal insulation measures and intelligent technology.
As we can see from table 2.3.22, energy economy begins with the choice of location. We have to assess the compass direction, the lie of the land, whether we are building on a north- or south-facing slope, the subsoil and groundwater.
The microclimate, which is affected by factors such as shading from trees or other buildings, driving rain, cold air dam or fog, can have an adverse effect on the heating energy requirement.
When aligning the building, small areas should face north or the prevailing wind direction. Compact building forms with a low ratio between heat-exchanging external surfaces and the volume are advantageous.
When planning the layout, rooms with a high heating requirement, e.g. living rooms, should be placed on the southern side and those with a low heating requirement, e.g. storage rooms, on the northern side.

The aim should be to optimise thermal insulation in economic terms. In Germany this should lie between 0.4 and 0.6 W/m²K for external walls, between 0.2 and 0.3 for roofs, and between 0.3 and 0.4 for floor slabs over basements, taking into account currrent prices of building and insulating materials, and of heating energy.
Temporary thermal insulation for windows (e.g. sliding or roller shutters) are intended to prevent unnecessary thermal losses during the night.

In compact buildings with good thermal insulation, thermal losses due to ventilation can exceed those due to transmission through the components. However, the minimum air change rate must be maintained to satisfy the demands of hygiene and building science; but this minimum should not be exceeded by too much if energy economy demands are to be met. This means ensuring controllable ventilation facilities instead of uncontrolled changes of air via leaky joints. Systems for recovering the heat from the exhaust air can further reduce heat losses due to ventilation.

*Compensating for thermal losses*
The correct design and control of heating plant is another important factor in saving energy. The passive use of solar energy is still primarily via the windows. They function best as solar collectors when positioned facing south. Good designs have large window areas facing south and solid interior components to store the incoming heat temporarily. The use of solar energy with typical external wall and roof constructions is relatively low – between 2 and 12%. Table 2.3.25 contains details of solar gain factors for components depending on their compass direction.

However, if we attach transparent layers of insulation to the outside of external components, such surfaces can be transformed from thermal loss to thermal gain areas. Figure 2.3.23 illustrates the comparison between transparent and opaque thermal insulation.

The principle of transparent thermal insulation is simple:
A transparent layer of thermal insulation is attached to a solid external wall, e.g. of concrete. Both direct and diffuse solar radiation penetrate the insulation. This strikes the solid, dark-coloured external wall and is converted from short-wave light radiation into long-wave heat radiation. But the transparent thermal insulation is less permeable for the heat radiation and so it is stored in the solid external wall, which has good conducting properties, and released into the interior of the building.

One system that can be employed for active solar energy use in monovalent heating operations is the massive absorber heating system (figure 2.3.24). In this system, pipes carrying a liquid are embedded in external concrete components, e.g. external walls, balcony walls, retaining walls, roof surfaces. The circuit on the cold side of a heat pump cools down the absorber surfaces to just below the ambient temperature. This means that energy is absorbed and raised by means of a heat pump to a level suitable for space heating. Concrete is the ideal building material for both systems (transparent thermal insulation and massive absorber heating system) because its

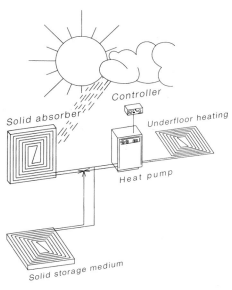

2.3.24 Massive absorber heating system [31]

thermal properties (heat storage capacity and thermal conductivity) can be custom-designed by choosing the right wall thickness, density and aggregates.

*Protecting against heat*
In summer the aim is to protect the building against incoming heat. One of the main ways of doing this is to shade the windows from solar radiation using plants, awnings or shutters. A low temperature-amplitude ratio (TAV), especially for roof slabs, has a beneficial effect on the interior temperature in summer; the TAV of a roof slab should therefore not exceed 0.10. This means that no more than 10% of the external temperature fluctuations are transmitted to the interior. Components with usable storage masses on the inside, e.g. concrete roof slab with layer of thermal insulation on the outside, fulfil this requirement. Furthermore, suitable internal components can store heat throughout the day and then release it again at night via the ventilation when cooler temperatures prevail (see "Requirements in summer").

*Developments in thermal insulation requirements*
All the measures for reducing the heating requirement specific to a building generally have a positive effect on the physical factors of thermal comfort. They therefore also create the right conditions for a building with a healthy interior climate.
The statutory provisions for thermal insulation in buildings contain minimum requirements for individual components; these are intended to guarantee their long-term serviceability. The provisions also contain enhanced requirements for the heat-exchanging envelopes of complete buildings. The aim here is to save as much heating energy as possible.

*Minimum requirements*
DIN 4108 lays down the foundation for thermal insulation and energy economy in buildings. It includes parameters for building materials and components, methods of analysis and advice concerning design and workmanship. The requirements represent minimum values for thermal insulation. The forthcoming harmonised European standard will also contain no more than the principles of thermal insulation and specify minimum requirements, e.g. according to hygiene aspects.

2.3.25 Solar gain factors for external walls and roofs after [32]

| Orientation | $\psi_w$-values for walls | | Collector wall |
|---|---|---|---|
| | typical external wall | | |
| | light colour | dark colour | translucency T = 0.5 |
| south | 0.04 | 0.12 | 1.14 |
| east, west | 0.03 | 0.07 | 0.93 |
| north | 0.02 | 0.06 | 0.85 |
| | $\psi_D$-values for roofs | | |
| general | 0.07 | | |
| horizontal | 0.12 | | |

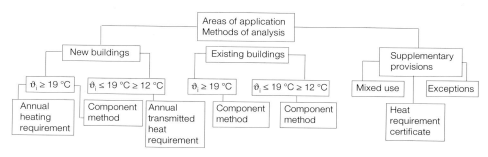

2.3.26 The Thermal Insulation Act: overview of areas of application and methods of analysis

2.3.27 Maximum values for the annual heating requirement related to the heated volume of the building V or the usable floor area of the building $A_N$ in relation to the ratio A/V

| A/V | max. annual heating requirement | |
|---|---|---|
| | related to V $Q'_H$ [1] | related to $A_N$ $Q''_H$ [2] |
| [m⁻¹] | [kWh/m³a] | [kWh/m²a] |
| 1 | 2 | 3 |
| ≤ 0.2 | 17.3 | 54.0 |
| 0.3 | 19.0 | 59.4 |
| 0.4 | 20.7 | 64.8 |
| 0.5 | 22.5 | 70.2 |
| 0.6 | 24.2 | 75.6 |
| 0.7 | 25.9 | 81.1 |
| 0.8 | 27.7 | 86.5 |
| 0.9 | 28.4 | 91.9 |
| 1.0 | 31.1 | 97.3 |
| ≤ 1.05 | 32.0 | 100.0 |

[1] Intermediate values are to be calculated according to the following equation:
$Q'_H = 13.82 + 17.32(A/V)$ [kWh/m³a]

[2] Intermediate values are to be calculated according to the following equation:
$Q''_H = Q'_H / 0.32$ [kWh/m²a]

2.3.28  Requirements for the thermal transmittance coefficients for individual external components in the heat-transmitting envelope "A" for new, small residential buildings

| Line | Component | Max. thermal transmittance coefficient $U_{max}$ [W/m²K] |
|---|---|---|
| Column | 1 | 2 |
| 1 | External walls | $U_W \leq 0.50$ [1] |
| 2 | Windows and glazed doors fitted externally, roof windows | $U_{m,F\ eq} \leq 0.70$ [2] |
| 3 | Slabs below unused roof spaces, slabs (including pitched sections) separating the interior from the outside air above and below rooms | $U_D \leq 0.22$ |
| 4 | Floor slabs over basements, walls and floor slabs adjacent to unheated rooms, floor slabs and walls adjacent to the soil | $U_G \leq 0.35$ |

[1] The requirement is deemed to have been satisfied when masonry walls 365 mm thk are constructed using materials having a thermal conductivity $\lambda \leq 0.21$ W/mK.

[2] The average equivalent thermal transmittance coefficient $U_{m,F\ eq}$ corresponds to a thermal transmittance coefficient averaged over all windows and glazed doors fitted externally, taking into account the solar heat gains.

2.3.29  Reasonable sound levels to VDI directive 2058, 1973

| Area | Reasonable sound level [dB(A)] | |
|---|---|---|
| | day | night |
| purely commercial complexes | 70 | |
| primarily commercial complexes | 65 | 50 |
| general residential areas | 55 | 40 |
| purely residential areas | 50 | 35 |
| rehabilitation centres, hospitals, nursing homes | 45 | 35 |

2.3.30  Basic noise levels in apartments (according to information from the Deutsche Gesellschaft für Wohnungsmedizin e.V., Baden-Baden, 1985)

| Type of room, area | Recommended noise level [dB(A)] | |
|---|---|---|
| | day | night 10 p.m. to 7 a.m. |
| bedrooms with open windows (irrespective of type of housing area) | 30 | 25 |
| living rooms | 45 | 35 |
| gardens, balconies etc. | 35 | 30 |

*Enhanced requirements*

In Germany the requirements of energy-saving thermal insulation will continue to differ from state to state and will be specified in special rules according to climatic conditions.

As early as September 1988, Germany's Minister of Building published recommendations for "Ways to Achieve Low-Energy Housing". The need for further savings in heating energy was also based on the urgent need to cut the level of atmospheric pollution. The most significant pollutants resulting from combustion connected with heating are sulphur dioxide, nitrogen oxide, carbon monoxide and, in particular, carbon dioxide. The latter is regarded by climate experts as being responsible for unpredictable changes in our climate.

The current Thermal Insulation Act came into force on 1 January 1995. It regulates the areas of application illustrated in figure 2.3.26. The building group that in its normal use is heated to at least 19°C represents by far the largest number of new buildings covered by the Act, e.g. housing, offices, hospitals, schools.

An upper limit is placed on the annual heating requirement $Q_H$ per cubic metre of heated volume V or square metre of heated surface area depending on the ratio of the heat-exchanging envelope A to the volume V of the building it encloses (table 2.3.27).

These requirements must be verified using a prescribed method of analysis and uniform boundary conditions stipulated for the whole of Germany. In doing so, the heat losses are balanced against the heat gains according to the equation:

$Q_H = 0.9 (Q_T + Q_L) - (Q_I + Q_S)$ kWh/a

where:
$Q_T$ = transmission heat requirement, i.e. the proportion of the annual heating requirement caused by thermal transmittance through the external components
$Q_L$ = the ventilation requirement, i.e. the proportion of the annual heating requirement caused by heating the interior air exchanged for cold outside air
$Q_I$ = the internal heat gains, i.e. the usable heat gains, e.g. from household appliances, lighting, heat given off by the occupants, occurring within a building used for its intended purpose
$Q_S$ = the solar heat gains, i.e. the usable heat gains due to the incidence of solar radiation while the building is being used for its intended purpose

The factor 0.9 takes into account the fact that, generally, not all of the building is heated all of the time, and that, if necessary, the temperature is lowered at night.

These requirements are deemed to be satisfied in the case of small residential buildings not exceeding two full storeys and three residential units when the maximum U-values for individual components do not exceed those given in table 2.3.28.

Requirements for individual components are a departure from the thermal balance method. They ignore a major factor affecting the heating requirement – the compactness of the building. The component method allows detached, semi-detached and terraced houses to be built with a high heating requirement despite the low U-values of their individual components. This means that the most important aim of the Act, namely to save heating energy, is not properly implemented.

The new edition of the Thermal Insulation Act, currently in draft form, will make the low-energy standard the norm. The heating requirement of new buildings will thereby be reduced by about 30% compared to current requirements. Based on new national and European standards, there will be a comprehensive analysis of the annual heating energy requirement Q.

The new equation is:

$Q = Q_h + Q_w + Q_t - Q_r$ kWh/a

where:
$Q_h$ = annual heating requirement
$Q_w$ = annual hot water heating requirement
$Q_t$ = thermal losses of heating and hot water systems
$Q_r$ = heat introduced by regenerative systems

In calculating the annual heating requirement $Q_h$, the losses through the components of the building, including thermal bridges and heat losses due to ventilation, are balanced against the direct solar and internal heat gains.

For the first time, the efficiency of heating and hot water systems can be taken into account when assessing the thermal quality of a building (variables $Q_w$ and $Q_t$). Active systems or additional facilities for using regenerative heat sources can be included in the calculation by way of the variable $Q_r$.

Behaviour in fire and fire protection

The sizing of components
There are two methods for verifying that the dimensions of a construction are adequate for the fire resistance class called for: DIN 4102 or a test certificate.

The latter method is useful when the savings gained by deviating from the standard justify the time and expense of carrying out tests. Table 2.3.41 lists the minimum dimensions of the most important types of concrete component. The information applies to fire resistance classes F 30, F 90 and F 180 because these are the classes in the building codes and the guidelines of property insurers which correspond to the designations "fire-retardant", "fire-resistant" and "highly fire-resistant".

In addition, the concrete cover to the reinforcement must be adequate for the fire resistance class called for to protect the steel against excessive heat and hence premature loss of its loadbearing capacity. As a rule, the concrete covers specified in DIN 1045 are adequate up to fire resistance class F 60; the figures in DIN 4102 apply to classes F 90 and higher.

If the maximum slenderness ratio causes a problem for a certain wall, this can be dealt with by assigning the vertical forces to concealed columns (figure 2.3.36). Columns in enclosing walls need only be 140 mm thick to satisfy fire resistance class F 90; there is no maximum slenderness ratio for the columns in this case. The sections of wall between the columns finished flush are then merely non-loadbearing infill panels.

Concrete sandwich panels are often used as non-loadbearing external walls. This also includes spandrel panels and aprons as well as combinations thereof. Such components are subject to the same requirements as non-loadbearing external walls. If this is not possible in a particular case, only the loadbearing layer of the concrete sandwich panel may be positioned in the plane of the construction; insulation and facing skin must lie outside in every case. Brackets carry, for example, the vertical loads from the spandrel panels. Horizontal wind loads are carried by additional, usually steel, fixings. The best and easiest way to incorporate these is by way of pockets subsequently filled with grout. Other claddings to provide fire protection are then unnecessary. Special attention must be paid to the careful detailing of joints between adjoining components. A mineral fibre seal (materials class A, melting point ≥ 1000 °C, density ≥ 30 kg/m³) is recommended here. It must also be compressed during installation (compression ≥ 10 mm). Figure 2.3.37 gives a practical example.

2.3.39  Building materials classes for insulating materials according to [21]

| Product group | Building materials class to DIN 4102 part 1 | | | | |
|---|---|---|---|---|---|
| | incombustible | | combustible | | |
| | A 1 | A 2 | B 1 | B 2 | B 3 |
| Wood-wool lightweight building boards | | | ☐ | | |
| Multi-ply lightweight building boards | | | | | |
| Mineral fibre multi-ply ltwt boards | | | ☐ | | |
| Rigid expanded foam multi-ply ltwt boards | | | ☐ | ☐ | |
| Cork products | | | | ☐ | |
| Phenolic resin rigid foam | | | ☐ | ☐ | Not permitted in buildings |
| Polystyrene particle foam | | | ☐ | | |
| Polystyrene extruded foam | | | ☐ | | |
| Polyurethane rigid foam | | | ☐ | ☐ | |
| In situ polyurethane foam | | | ☐ | ☐ | |
| Urea formaldehyde foam | | | ☐ | ☐ | |
| Mineral fibre insulating materials | ☐ | ☐ | ☐ | | |
| Foamed glass | ☐ | | | | |
| Perlite insulating batts | | ☐ | ☐ | ☐ | |
| Calcium silicate insulating batts | | ☐ | | | |
| Loose perlite | ☐ | | | | |
| Cellulose fibre insulating materials | | | | ☐ | |

2.3.40  Allocation of building authority and DIN 4102 part 2 designations to components

| Building authority designation | DIN 4102 part 2 designation | Code |
|---|---|---|
| Fire-retardant | Fire resistance class F 30 | F 30 – B |
| Fire-retardant, and with loadbearing parts made from incombustible materials | Fire resistance class F 30, and with essential parts made from incombustible materials | F 30 – AB |
| Fire-retardant, and made from incombustible materials | Fire resistance class F 30, and made from incombustible materials | F 30 – A |
| Fire-resistant | Fire resistance class F 90, and with essential parts made from incombustible materials | F 90 – AB |
| Fire-resistant, and made from incombustible materials | Fire resistance class F 90, and made from incombustible materials | F 90 – A |

# Building science

2.3.41 Minimum thickness d or minimum width b and edge distances u[1] of various unclad concrete components for fire resistance classes F 30, F 90 and F 180 according to [30]

| Type of component | | | min. thickness d or min. width b and edge distance u [mm] for fire resistance class | | | | | |
|---|---|---|---|---|---|---|---|---|
| | | | F 30-A | | F 90-A | | F 180-A | |
| | | | d or b | u | d or b | u | d or b | u |
| Floors | Solid slab w/o screed | static. determinate supports | 60 | | 100 | | 150 | |
| | | static. indeterminate supports | 60 | | 100 | | 150 | |
| | Solid slab with bonded screed Slab thickness | | 50 | | 50 | | 75 | |
| | Total floor thickness (slab + screed) | static. determinate supports | 60 | | 100 | | 150 | |
| | | static. indeterminate supports | 80 | | 100 | | 150 | |
| | Solid slab with floating screed Slab thickness | static. determinate supports | 60 | | 60 | | 80 | |
| | | static. indeterminate supports | 80 | | 80 | | 80 | |
| | Screed thickness | | 25 | | 25 | | 40 | |
| | Edge distance of span reinforcement | | | | | | | |
| | spanning 1 way | Steel grade BSt 420 S Steel grade BSt 500 S | | 12 | | 35 | | 60 |
| | spanning 2 ways supported on 4 sides | Steel grade BSt 420 S Steel grade BSt 500 S | | 12 | | 15 | | 30 |
| Walls | Non-loadbearing | | 80 | | 100 | | 150 | |
| | Loadbearing | $\sigma \leq 0.5\beta_R/2.1$ | 120 | | 140 | | 200 | |
| | | $\sigma \leq 1.0\beta_R/2.1$ | 120 | | 170 | | 300 | |
| | Edge distance at crit T = 500 °C | | | | | | | |
| | | $\sigma \leq 0.5\beta_R/2.1$ | | 12 | | 25 | | 55 |
| | | $\sigma \leq 1.0\beta_R/2.1$ | | 12 | | 35 | | 65 |
| Beams | Exposed to fire on 3 sides crit T ≥ 450 °C | static. determinate supports | 80 | 25 | 150 | 55 | 240 | 80 |
| | | | 120 | 15 | 200 | 45 | 300 | 70 |
| | | | 160 | 12 | 250 | 40 | 400 | 65 |
| | | | ≥ 200 | 12 | ≥ 400 | 35 | ≥ 600 | 60 |
| | | static. indeterminate | 80 | 12 | 150 | 35 | 400 | 60 |
| | | | ≥ 160 | 12 | ≥ 250 | 25 | ≥ 400 | 50 |
| Columns | Exposed to fire on several sides | | 150 | 18 | 240 | 45 | 400 | 70 |
| | | | 150 | 18 | 300 | 35 | 500 | 60 |

Please also refer to table 2.1.12, p. 51, and table 2.1.19, p. 57

[1] u = distance from outer face of component to centre of reinforcing bar

## An overview of building science requirements

Many of the building science relationships and influences on the interior climate given above cannot be expressed as a single figure in the form of a minimum or maximum value. However, as in the case of the thermal comfort zones, they can be specified as ranges that then serve as targets for the design. Other influences on the interior climate and the well-being of the occupants depend on the behaviour of users, their individual needs and the circumstances of each individual case.

For these reasons, not all the issues of "healthy living" can be incorporated into building codes. Other basic demands on hygiene have long been taken into account in statutory instruments and technical specifications. For instance, the minimum requirements for thermal insulation and moisture control are intended to prevent condensation forming on interior surfaces and hence the growth of mould. The enhanced thermal insulation requirements designed to save energy lead to higher interior surface temperatures and hence improve thermal comfort. The regulations and requirements for sound insulation and fire protection also take hygiene aspects into account.

The most important requirements from the regulations concerning components are summarised in tables 2.3.42 and 2.3.43.

The minimum values for thermal insulation are taken from DIN 4108 part 2. The values for enhanced thermal insulation correspond to the Thermal Insulation Act of January 1995. Both the minimum requirements for the airborne sound insulation index R'w and the proposals for enhanced airborne and impact sound insulation are derived from DIN 4109 "Sound insulation in buildings". Requirements from the Protection Against Aircraft Noise Act are also included. The fire protection requirements correspond to those of the Building Code of the Federal State of North Rhine-Westphalia. The building codes of other German federal states may deviate from this.

Therefore, the use of tables 2.3.42 and 2.3.43 does not relieve the designer of his responsibility to study the respective statutory instrument because this contains further details for special cases and important information on the design of components.

## 2.3.42 Walls – requirements and recommendations

| Component | Thermal insulation | | | | Sound insulation | | | | Fire protection | | | |
|---|---|---|---|---|---|---|---|---|---|---|---|---|
| | Residential buildings | $m_d$ [kg/m²] | $1/\Lambda$ [m²K/W] | $U_w$ [W/m²K] | Noise level range | Hospitals | Apartments | Offices | Type of bldg | Low-rise resid. bldg with ≤ 2 housing units | Low-rise bldg | Other bldgs except high-rise bldgs |
| | | | | | | $R'_w$ [dB] | | | | | | |
| External walls (AW) | Enhanced thermal insulation to WVO[2] 1993 edition | | | | Sound insulation of wall/ window against external noise | | | | Loadbearing, stiffening, to BauONW[3] | | | |
| | where ≤ 0.70 W/m²K | | | | Window area proportion where $R_{w,res}$ = 40 dB | | | | | | | |
| | ≤ 2 storeys ≤ 3 apartments | – | ≥ 1.83 | ≤ 0.50 | 10% 20% | | | 45/30 40/35 | Resid. bldg | F 30 – B | F 30 – AB | F 90 – AB |
| | Next to soil | – | ≥ 2.73 | ≤ 0.35 | 30% 40% | | | 45/35 45/35 | In basements | F 30 – AB | F 90 – AB | F 90 – AB |
| | Min. values for thermal bridges to DIN 4108 pt 2 | | | | Min. values to DIN 4109 | | | | Non-loadbearing or parts thereof to BauONW[3] | | | |
| | – | 0 | 1.75 | 0.52 | I | 35 | 30 | – | All bldgs | – | – | A or F 30 |
| | | 20 | 1.40 | 0.64 | II | 35 | 30 | 30 | | | | |
| | | 50 | 1.10 | 0.79 | III | 40 | 35 | 30 | | | | |
| | | 100 | 0.80 | 1.03 | IV | 45 | 40 | 35 | | | | |
| | | 150 | 0.65 | 1.22 | V | 50 | 45 | 40 | | | | |
| | | 200 | 0.60 | 1.30 | VI | [1] | 50 | 45 | | | | |
| | | ≥ 300 | 0.55 | 1.39 | VII | [1] | [1] | 50 | | | | |
| Party walls between apartments (IVW) | – | | ≥ 0.25 | ≤ 1.96 | Enhan. Min. | 55 53 | 55 53 | 55 53 | – | F 30 – B | F 60 – AB | F 90 – AB |
| Walls to staircases (IW) | Enhanced Min. | | ≥ 1.56 ≥ 0.25 | ≤ 0.55 ≤ 1.96 | Enhan. Min. | 55 47 | 55 52 | 55 52 | All bldgs | F 90 – AB | Constructed as fire walls | |
| End walls of buildings, party walls (IW) | – | | – | – | Enhan. Min. | 67 57 | 67 57 | 67 57 | | F 90 – AB | Fire wall or F 90 – AB | Fire wall |

[1] Requirements should be specified according to local conditions
[2] Thermal Insulation Act
[3] Building Code of North Rhine-Westphalia

# Building science

### 2.3.43 Floors and roofs – requirements and recommendations

| Component | Thermal insulation to DIN 4108 pt 2 (min.), WVO 93[1] (enhanced)[2] | | | | | Sound insulation to DIN 4109 | | | | Fire protection to BauONW[3] | | |
|---|---|---|---|---|---|---|---|---|---|---|---|---|
| | $m_d$ | $1/\Lambda$ [m²K/W] | | U [W/m²K] | | $R'_w$ [dB] | | TSM [dB] | | Low-rise resid. bldgs with ≤ 2 housing units | Low-rise bldgs | Other bldgs except high-rise bldgs |
| | [kg/m²] | min. | enhan. | min. | enhan. | min. | enhan. | min. | enhan. | | | |
| Floors in detached & terraced houses | – | – | | – | | 55 | 15 | 25 | | F 30 – B | F 30 – AB | F 90 – AB |
| Separating floors (DE) with heat flow up (DE0) down (DEU) | – – | 0.35 0.35 | – – | 1.64 1.45 | – – | 54 | 55 | 10 | 17 | No requirement | | |
| Floors next to soil (FB) | – | 0.90 | ≥ 2.86 | 0.93 | ≤ 0.35 | – | | – | | – | | |
| Floors below unused roof spaces (DD) | 0 20 50 ≥ 83 therm. | 1.75 1.40 1.10 0.90 0.45 bridge | ≥ 4.51 | 0.51 0.62 0.76 0.90 1.52 | ≤ 0.22 | 53 | 55 | 10 | 17 | F 30 – B | F 30 – AB | F 90 – AB |
| | | | | | | | | | | No requirement, provided no occupancy possible | | |
| Floors over basements (DK) | – therm. | 0.90 0.45 bridge | ≥ 2.86 | 0.81 1.27 | ≤ 0.35 | 52 | 55 | 10 | 17 | F 30 – AB | | F 90 – AB |
| Floors over driveways or sim. (DL) | – 1.30 therm. bridge | 1.75 | ≥ 4.51 0.66 | 0.51 | ≤ 0.22 | 55 | – | 10 | 17 | | F 90 – AB | |
| Floors over occupied rooms separating them from the outside air (DA) | 0 20 ≥ 50 therm. bridge | 1.75 1.40 1.10 0.80 | ≥ 4.55 | 0.52 0.64 0.79 1.03 | ≤ 0.22 | as AW tab. 47 | Aircraft noise: Z1 50 Z2 50 | – | | F 30-B | F 30-AB | F 90-AB |
| Stairs general terr. houses | – | – | | – | | – | | 5 5 | 17 17 | F 30 – AB | F 30 – AB | F 90 – AB |
| Roofs (unused) | – | – | | – | | – | Aircraft noise: Z1 30 Z2 25 | – | | Roof covering resistant to flying sparks & radiated heat | | |

[1] Thermal Insulation Act
[2] In residential buildings with ≤ 2 storeys and ≤ 3 apartments
[3] Building Code of North Rhine-Westphalia

# Part 3 · Reinforced concrete in buildings

Stefan Polónyi · Claudia Austermann

Multistorey structures

    Aspects of industrial production
    Joints in structures
    Roofs
    Floors
    Walls
    Columns
    Service cores, stairs
    Facades
    High-rise buildings
    Suspended high-rise buildings

Single-storey sheds

    Single-storey sheds of trusses and frames
    Stressed skin structures
    Suspended roofs

Foundations

    Shallow foundations
    Deep foundations, pile foundations
    Securing excavations
    Building below the water table, ground water control

References

This section demonstrates the various structural possibilities of reinforced concrete. It is essentially based on lectures on structural design for the architecture and construction engineering courses – including structural engineering, production and management in the building industry – at Dortmund University. It is intended to provide guidance for planners and engineers during the preliminary and detailed design stages of the HOAI (fee tariffs for architects and engineers). Therefore, special attention is paid to the factors that influence the design and dimensions of reinforced concrete structures, e.g. thermal insulation, sound insulation, fire protection, building services. Aids for decision-making are provided for general cases.

Concrete technology processes are less interesting during these planning phases. Reinforcement (layout, calculating the cross-sections) is thus not dealt with in detail, if at all. Problems that are easily solved by referring to the *Beton-Kalender* are also not mentioned here. Exceptions to this are aids that can only be found in the wider literature and those that considerably ease the engineer's workload.

The reader is asked not to use the tables and diagrams without first referring to the associated passages of text.

Part 3 of this book is primarily based on Eurocode 2. Reference to German DIN standards is restricted to situations not covered by the Eurocode.

This book has been conceived as a design aid. The arrangement therefore differs from the usual approach to building design, in which the individual components are discussed in isolation. Contrary to usual practice, reinforced concrete construction will be shown in context; the reader will not be distracted by forms of construction irrelevant to the task in hand.

This section of the book has been arranged based on the utilisation of the buildings. This determines the type of structure – multistorey block or single-storey shed. Of course, the top storey of a multistorey building may also be built like a single-storey shed.

The chapter on multistorey buildings is further subdivided according to type of use. Flooring systems supported on walls are primarily suited to residential and office buildings, whereas floors supported on columns are better suited to commercial buildings and multistorey car parks.

The main loadbearing structures of single-storey sheds are divided into those composed of trusses and frames and those made up of stressed skin structures. The former are primarily used for industrial buildings, the latter for exhibition halls, railway canopies, churches etc.

The chosen allocation of types of construction is not intended to produce a strict classification; cross-references are therefore included. Allocating the foundations according to the multistorey and single-storey scheme does not make much sense and so this approach is not employed in the chapter on foundations.

Loadbearing structures

3.0.1 Structural systems

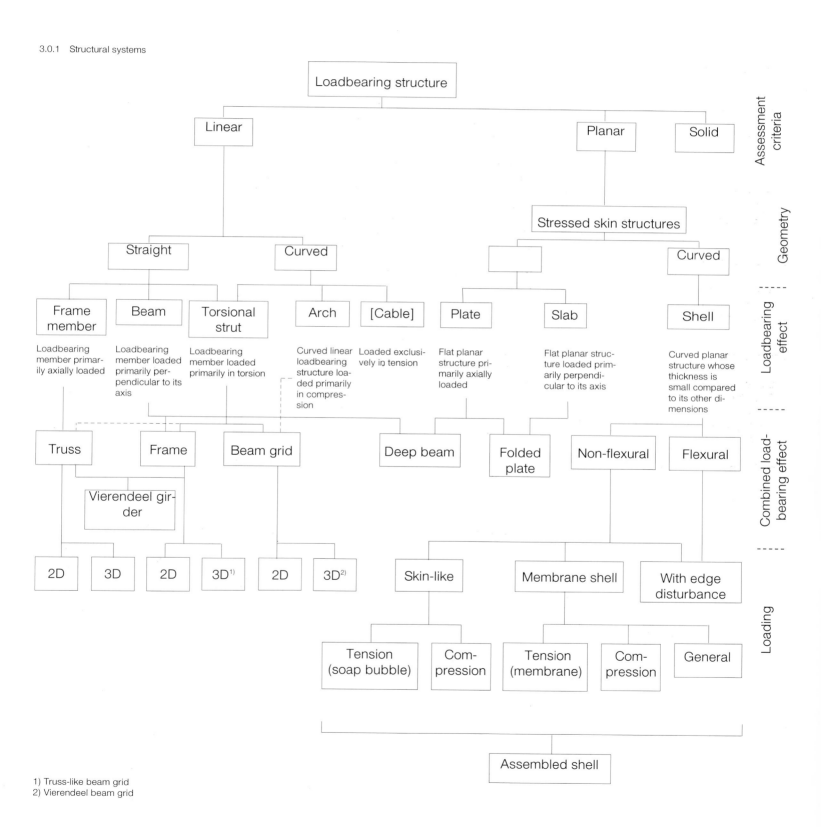

1) Truss-like beam grid
2) Vierendeel beam grid

# Multistorey structures

**Aspects of industrial production**

The industrial production of structures was made possible by the development of "easy-to-use" cranes. At first, cranes were used for lifting building materials (bricks, mortar, concrete, formwork) on the building site. Later, formwork and scaffolding were combined to form larger units. Parallel with these developments came the design of taller, heavier components, blocks, and finally, wall and floor panels, which in turn called for cranes capable of lifting heavier loads.

These improved methods of producing the loadbearing structure have consequences for the interior finishes. As these large formwork panels can be finished smooth and reused many times economically, plaster is not necessarily required. The interface between the finishes and the structure must adapt to the new methods. But the new methods of production also have consequences for the layout of the building:

- For in situ concrete, the layout should allow the formwork and scaffolding units to be reused many times; they should be able to be relocated without having to be dismantled first.
- For precast (prefabricated) concrete components, the aim is to minimise the number of different elements.

Initially, an attempt was made to use the new techniques to construct buildings designed according to conventional methods. The desired rationalisation effect was not forthcoming. This was not surprising, as the conditions that enable sensible application and promotion of the new building methods were recognised only gradually.

What we understand by industrial production is the manufacture of commodities more or less without manual work. The identity of industrial production and prefabrication cannot be derived from this definition. In terms of the loadbearing structure, production on site can be either an industrial or semi-industrial process. In many cases, however, prefabrication is the prerequisite for industrialisation.

The word "industrial" should not be simply regarded as a positive attribute without any drawbacks. It does not necessarily represent the most sensible method of production. However, we can assume that under certain conditions industrial production is the most economic, taking into account current costs of materials and labour in the industrialised world.

Semi-industrial production means that the loadbearing structure is concreted in situ but with the help of large scaffolding and formwork elements.

Construction with the industrial building methods described below presupposes prefabrication in a permanent factory off site. The use of a temporary, on-site plant enables larger elements to be produced because the dimensions are not limited to those possible for road transport. But considering the distribution of precasting plants in Germany, on-site plants are set up only in exceptional circumstances because the conditions in a factory are more favourable. Therefore, only elements suitable for fabrication in a permanent factory are discussed here.

The most important conditions for the industrial and semi-industrial production of buildings have already been mentioned above. A further distinction results from the following explanations of the building methods. However, the following general planning criteria apply:

- simplified layouts
- the avoidance of changes of level within individual storeys
- a strict architectural form

**The planning sequence**

In manual methods of production, the steps in the planning of the building can be summarised as shown in table 3.1.1. The adaptability of manual production makes feedback between the two planning phases unnecessary.

But feedback is indispensable in the planning of system buildings. This is why it is impossible to separate the individual planning areas. The term "system building" is intended to describe structures produced either industrially or semi-industrially. Table 3.1.2 attempts to clarify the sequence of operations.

# Aspects of industrial production

The choice of a loadbearing construction comprises the following steps:

- Drawing up a list of the potential types of construction
- Checking whether these fulfil utilisation requirements
- Appraising the types of construction that do justice to the utilisation

One of the most important aspects in appraising building systems is a comparison of costs. The cost of a structure can only be determined by planning it completely, working out the quantities of materials required and finally applying prices to those materials and the labour input. The amount of work required to compare a large number of potential solutions cannot be justified. The appraisal is therefore carried out in two stages:

- outline analysis
- detailed analysis

In the outline analysis the choice is made according to unquantified criteria (table 3.1.3). This involves drawing up a list of all the factors that affect cost in the form of a list of criteria. Only the forms of construction that fulfil the majority of, and most important, criteria are included in the detailed analysis, where a complete or partial cost comparison is carried out.

**Building services**

The services for a building must be taken into account when designing the loadbearing construction. The loadbearing structure and the building services form a constructional unit characterised by the fact that they do not obstruct each other. Close cooperation is therefore necessary between architect, structural engineer and building services engineer. In doing so, the individual details should be specified during the detailed design stage and not left to be sorted out on site.

Vertical runs of services are grouped together in layers. The horizontal runs do not disrupt the loadbearing structure in the case of flat slabs. It is sensible to concentrate services in certain areas (e.g. in a corridor) in which a lower clear ceiling height is permissible and therefore a suspended ceiling can be incorporated, or positioned lower than in other rooms. If downstand beams are required for longer spans, their positions should be coordinated with the service runs (see p. 120). The building services should be arranged in such a way that they do not disrupt or weaken the loadbearing construction, nor introduce any problems during erection. Slots should be avoided. Conduits for electric cables should not be cast into the concrete.

3.1.1 Planning sequence for loadbearing construction – manual production methods

| 1st planning phase | Needs –> Wish to build –> Utilisation demands, room schedule | | |
|---|---|---|---|
| Preliminary and detailed design | Plot of land, permitted building density | | |
| | Infrastructure provision, room allocation, plans, elevations | | |
| 2nd planning phase Construction scheme | Loadbearing construction<br>• selecting load-carrying components (walls, columns)<br>• selecting flooring system<br>• selecting foundation | General interior finishes and fittings | Building services |
| | Construction specifications (tender) | | |
| | Evaluating tenders, awarding contract | | |

3.1.2 Planning sequence for loadbearing construction – system buildings

| 1st planning phase | Needs –> Wish to build –> Utilisation demands, room schedule | | |
|---|---|---|---|
| Preliminary design | Plot of land, permitted building density | | |
| | Room allocation, plan and elevation concept | | |
| 2nd planning phase Detailed design | Selecting loadbearing construction | Selecting general interior finishes and fittings | Selecting building services |
| | Plans, elevations | | |
| 3rd planning phase Construction scheme | Construction specifications | | |

3.1.3 Design criteria for industrialised building methods

- Materials requirements (low materials requirements)
    - design favourable in structural terms (no transfer structures)
    - spans
    - structural depth (e.g. storey-height girders) structural function of other components
    - no additional reinforcement for transport and erection
- Production (low labour costs)
    - transfer as many operations as possible to the factory
    - small number of different elements
    - small number of parts (fewer but larger)
    - simple form
    - few operations
    - mechanised and automated production
- Storage
    - stackable
    - no bulky units
- Transport
    - dimensions for transport
    - no bulky units
- Erection
    Placing
    - can be placed directly from the vehicle
    - parts of equal size
    - parts of equal weight
    - same size of crane jib
    - crane jib as short as possible
    - no scaffolding
    - no temporary supports
    - few connections
    - simple adjustment (no lateral propping, e.g. for wall panels)
    - adjustment and completion of connections depends on lifting equipment
    Connections
    - few connections (only two units meeting at one point wherever possible)
    - identical types of connection
    - no extra parts for connections
    - no bending moment capacity in precast concrete units
    - generous tolerances
    - no restraints

When gas lamps and, later, electric lights were first introduced, the pipes and cables were mounted on the plaster; only later were the pipes and cables concealed beneath the plaster. In masonry construction this is not unreasonable because slits can simply be cut into the wall. (For further information see "Masonry Construction Manual" pp. 148-49.) But this is not advisable with reinforced concrete, particularly when the formwork is so smooth that a skim coat of plaster suffices to finish the concrete surface. Electric cables should be routed in the general fittings: in skirting boards, in door frames, behind mouldings in the corners between the walls and the ceiling if necessary. The cables can be taken from here to ceiling-mounted lighting units. The outlets from conduits fitted in the floor slabs seldom suit the layout of the furnishings, so the cable has to be suspended exposed in order to reach the light; this could also be done from a ceiling corner moulding. In any case, it must be ensured that the reinforced concrete construction is disrupted as little as possible by the building services.

### General interior work

The term "structure" implies the loadbearing part of a building before it is completed by adding plaster, claddings or other finishes. This concept does not apply to modern reinforced concrete construction and certainly not to precast concrete construction. In order to maximise the usable floor area, the walls should be built no thicker than necessary.

The loadbearing walls should be positioned so that the floor spans lie within an economic range. In doing so, the function of the walls should not be forgotten. As a rule, party walls between apartments are constructed in reinforced concrete (d = 180 mm) for reasons of airborne sound insulation. If required, the wall between the master bedroom and a child's room can be built of reinforced concrete, too. The remaining walls are lightweight partitions (plaster, plasterboard on timber studding, aerated concrete). It is unwise to load the floor slab with heavy partitions. The aforementioned lightweight partitions can accommodate the deformations of the floor slab. This is why the more stringent deflection limits do not need to be adhered to (see p. 113).

A floor slab supported on two edges deforms like the surface of a cylinder. In doing so, the walls transverse to the main loadbearing direction are not adversely affected by the deflection of the slab. Walls parallel to the primary loadbearing direction are not at risk either when they are interrupted, e.g. by corridors and doors in which the door openings reach to the ceiling. Non-loadbearing walls that reach to the ceiling can be protected against being loaded by means of suitable joint details.

## Joints in structures

### Settlement joints

Settlement joints are included in order to prevent damage to components as a result of the differential settlement of different parts of the structure. Differential settlement can occur when the subsoil is subjected to different loads, e.g. different parts of the building are of different heights or are founded on different subsoil conditions. Differential settlement can also be expected when different types of foundation are used, e.g. when one part of the building has a shallow foundation and an adjoining segment is founded on piles. Settlement joints are also advisable when parts of a structure are built at different times, as the subsoil under the section built first will consolidate before the next section is constructed. As differential settlement leads to an offset at the joint, the finishes must take this into account. To avoid a step at this point, it is often helpful to provide coupling plates between the two parts of the structure. Settlement joints should be included in the design at an early stage (scale 1:200). They must continue through to the underside of the foundations.

Wherever possible, a settlement joint should not simply be a slice through the building but should be harmonised with the form of the structure, e.g. coinciding with a recess or projection on the facade.

### Movement joints

Movement joints are intended to prevent damage caused by temperature fluctuations, the use of materials with different coefficients of thermal expansion, shrinkage and, if necessary, creep of the concrete.

The regulations recommend including movement joints in structures at a spacing of about 35 m. This recommendation is correct for masonry in combination with reinforced concrete components because masonry and concrete exhibit different coefficients of thermal expansion; the shrinkage of the concrete acts like a negative temperature change. Displacements between walls and floor therefore occur in longer sections and these cause cracks in the walls.

The movement of the structure as a result of temperature fluctuations is reduced considerably by the thermal insulation. Therefore, insulated and non-insulated buildings need to be approached differently.

For insulated structures, the critical phase lasts from the completion of construction to when the structure reaches its operating temperature for the first time. Afterwards, temperature fluctuations are usually so small that the tensile strength of the concrete is able to absorb the restraint stresses. Movement joints are not usually necessary in insulated structures, even over longer lengths, if the shrinkage stresses are substantially reduced by means of appropriate technology. Shrinkage is deformation caused by the declining temperature due to the slowing hydration process and is not dependent on the loading. A low temperature during hydration is desirable. This can be achieved by using a low-heat cement and, during periods of hot weather, by cooling the fresh concrete and/or pouring the concrete at night. It is known that the shrinkage strain is small for low water/cement ratios. The concrete should not be placed starting at one end of the structure but instead at the middle (in the form of a spiral) whenever possible by two or more concreting gangs so that it is always possible to join on to concrete that has not yet hardened.

The curing of the concrete has a special significance. The use of sheeting or sprayed films is intended to ensure that the water cannot evaporate. This guarantees that

- the concrete has sufficient water for crystallisation, and
- the concrete does not cool down too quickly by evaporation.

Besides the sheeting acting as a vapour barrier, thermal insulation should ensure that the concrete cools down slowly and as evenly as possible. Spraying the newly placed concrete with water, common in the past and still encountered today, is wrong. This cools the concrete suddenly and unevenly, which leads to the formation of cracks.

Cracks cannot be ruled out completely even in concrete produced and cured with great care. Many authors recommend including heavy shrinkage reinforcement to minimise the crack widths. The amount of steel required is usually great because the cross-section of anti-crack reinforcement is proportional to the thickness of the component. The more economic solution appears to be to include only a small amount of shrinkage reinforcement and to grout up any cracks with a width > 0.3 mm. This grouting should not be regarded as a repair but rather as an additional stage in the production process. Grouting is carried out after the building has been insulated and the openings (windows, doors) closed off, but before heating the building to its operating temperature.

In basement car parks temperature fluctuations are limited; movement joints are therefore unnecessary. Any cracks that appear are grouted up at a low temperature.

When placing concrete against components with a lower temperature (e.g. walls at ground slabs and floor slabs), it is advisable to heat these components in order to avoid restraint stresses in the newer components caused by the declining temperature due to the slowing hydration process [24].

The construction of structures subjected to severe daily and seasonal temperature fluctuations, e.g. multistorey car parks, should be planned so that the temperature changes do not lead to any appreciable restraints. This is achieved by supporting the floors on the walls; these are designed in such a way that the plate effect merely serves to accommodate wind forces and does not resist the deformation of the floors. Wherever possible, the columns should be slender so that they do not act as restraints. The restraint stresses should be analysed and reinforcement included to accommodate them, or the sections prestressed (see p. 120). This is especially advisable on vehicle parking decks (without finishes to the slab) to ensure that the floor remains free from cracks. In order to avoid restraints, it can be necessary to install elastomeric bearings, e.g. to form hinged columns. One example of this is the Bahnhofstraße multistorey car park in Münster, Germany (figure 3.1.4). The floors are supported in such a way that they can move without restraint outside the service core.

The fire behaviour of structures without joints is assessed according to a fire report that incorporates various scenarios. In the case of a fire confined to one storey, the construction is investigated theoretically using the temperature difference between the floors exposed to the fire and the neighbouring ones. In doing so, steel is incorporated with the yield stress associated with the respective temperature, and concrete with the corresponding modulus of elasticity. In this way the restraint stresses quickly diminish. The loads on the stiffening core and/or plates are critical. They may not fail in shear during the first 90 minutes of the fire. This is where the "full fire" scenario applies. Studies of fire behaviour are necessary only in structures with an increased risk of fire.

The preceding concepts may not always apply. When the plan layout of the building has a constriction, movement joints will probably be provided at these points. In the case of a precast concrete structure, shrinkage and creep have mostly been completed by the time the elements are erected. The relationships are therefore more favourable in this case. Apart from that, with appropriately designed bearings,

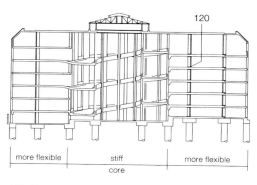

120 Elastomeric bearing

3.1.4 Bahnhofstraße multistorey car park, Münster, Germany

# Multistorey structures

## 3.1.5 Loads on floors and stairs

| Permanent loads | calculated according to EC and DIN 1055 part 1 | | | |
|---|---|---|---|---|
| Variable loads | DIN 1055 | | EC | * |
| | $q_K$ [kN/m²] | $Q_K$ [kN] | $q_K$ [kN/m²] | $Q_K$ [kN] |
| Imposed loads | | | | |
| Residential accommodation | | | | |
|   Floors | 1.5 | | 2.0 | 2.0 |
|   Stairs | 3.5 | 1.5 | 3.0 | 2.0 |
|   Balconies | 3.5 (5.0) | | 4.0 | 2.0 |
| Offices | 2.0 | | 3.0 | 2.0 |
| Places of assembly | 5.0 | | 4.0 | 4.0 |
| | | | 5.0 | 5.0 |
| Retail premises | 5.0 | | 5.0 | 7.0 |
| Garages, multistorey car parks | 3.5 | | 2.0 | 10.0 |

\* Contact area 50 × 50 mm

The surcharge for non-loaded lightweight partitions is also taken from DIN 1055 part 3 for calculations according to the EC.

| For wall load | ≤ 3.0 kN/m | ≤ 4.5 kN/m |
|---|---|---|
| $q_{KW}$ | 0.75 kN/m² | 1.25 kN/m² |

No surcharge is necessary for partition loads ≤ 4.5 kN/m when $q_K ≥ 5.0$ kN/m².

## 3.1.6 Partial safety factors for actions

| Effect | Type of action | |
|---|---|---|
| | permanent ($\gamma_G$) | variable ($\gamma_Q$) |
| unfavourable | 1.35 | 1.50 |
| favourable | 1.00 | 0 |

Combined factors may also be used. When designing floors, this has an effect only on the surcharge for partitions:

$$S_d = 1.35 G_K + 1.50(Q_K + 0.7 Q_{KW})$$

02 Reinforced in situ concrete
29 Screed
30 Plaster
42 Impact sound insulation

3.1.7 Floor construction for residential buildings

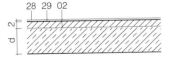

02 Reinforced in situ concrete
28 Resilient floor covering (textile)
29 Screed

3.1.8 Floor construction for office buildings

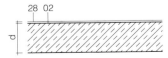

02 In situ concrete
28 Resilient floor covering (textile)

3.1.9 Floor construction with surface of concrete slab finished smooth

---

some of the movement can be accommodated at these points. Joints are weak points, and are also expensive. They should be included only where absolutely essential. They must be planned thoroughly, in terms of both structure and finishes, and installed with the utmost care.

### Construction joints

The size of a concrete pour is limited by construction methods and other aspects; construction joints separate the individual pours. The reinforcement passes through the construction joints without interruption. In the first pour a stop board prevents the concrete from flowing beyond the intended construction joint. In the next pour the concrete is cast directly against the rough end of the previous pour. Construction joints are established by the contractor in consultation with the designer and the site management. The construction joint is not visible in the finished structure.

To limit the effect of shrinkage, gaps 300-500 mm wide are often left between concrete pours. These are then filled in with concrete later after some of the shrinkage process has already been completed. The joints between the new and earlier pours are formed as construction joints.

### Methods of constructing reinforced concrete structures

Components cast in formwork at their final position are known as in situ or cast-in-place components.

In precast concrete construction, the components can be cast on the building site, but are preferably produced in a precast concrete works. They are then transported to the site, lifted into position by crane and the connections completed. In such cases the force transfer is guaranteed by means of bolts, welding of cast-in steel plates or starter bars (by way of bond, seldom with screwed connections).

The advantages of precasting:
- savings in terms of scaffolding and formwork
- production unaffected by the weather
- a more constant quality
- better organisation and control of the production
- a part of the workforce does not need to migrate between building sites

The disadvantages of precasting:
- connections are expensive
- moments are not transferred across connections (see table 3.1.3) and hence favourable continuity effects are unavailable, which may lead to thicker floors
- more loads sent by road

For semi-precast construction, the disadvantages of forming the connections are overcome by ensuring that no precast components need to be joined together. There is always a piece of in situ concrete between them, with the force transfer accomplished by overlapping the reinforcing bars.

## Roofs

At this point we will only consider roofs with reinforced concrete loadbearing structure. For other types of structures the reader is referred to [8] and [23].

### Pitched roofs

Pitched roofs are often built in reinforced concrete these days, especially when the roof space is to be fully utilised (mansard roofs). Precast concrete units are very common in larger structures or housing developments (see p. 186).

### Flat roofs

Flat roofs are constructed using reinforced concrete planks in conjunction with thermal insulation and bitumen or plastic (PVC) roofing felt. If the waterproofing is damaged, it is difficult to find the exact location. It is therefore recommended to divide the roof into smaller bays, each with its own fall, so that the search for the leak can be localised.

The examples in part 4 (p. 178) illustrate non-ventilated roofs (warm decks) only. Ventilated roofs (cold decks) require a twin loadbearing construction when the second layer also has a smaller span. As the warm deck can be constructed reliably, the extra cost of the cold deck is only justified in special circumstances.

Reinforced concrete roofs with bitumen roofing felts and gravel chippings or planting are heavy and advisable only for smaller spans (residential and office buildings). They are not suitable for the roofs to single-storey sheds.

If a waterproof concrete is used, the waterproofing task can be assigned to the reinforced concrete slab (upside-down roof). To avoid cracks that pass right through the concrete, the slab must be supported without restraint (slip layers). Where necessary, prestressed concrete can be used for the slab (see "Prestressed concrete floor slabs", p. 120). In all other respects, reinforced concrete roofs are designed like floor slabs. Generally, loading class I can be assumed (see table 3.1.22 and p. 185, figure 4.7).

# Floors

### Requirements

*Loading*
Tables 3.1.5 and 3.1.6 give the loads and partial safety factors.

*Sound insulation*
The requirements for sound insulation are stipulated in DIN 4109 and its supplement dating from November 1989 (see p. 93). A distinction should be made between residential buildings and general multistorey buildings when designing floors to meet the requirements of DIN 4109. Figure 3.1.7 shows the acoustic requirements for residential buildings; the plaster may be omitted and compensated for by employing a deeper concrete slab.

Although the November 1989 edition of DIN 4109 confirms that a reinforced concrete slab with a resilient floor covering offers adequate impact sound insulation, it prohibits this type of construction, obviously realising that some users could exchange fitted carpets for a different floor covering not complying with the requirements. This rule has resulted in a considerable increase in building costs. But without the rule, the consequences of the general application of this principle would be unpredictable.

Figure 3.1.8 shows the floor construction of an office building with a solid floor slab. Dimensional discrepancies in the reinforced concrete slab due to its production can lead to the levelling screed being just 20 mm thick in some places. We therefore assume this minimum figure when assessing the sound insulation.

There are various methods for finishing the surface of the concrete slab so smooth that a resilient floor covering can be laid directly onto it without the need for a levelling screed (figure 3.1.9 and "Surface finish", p. 114). Office buildings are increasingly employing cavity floors to cope with the constant increase in the number of cables required by office equipment. However, such floors may not be used to improve the airborne sound insulation.

*Fire protection*
The building codes of the federal states regulate fire protection requirements. In most cases the floors must comply with fire resistance class F 90, i.e. minimum thickness d = 100 mm. The concrete cover required for various boundary conditions, e.g. support, type of construction, type of screed, plaster etc., are described in DIN 4102 part 4, 1994 edition (see p. 94).

The requirements are less rigorous for continuous floor slabs when, for example, 20% of the reinforcement required at the supports is included in all spans as the minimum reinforcement in the top (figure 3.1.10). The effectiveness of this reinforcement can be explained as follows:
In a fire the yield point of the bottom reinforcement decreases. This leads to the formation of a plastic hinge acting on one side. The top reinforcement, which is exposed to the fire to a lesser extent, does not yield and prevents the creation of a kinematic chain (figure 3.1.11 A).

Figure 3.1.11 B shows the kinematic chain of a four-span floor without top reinforcement as a result of the formation of plastic hinges. DIN 4102 part 4, 1994 edition, requires 20% of the top reinforcement to be carried through only for floors supported on columns. For floors supported on beams and walls, the reinforcement at the supports merely needs to be extended by 0.15 l. This is, however, implausible.

*Limitation on deflection*
Limiting the deflection is intended to guarantee the serviceability of the structure [22]. In doing so, care is taken to ensure that the non-loadbearing walls supported on the floor are not cracked due to the deflection of the floor; also that the non-loadbearing walls below the floor are not loaded by the floor.

EC 2 recommends limiting the deflection f of components subjected to bending to $f = l_{eff}/250$ (where $l_{eff}$ = effective span) under the action of quasi-permanent loads. For components whose excessive deformation could lead to damage, deflection should not exceed $f = l_{eff}/500$.

The following influences must be known in order to determine the deflection accurately [15]:

- Parameters of the supporting member
  shape of cross-section
  amount of tension and compression reinforcement
  grade of concrete
  internal forces
- Concrete mix
  type of cement
  quantity of cement
  water/cement ratio
  type of aggregate
- Support conditions
  curing
  humidity
  average temperature during the first 7 days
- Loading history
  concreting, date ...
  side forms removed after ... days
  temporary supports removed after ... days
  all changes to the permanent loading

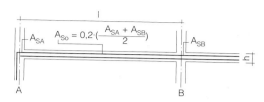

3.1.10  Continuous top reinforcement for fire protection purposes

A  Plastic hinge acting on one side

B  Effect without top reinforcement after formation of plastic hinges

3.1.11  Kinematic chain for a 4-span slab

3.1.12  Permissible ratio of effective span to effective depth (span/depth ratio) for simplified analysis of deflection limit to Eurocode 2

| System Linear member | Slab | Permissible ratio $l_{eff}/d$ or $l_1/d$ or $l_2/d$; component is: heavily loaded | lightly loaded | Applies |
|---|---|---|---|---|
| 1 |  | 2 | 3 | 4 |
| 1 |  | 18 | 25 | $l_1$ |
| 2 |  | 23 | 32 |  |
| 3 |  | 25 | 35 |  |
| 4 | flat slab | 21 | 30 | $l_2$ |
| 5 |  | 7 | 10 | $l_1$ |

# Multistorey structures

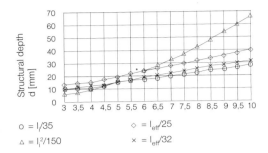

o = l/35  
△ = $l_i^2/150$  
◇ = $l_{eff}/25$  
× = $l_{eff}/32$

3.1.13 Structural depth of floor slab

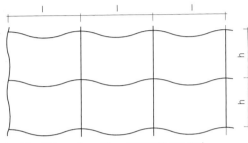

3.1.14 Deformation of floor slabs with one end restrained by reinforced concrete wall

| Structural system | Span l [m] | |
|---|---|---|
| | EC | DIN |
| (simply supported) | 3.12 | 4.35 |
| (one end fixed) | 4.00 | 5.40 |
| (both ends fixed) | 4.35 | 7.25 |

3.1.15 Maximum spans of floor slabs with minimum thickness for sound insulation

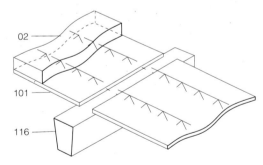

Detail of end support (during erection)

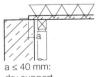

a ≤ 40 mm: dry support  
a > 40 mm: mortar bed

Temporary propping at support not required when a ≥ 35 mm and one node of the lower chord of every second lattice girder in the precast plank is located above the support.

02 In situ concrete  
101 precast floor plank  
116 Beam

3.1.16 Semi-precast (hybrid) floor slab

Apart from the parameters of the supporting member, none of these influences can be determined at the design stage. It is therefore advisable to limit not the deflection but rather the span/depth ratio $l_{eff}/d$ (where d = structural depth, i.e. distance from centre of gravity of reinforcement to edge in compression; and $l_{eff}$ = effective span) (table 3.1.12). Figure 3.1.13 compares the structural depths of various regulations. The designation in EC 2 regarding heavily or lightly reinforced components is defined by the reinforcement ratio (ratio of cross-section of span reinforcement to concrete cross-section). In buildings the components are generally lightly reinforced. Intermediate values may be interpolated. In determining the slenderness of the floor slab, the end span can be assumed to be restrained by a reinforced concrete wall (figure 3.1.14) because the stiffness of the two adjoining walls, in relation to the larger span of the floor slab, does not permit any appreciable angular rotation.

Table 3.1.15 gives the maximum spans of floor slabs with a minimum thickness to suit acoustic requirements, and according to type of support.

The figures in the table should be regarded as recommendations. Greater slenderness ratios can be allowed "when these are in agreement with the requirements placed on the structure or when excessive deflection can be ruled out because of structural measures employed (e.g. cambering, prestressing), a structural analysis and/or experience" [20]. However, $l_i/35$ should be regarded as an absolute minimum.

The details that prevent damage to non-loadbearing walls supported by the floor slab depend on the type of wall and its construction. Masonry (brittle) walls are more sensitive to deflection of the floor slab than timber-stud or demountable partitions see "General fitting-out", p. 110).

### Surface finish
The surface of the concrete slab cannot be finished smooth without further treatment. A levelling screed (bonded screed) is therefore usually provided. The task of the screed is to compensate for unevenness due to the building conditions and possibly also deviations from the horizontal. Levelling screeds usually have a nominal thickness of 40-50 mm. As the screed represents a dead load and increases the thickness of the floor, various methods have been developed for attaining a better surface finish to the structural slab, e.g. vacuum dewatering.

Even low-slump concrete contains roughly twice as much water in the mix as is necessary for crystallisation. The compacted concrete is covered with vacuum mats incorporating filter layers, and the excess water is extracted. Concrete hardened in this way can be finished smooth and exhibits a high strength at the surface. This method is therefore very popular for floor slabs carrying vehicular traffic, e.g. multi-storey car parks. If required, an angular-grained sand can be worked into the top surface. Large uninterrupted areas are necessary if this method is to be used economically.

In the Quintling method the hardening process of the concrete is controlled by adding a retarder so that the surface can be trowelled and floated a long time after placing the concrete.

### Flooring systems
*In situ concrete*
In situ concrete floor slabs are cast on site in the formwork. The use of suitable formwork can make it possible to produce such a smooth soffit that plaster or other claddings are no longer required. Modern formwork and scaffolding systems permit short turnaround times.

Floors supported on beams and walls
These floors may be simply supported on masonry walls or given a degree of fixity by being cast monolithically with reinforced concrete walls. The thickness of the floor depends on the sound insulation and fire protection requirements. With lower imposed loads the deflection limit or the slenderness governs, with higher loads it is the stress in the concrete.

Precast concrete floor planks and similar systems are classed as partly prefabricated or hybrid constructions. This involves using a reinforced concrete unit 50-60 mm thick precast at the works and already containing the reinforcement for the primary loadbearing direction. These lightweight planks are easily lifted into position on site by a rotating tower crane. They serve as permanent formwork for the subsequent in situ concrete topping (figure 3.1.16). Lattice girders, partly cast in and partly projecting from the planks, guarantee the necessary stiffness during transport and erection. These also render intermediate temporary supports unnecessary up to a certain span; however, supports are still required for longer spans. In the final condition the projecting lattice girders help to achieve a connection between precast and in situ concrete. These floor planks carry loads primarily in one direction. However, fixing additional reinforcement on the precast planks, which thus has only a small effective structural depth, can achieve a two-way-spanning action and so also guarantee a biaxial plate effect.

## Floors and roofs

*Crosswalls*

Floors with shorter spans, as is usually the case in residential and office buildings, are formed as reinforced concrete slabs without beams or ribs.

In residential buildings the crosswall arrangement shown in figure 3.1.17 is advantageous. Providing the crosswalls in the form of party walls is a sensible concept. Two compartments can be linked together by way of an opening (figure 3.1.23). So-called duplex, triplex and semi-duplex types etc. can be designed by incorporating openings in the floor and providing internal stairs. Access is by way of a central corridor, outer corridor or covered external passageway. Different (alternating) distances between crosswalls enable apartments of different sizes to be realised.

The arrangement of the crosswalls enables scaffolding and formwork units to be produced to match the size of the compartment and to be relocated in one piece. A central longitudinal wall does not present an obstruction to this method. After repositioning the scaffolding and formwork units, the reinforcement, predominantly in mesh form, is laid and afterwards the concrete placed, normally using a concrete pump. To be able to remove the scaffolding and formwork units without dismantling them, there should be no projecting columns or piers, even if these are desirable from a structural point of view. If this rule is infringed, the extra cost of dismantling can be higher than that for the extra reinforcement required to strengthen strips of slab and to provide thicker walls.

The facade consists of lightweight elements forming a unit together with windows and, possibly, doors.

The escape route from any point on a floor to the stairs may not exceed 30 m. The maximum centre-to-centre spacing of two stairs is therefore approx. 56 m.

In multistorey structures of reinforced concrete it is possible to support walls on just one column; in this case the position of the wall must be secured by the floor plates. The forces transferred to the floor must be able to reach the subsoil. In doing so, it must be ensured that scaffolding is removed only after all components necessary for stability of the structure have attained their intended load-carrying capacity.

In residential buildings with retail premises on the ground floor and/or parking in the basement, in hotels with restaurants and conference rooms below ground level, and in hospitals in which patient care is carried out on the upper floors and treatment rooms are located

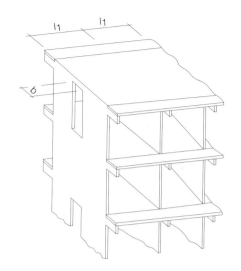

3.1.17 Crosswalls in in situ concrete

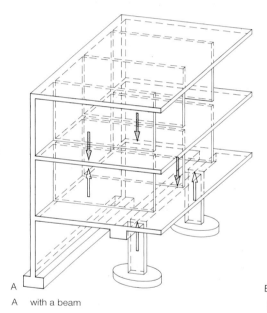

A with a beam

3.1.18 Supporting a crosswall

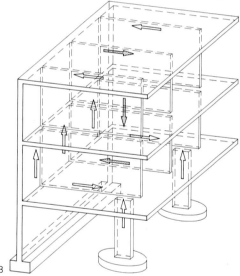

B by means of plate action

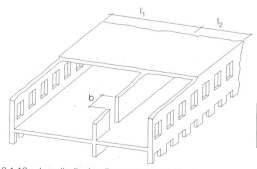

3.1.19 Longitudinal wall arrangement with one central wall

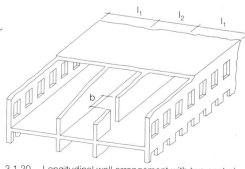

3.1.20 Longitudinal wall arrangement with two central walls

3.1.21  Loading combination p* [kN/m²]

| $G_K^*$ \ $Q_K$ | 0.75 | 2.00 | 3.00 | 5.00 |
|---|---|---|---|---|
| 0 | 1.125 | 3.00 | 4.5 | 7.50 |
| 0.5 | 1.80 | 3.675 | 5.175 | 8.175 |
| 1.0 | 2.475 | 4.35 | 5.85 | 8.85 |
| 1.5 | 3.15 | 5.025 | 6.525 | 9.525 |
| 2.0 | 3.825 | 5.70 | 7.20 | 10.20 |
| 2.5 | 4.50 | 6.375 | 7.875 | 10.875 |

The loads are combined according to the Eurocode. The values listed in the table have been multiplied by load factors (without self-weight). Intermediate values may be interpolated.

3.1.22  Loading classes

| Loading class | I | II | III | IV |
|---|---|---|---|---|
| p* [kN/m²] | 3.15 | 5.25 | 7.50 | 10.00 |

Based on the data given in Eurocode 2, the loading group is calculated according to the following equation:
$$p^* = \gamma_G G_K + \gamma_Q Q_K$$
$$p^* = 1.35 G_K + 1.5 Q_K$$
where:
- $G_K^*$ = permanent action (excluding self-weight of reinforced concrete slab)
- $Q_K$ = variable action
- $\gamma_G$ = safety factor of 1.35 for dead loads
- $\gamma_Q$ = safety factor of 1.5 for imposed load

Determining the possible door opening widths for crosswalls and longitudinal walls

Example 1:
Given parameters:
- crosswalls at centre-to-centre spacing l = 6.00 m = a'
- slab thickness h = 160 mm
- dead loads $g_K$ = 1.00 kN/m²
- imposed loads $q_K$ = 2.00 kN/m²

According to tables 3.1.21 and 3.1.22:
loading class II, where p* = 5.25 kN/m²
From table 3.1.23 loading class II read off opening width b = 3.05 m

Example 2:
Given parameters:
- longitudinal walls at centre-to-centre spacing $l_1$ = 5.70 m and $l_2$ = 3.90 m, floor slab is restrained by external walls
- slab thickness h = 180 mm
- dead loads $g_K$ = 1.00 kN/m²
- imposed loads $q_K$ = 3.25 kN/m²

According to tables 3.1.21 and 3.1.22:
loading class III, where p* = 7.50 kN/m²
floor span a' = $(l_1 + l_2)/2$ = (3.90 + 5.70)/2 = 4.80 m
From table 3.1.23 loading class III read off opening width b = 3.75 m

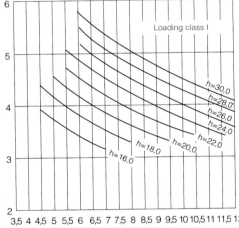

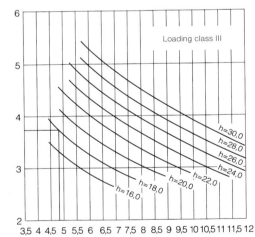
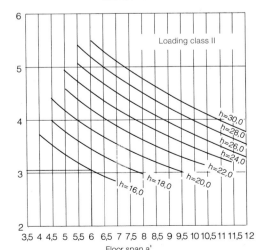
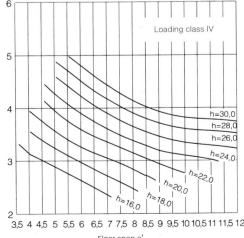

3.1.23  Determining the opening widths in walls without beams

The critical floor span a' is calculated from the average value of adjacent floor spans, i.e. for a crosswall with identical wall spacing a' = floor span l, for a longitudinal wall a' = $(l_1 + l_2)/2$.
The loading classes are given in table 3.1.22.
The diagrams have been prepared for the following conditions:
- Concrete grade C 30/37
- No shear reinforcement has been included
- Max. deflection of floor slab d = 0.6l/35
- Opening widths greater than the span have been ignored because this is then a type of column and no longer a longitudinal wall or cross wall.

on the lower floors, it is beneficial to omit one column below every second crosswall (figure 3.1.18). This means that the load on this wall is transferred to the outer wall or the columns in the outer wall and the secondary moment from the couple of forces in the floors is transferred into the neighbouring walls. A beam is totally unnecessary here because the plate effect of the walls and floors is stiffer than the flexural stiffness of the beam. A beam in this position will carry the load attributed to it only after the plate effect has failed. It is therefore wrong to include a beam. Exploiting the plate effect is also more economic because it saves the cost of formwork for the beam. The additional reinforcement required for the plate effect is considerably less than that required to reinforce the beam.

*Longitudinal walls*
In office buildings (figures 3.1.19 and 3.1.20) it is convenient to use one wall to a corridor as a loadbearing wall for supporting the floors. Longitudinal walls are primarily suitable for layouts with a central corridor. The crosswalls are incorporated later as non-loadbearing lightweight partitions. Party walls in the transverse direction between apartments will usually need to be built with two leaves and appropriate sound insulation.

Flat slabs
Only a limited number of walls are desirable in commercial buildings and, in particular, multi-storey car parks. In these cases the floor slabs are supported directly on columns. Such floor slabs result in a low overall depth of construction for the floor and this reduces the volume of the building. Furthermore, in buildings with a prescribed eaves height, the number of storeys can be increased.

Flat slabs are those supported directly on columns and are characterised by their low overall thickness. Punching shear around the column often governs the thickness of the slab.

Flat slabs with studrails
Studrails have been developed by various manufacturers to accommodate the punching shear force. These permit thinner slabs than is the case with standard punching reinforcement. Studrails consist of headed studs welded to a steel flat. This form of construction exhibits very little slip and can be used up to yield stress even in thin components. It is easier to incorporate a studrail than shear links because it can be inserted before fixing the bending reinforcement. It is also possible to incorporate studrails in semi-precast floors.

The manufacturers provide design programs for sizing their studrails.

The advantages of flat slabs are that there are no obstructions to services routed beneath the slab and non-loadbearing partitions can be built identically in any direction because there are no beams or column heads in the way (figure 3.1.24). See p. 126 for openings in flat slabs.

Flat slabs with column heads
The punching shear resistance can also be improved by enlarging the cross-section of the column or by forming a column head (figure 3.1.25a-d). The column head can also be built above the slab, e.g. like a protective bollard in car parks (figure 3.1.25 e). The column heads reduce the moments and so permit a larger column grid (see pp. 17, 21).

Hollow core slabs
These are slabs into which void formers (e.g. tubes of metal or cardboard) have been cast (figure 3.1.26). During concreting, these void formers must be anchored to prevent them floating on the concrete. The purpose of the void formers is to reduce the self-weight of the slab and save concrete. These are becoming less and less common because of the high cost of labour required to position the void formers and the need to reinforce the webs in between. Solid slabs with d = 350 mm are therefore no longer uncommon.

Ribbed slabs
The idea of omitting part of the concrete in the tension zone while still maintaining the structural depth, reducing the self-weight and increasing the load-carrying capacity for imposed loads, led to the development of the ribbed slab. However, the higher cost of production means this type of slab is only advisable for longer spans and/or larger areas.

Figure 3.1.27 shows the geometrical conditions for ribbed slabs according to DIN 1045, section 21.2.2. Other conditions are:
- imposed load $p \leq 5.0$ kN/m²
- not permitted for slabs subjected to vehicular traffic heavier than private cars
- point loads on the slab must be investigated for their effect on both the slab and the ribs (table 3.1.5, EC $Q_K$)
- ribbed slabs are economical for spans of 8-15 m

Ribbed slabs are supported on walls or beams. With smaller spans the beam can also be accommodated within the depth of the slab. The ribs must be widened in the region of the negative moments of continuous slabs (figure 3.1.28).

These slabs are constructed using reusable formwork elements made from sheet steel or plastic, or with permanent formwork, with fillers of concrete with pumice or granulated slag

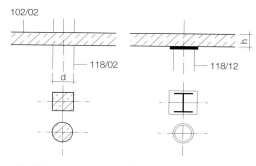

02 Reinforced in situ concrete
12 Steel
102 Flat slab
112 Column

3.1.24 Flat slab without column heads

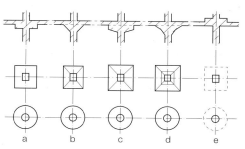

3.1.25 Flat slab with column heads

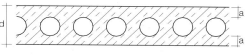

3.1.26 Hollow core slab    a ≥ 50 mm

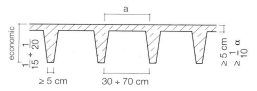

3.1.27 Ribbed slab to DIN 1045

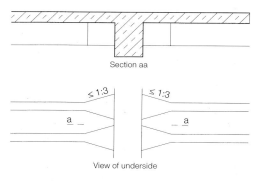

3.1.28 Detail of junction between ribbed slab and beam

# Multistorey structures

3.1.29 Guide figure for reinforcement cross-section
for beam with a

| structural depth (d) of 500 mm | |
|---|---|
| M [kNm]/10 | erf $A_s$ [cm²] |
| is h < 500 mm | erf $A_s$ > M/10 |
| is h > 500 mm | erf $A_s$ < M/10 |

If compression reinforcement is included, a moment max. 30% higher than that without compression reinforcement can be accommodated. (This corresponds to the condition that there is not more compression reinforcement than tension reinforcement.)

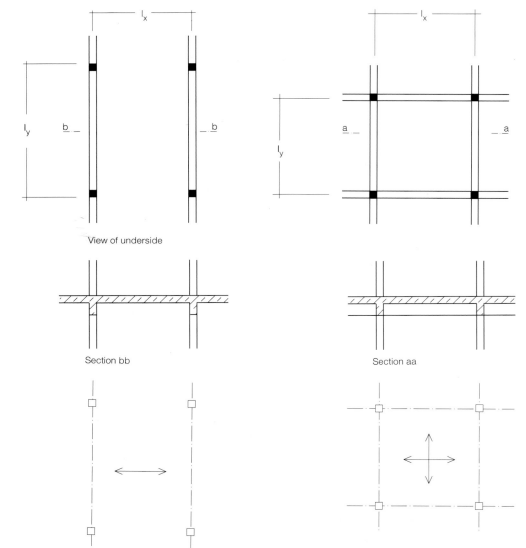

3.1.30 Beam and slab floor spanning one way

3.1.31 Beam and slab floor spanning two ways

aggregate, hollow clay blocks or lightweight wood-wool boards. Ribbed slabs are used for floor spans with one primary loadbearing direction.

Waffle slabs
Floors with square or nearly square bays ($l_y/l_x$ < 1.5) can be formed as waffle slabs, with the ribs arranged at right angles to each other. In principle, the comments regarding ribbed slabs also apply to waffle slabs. However, they can be supported directly on columns without the need for any intermediate beams. The floor is cast as a solid slab around the columns in order to withstand negative moments and punching shear. These slabs are constructed similarly to ribbed slabs. Appropriate formwork elements can be hired from special companies.

Waffle slabs can also be formed with a triangular instead of square structure. Pier Luigi Nervi wanted to arrange the ribs corresponding to the diagram of the primary bending moments. He used ferrocement elements (wire-reinforced fine-aggregate concrete) as permanent formwork for producing a number of roofs (see p. 31).

Beam grid slabs
If we deviate from the prescribed conditions for ribbed slabs as stated in DIN 1045, the simplifications for analysis and construction cannot be employed. The waffle slab with greater rib pitches or, in other words, a beam and slab arrangement in which the beam intersections are not supported is called a beam grid. An example can be seen in Frank Lloyd Wright's Unity Church in Oak Park near Chicago (p. 21). So the beam grid consists of the slab and the beams that act together with the slab. The beams are positioned at right angles to each other but other, angles are also feasible.

Beam and slab arrangements
We speak of a beam and slab system when the floor slab is supported on beams and these beams act together with the slab (T-beams). This is also the case when the maximum rib pitch is exceeded on a ribbed slab. The beams are supported on walls or other – main – beams. The slabs span in one or two directions depending on the ratio of the spans $l_y/l_x$ (figures 3.1.30 and 3.1.31). Beams are only required transverse to the direction of span in the case of a one-way-spanning slab. Beams at angles other than 90° to each other are also possible.

Imposed loads can be reduced by the factor $\alpha_A = 0.5 + 10/A$ for beams carrying an area (A) larger than 20 m².
Flat slabs and flat slabs with column heads present no obstructions to building services. But if a floor slab with beams is chosen, the interdependence of the arrangement of the beams and the services must be considered.

Systems in which the beams are not continuous but are omitted at certain points so that the building services can be routed more easily are recommended. These systems are preferable to beams with openings because they offer more freedom for the layout of the services (figures 3.1.32 to 3.1.34).

Rectangular slabs with a span ratio $l_y/l_x$ of about 1.0 exhibit considerably smaller moments than slabs that carry the load in just one direction. So it would seem that beams in both directions are desirable. However, comparative studies have shown that the reinforcement in slabs spanning in two directions and the additional beams generate costs that are difficult to justify, especially considering the effects of the additional beams on the interior fittings and the services. At this juncture it is worth pointing out again that the loadbearing structure can only be designed in conjunction with the building services.

Upstand beams
Upstand beams are generally used at the edges of slabs to form spandrel panels or parapets above openings in the walls. These act together with the slab. The slab lies in the compression zone in the case of support moments. Deep beams (h = 1.30 m) serving as spandrel panels can span distances of up to 15 m. The upper edge of the spandrel panel is strengthened to accommodate the span moments in longer spans (see p. 182).

Upstand beams are costly to produce because they can only be cast in a second concreting operation after the formwork to the sides has been set up on the concrete slab after it has hardened.

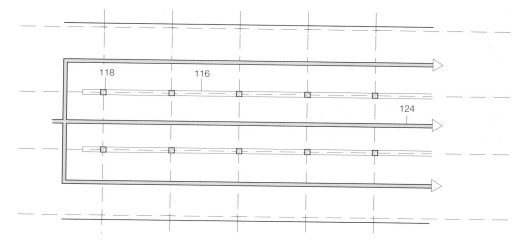

3.1.32  3-span floor slab spanning one way
continuous beams
service runs in one direction possible,
openings in beams unnecessary

116  Beam
118  Column
124  Services

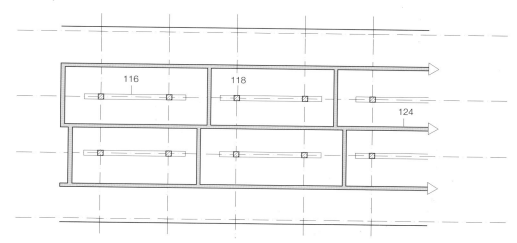

3.1.33  3-span floor slab spanning one way
with areas of flat slab
beams not continuous
service runs in both directions possible owing
to gaps in beams

116  Beam
118  Column
124  Services

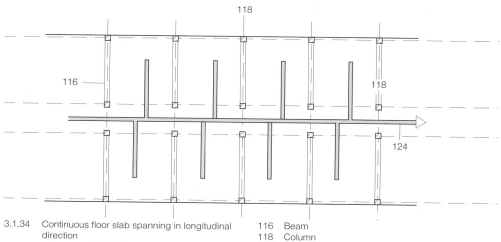

3.1.34  Continuous floor slab spanning in longitudinal
direction
with areas of flat slab
service runs: main leg with branches

116  Beam
118  Column
124  Services

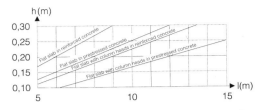

3.1.35  Guide figures for slab thickness depending on span

3.1.36  Arrangement of prestressing tendons around top of column [26]

3.1.37  Prestressing tendons with intermediate anchorages [26]

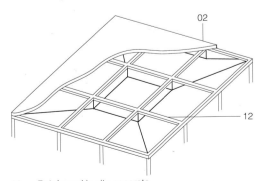

02  Reinforced in situ concrete
12  Steel beam

3.1.38  Trussed floor slab

Prestressed concrete floor slabs[1]
Compressing the concrete by means of prestressed steel tendons
- prevents cracks in the concrete (which exceed the tensile strength of the concrete),
- allow the use of thinner slabs, and
- create a more favourable condition around punching shear in flat slabs.

Here, besides conventional reinforcement, steel tendons in sheaths are cast in at a varying depth, corresponding to the bending moment diagram. The varying depth at which the tendons are cast in generates lateral forces in slab-like components similar to the bending moment curve due to loads in the loadbearing structure, and these counteract the forces generated by the loads. Depending on this, either the thickness of the slab can be reduced or the span enlarged considerably (figure 3.1.35). The risk of cracking is also decreased substantially.

Prestressing leads to another advantage compared to slabs reinforced conventionally. In the latter case there is a considerable concentration of reinforcement to resist bending and punching shear around the columns. This is often very difficult to accommodate sensibly and frequently means additional studrails or cast-in steel inserts have to be provided.
The tendons are lifted as they pass over the columns. The shear force in the slab is partly directed into the column by the vertical component of the force in the tendons, which means that the concrete is subjected to a much smaller punching shear load. The magnitude of the lateral forces is primarily determined by the depth of the tendons. In the span they should generate upward lateral forces, and near the columns, downward lateral forces (figure 3.1.36).

The prestressing tendons common in bridge-building are not suitable in this situation, both in terms of the prestressing force they can carry and their overall dimensions. Floor slabs require small tendons. To achieve optimum lateral forces, they should be positioned with an eccentricity as large as possible. Instead of bundling them, the tendons are therefore laid next to each other as bonded or unbonded tendons.

Bonded tendons are stressed wires or strands placed in a sheath, which is then filled with grout before the prestress is transferred to the surrounding concrete. The injected grout is an excellent means of protecting the tendons against corrosion.

[1] Co-author: E. Conrads

Unbonded tendons are given a coating of grease at the works to protect against corrosion and cast into the concrete in a polyethylene sheath. They are stressed after the concrete has hardened, whereby the prestressing force is transferred to the concrete via the anchorages only.

The tendons are usually unrolled within the column strip and fixed at the desired depths to chairs or "hold-downs". The section of floor, the size of which is determined by the amount of concrete and the amount of formwork, can be prestressed after the concrete has reached a compressive strength of just 25 N/mm$^2$. The slab formwork can be struck immediately after prestressing because the self-weight of the concrete is carried by the lateral forces. The formwork is then already available for the next pour.

At construction joints the tendons are connected to the previous section by special couplers. This enables construction to proceed rapidly with a possible turnaround time of one week.

If end anchorage points on the edges of the slabs are not accessible for prestressing, as is the case, for example, when securing excavations without any working space, the tendons are prestressed via special intermediate anchorages positioned within the section to be concreted (figure 3.1.37).

Care must be taken when designing prestressed concrete slabs to ensure that the slabs can deform unhindered during the prestressing operation; they should not be restricted significantly by walls or stiffening cores. Columns are therefore planned as hinged members with an elastomeric bearing. The bearings on the walls and cores include a slip layer. After introducing the prestressing force, the connections to walls for transferring horizontal forces are completed to resist shear.

Slabs subjected to direct vehicular traffic (multistorey car parks)
Damage to the floor coverings of parking decks has led to solutions with reinforced concrete slabs without any finishes. To do this it is necessary to finish the slab with a smooth surface by means of vacuum dewatering. As the vehicles bring de-icing salt onto the parking decks in winter and even with crack widths < 0.2 mm the concrete does not possess adequate protection against the chlorinated water, a surface without any cracks is desirable. Prestressed concrete can achieve this.

Trussed slabs
Reinforced concrete slabs spanning longer distances can be supported by cables and struts. The force in the struts is transferred into the slab via a strengthened head, e.g. cast

steel capital. This type of structure loses its elegance when the trussing has to be cladded to achieve F 90 fire protection.

The floors of the Museum for Work and Technology in Mannheim, Germany, employ an intermediate steel beam grid, which saved the cost of scaffolding (figure 3.1.38).

*Precast concrete floors*
Solid slabs supported on beams and walls
Solid slabs may be supported on beams, walls or columns. They are seldom used for spans exceeding 6 m. The floor units can be up to 3.6 m wide, which means they can still be transported tilted up at an angle on a low-loader (see table 3.1.3).

See "Precast walls", p. 129, for details of the manufacture of the floor units.

The exposed edges of the floor units are formed with a longitudinal groove which is filled with grout (figure 3.1.39). This filling guarantees that when one floor unit is loaded, the neighbouring panels deflect by an equal amount. To achieve a plate effect, the edge of the floor unit is given an interlocking profile. To ensure that horizontal forces (wind, unintended inaccurate positioning of load-carrying components) are transferred through the floor plate to the stair core and stiffening elements, the floors must be provided with continuous tension reinforcement. Reinforcement projecting from the side of a precast concrete floor unit should be avoided because this means that the reinforcement has to be passed through metal formwork cut like a comb to suit. Therefore, either a continuous reinforcing bar is laid in the joint or cast-in parts welded to the reinforcement within the element are connected by welding steel plates together.

Solid slabs are also produced in prestressed concrete. In these the prestressing wires are wedged into a steel frame; displacing the frame transom by means of hydraulic jacks induces a tensile force in the prestressing wires, which acts as a compressive force on the concrete after removing the jacks once the concrete has hardened. The force in the prestressing wires is transferred to the concrete via the bond.

Walls
Residential and office buildings can be constructed from solid precast concrete panels. Figure 3.1.40 to 3.1.42 show crosswalls and longitudinal walls made from wall panels and floor units.

We avoid rigid connections in reinforced concrete because they are very costly to produce (see table 3.1.3). This lack of continuity results in thicker floor slabs and heavier reinforcement

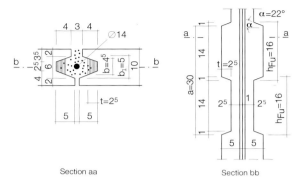

3.1.39 Detail of joint between precast concrete units

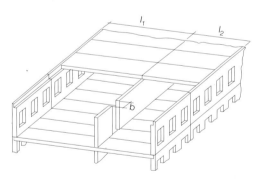

3.1.40 Precast concrete crosswalls

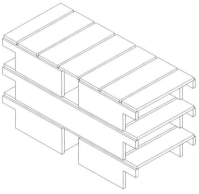

3.1.41 Precast concrete longitudinal wall arrangement with one central wall

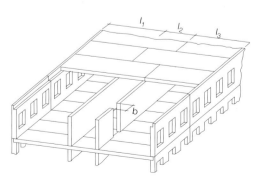

3.1.42 Precast concrete longitudinal wall arrangement with two central walls

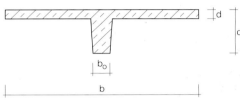

3.1.43 Single-tee floor unit

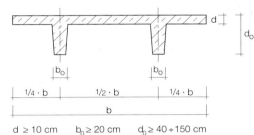

$d \geq 10$ cm  $b_0 \geq 20$ cm  $d_0 \geq 40 \div 150$ cm
3.1.44 Double-tee floor unit

3.1.45 Formwork inserts for double-tee floor units in various sizes

3.1.46 Double-tee floor unit with intermediate single-span unit

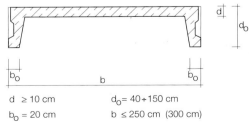

$d \geq 10$ cm     $d_0 = 40 \div 150$ cm
$b_0 = 20$ cm     $b \leq 250$ cm (300 cm)

3.1.47 Channel section floor unit

3.1.48 Double-tee floor units with conventional reinforcement

Dimensions [mm]

| $d_U$ | 200 | 300 | 400 | 500 | 600 | 700 | 800 |
|---|---|---|---|---|---|---|---|
| $b_U$ | 190 | | | | | | |
| $b_0$ | 210 | 220 | 230 | 240 | 250 | 260 | 270 |

All dimensions are adequate for fire resistance class F 90-A to DIN 4102.

| $d_0$ | ≥ 60 | F 30-A |
|---|---|---|
| | ≥ 100 | F 90-A |

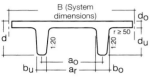

10 mm chamfers to bottom edges of webs

Note:
Deeper elements are not feasible.
Flange depth with concrete topping
min. $d_0 = 50 + 70 = 120$ mm
Depth adjusted by means of screed or concrete topping

Determining the depth of double-tee units

Loads:
- included in the table: self-weight $g_1$ of double-tee unit
- select as required: imposed load q (service load)

Coordinating size B = 2.50 m

| L System [m] | Floor depth d [mm] for imposed load q | | | | | | |
|---|---|---|---|---|---|---|---|
| | 3.5 | 5.0 | 7.5 | 10 | 15 | 20 | 25 |
| 6.00 | 320 | | 350 | | 450 | 500 | |
| 7.50 | | 420 | 450 | | 550 | 600 | |
| 10.00 | 520 | | 650 | | | 700 | |
| 12.50 | 720 | | 750 | 850 | | 900 | |
| 15.00 | 820 | | 850 | 950 | | 1000 | |
| 17.50 | 920 | | 950 | | | | |
| 20.00 | | | | | | | |
| Flange depth | $d_0 = 120$ | | $d_0 = 150$ | | $d_0 = 200$ mm | | |

Note: Additional measures may be required for supporting partitions on the floor.

Example:
imposed load   q = 10 kN/m²
coordinating size   L = 12.50 m
read off:   d = 850 mm
   $d_0$ = 150 mm
   $d_U$ = 850 - 150 = 700 mm

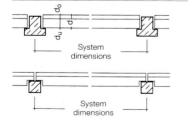

▨ Select a different beam

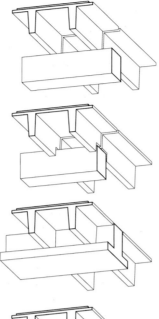

3.1.49 Support conditions for double-tee floor units on beams

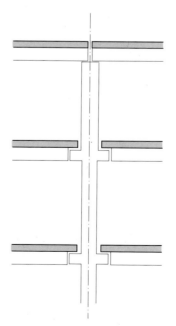

3.1.50 Support conditions for beams on multistorey columns

than with in situ concrete (figures 3.1.12 and 3.1.15). The extra work and extra cost must be offset by rational production, transport and erection. Moreover, savings are achieved in the first place through suitable design.

Hollow core floor units
To reduce the weight of the floors, hollow core units are produced in reinforced and prestressed concrete. The units available are 625 mm wide. The reason for this relatively narrow width is to allow the units to be adapted easily to the dimensions of the building. The edges of hollow units are similar to those of solid units (figure 3.1.39).

Tee units
These units are employed for longer spans (≥ 7.2 m).

Single-tee units
These floor units (figure 3.1.43) are characterised by their ease of production but more difficult erection owing to their inherent instability. The webs taper slightly to enable the units to be easily lifted out of the forms after the concrete has hardened. A taper of 1:20 for steel forms and 1:10 for timber forms is recommended (this applies to other types of tee units as well).

Double-tee units
These units (figure 3.1.44) are the most widely used form of floor unit. The principles mentioned for the single-tee unit apply here as well. Precasting companies have agreed on the dimensions given in table 3.1.48. This means that the majority of manufacturers stock steel forms with these dimensions.
The basic form can be varied by using inserts. Both the depth and the width of the web can be altered. To maintain the pitch of the webs, the inserts are not used symmetrically (figure 3.1.45).
The length of the cantilevering section is variable. The limit depends on the width for transport (3 m with special permit). A width deviating from 2.4 m results in a periodic web spacing.

Table 3.1.48 lists depths of double-tee units for different loads.

Floors and roofs

3.1.51 Dimensions of precast concrete beams with ⊥ cross-section [4]
Loads: included in the table: self-weight of double-T unit + self-weight of ⊥ beam; select as required: imposed load

| Span Beam l (m) | Pitch of beams L (m) | Beam cross-section $d/b_0$ [mm] for imposed q [kN/m²] | | | | | | | | | |
|---|---|---|---|---|---|---|---|---|---|---|---|
| | | 1,0 | 2,0 | 3,0 | 4,0 | 5,0 | 7,5 | 10,0 | 15,0 | 20,0 | 25,0 |
| 5,0 | 6,0 | | | | | | | | 500/600 | 500/600 | 600/600 |
| | 7,5 | | 400/300 | | | | | | 500/600 | 600/600 | |
| | 10,0 | | | | | | | | 600/600 | | 700/600 |
| | 12,5 | | | 500/400 | | | | | 700/600 | 800/600 | |
| | 15,0 | | | | | 500/600 | 600/600 | | | | |
| | 17,5 | | | | | | | | 800/600 | | 1000/600 |
| | 20,0 | | | | | | | 700/600 | | | |
| | 25,0 | 500/600 | 600/600 | | 700/600 | | 800/600 | | 1000/600 | | 1000/800 |
| 6,25 | 6,0 | | 400/400 | | | | | | 600/600 | | |
| | 7,5 | | | | | | 600/600 | | | | |
| | 10,0 | | | 500/500 | | | 700/600 | | | | |
| | 12,5 | | | | | | | | | 1000/600 | |
| | 15,0 | | | | 600/600 | | | | 1000/600 | | |
| | 17,5 | | | | | | 900/600 | | 1000/800 | 1200/800 | |
| | 20,0 | | | 700/600 | | | | 1000/600 | | | |
| | 25,0 | | | | | | 1000/600 | | | 1200/800 | |
| 7,50 | 6,0 | | 500/400 | | | 500/600 | | | | | |
| | 7,5 | | | | | | | | | 1000/600 | |
| | 10,0 | | | | 700/500 | | | 1000/600 | | | |
| | 12,5 | | 500/600 | | | | 1000/600 | | | | |
| | 15,0 | | | | 800/600 | | | 1000/800 | 1200/800 | | |
| | 17,5 | | | | | | 1000/600 | | | | |
| | 20,0 | | | | | | 1000/800 | 1200/800 | | | |
| | 25,0 | | | | 1000/600 | 1000/800 | | | 1400/800 | | |
| 8,75 | 6,0 | | | 600/400 | 600/600 | | | 900/600 | | | |
| | 7,5 | | | | 600/600 | | | 900/600 | | | |
| | 10,0 | | | 600/600 | | | 900/600 | | | | |
| | 12,5 | | | | 800/600 | | 900/600 | 1000/800 | | 1200/800 | |
| | 15,0 | | | | | | | | 1200/800 | | |
| | 17,5 | | | | 900/600 | | | 1200/800 | 1400/800 | | |
| | 20,0 | | | | | 1000/800 | | | | | |
| | 25,0 | 900/600 | 1000/800 | | | 1200/800 | | | – – | | |
| 10,0 | 6,0 | | 700/400 | | 700/600 | | | 800/600 | | | 1000/800 |
| | 7,5 | | | | | | 800/600 | | 1000/800 | | |
| | 10,0 | | 700/600 | | | | | 1000/800 | 1200/800 | | |
| | 12,5 | | | | | | 1000/800 | 1200/800 | 1400/800 | | |
| | 15,0 | | 800/600 | | | | 1000/800 | 1200/800 | | | |
| | 17,5 | | | 1000/600 | | | 1200/800 | 1400/800 | | | |
| | 20,0 | | | | | 1000/800 | | | | | |
| | 25,0 | 1000/600 | 1000/800 | | 1200/800 | 1400/800 | | | | | |
| 11,25 | 6,0 | | | | | | | 1000/600 | 1000/800 | | |
| | 7,5 | | 800/500 | | | | 1000/600 | 1000/800 | | 1200/800 | |
| | 10,0 | | | | | 1000/600 | 1000/800 | 1200/800 | | | |
| | 12,5 | | 900/600 | | 1000/600 | 1000/800 | | 1400/800 | | | |
| | 15,0 | | | 1000/600 | | | 1200/800 | 1400/800 | | | |
| | 17,5 | | 1000/600 | 1000/800 | | 1200/800 | 1400/800 | | | | |
| | 20,0 | 1000/600 | 1000/800 | 1200/800 | 1400/800 | | | | | | |
| | 25,0 | 1000/800 | | 1200/800 | 1400/800 | | | | | | |
| 12,50 | 6,0 | | | | | | 1000/600 | 1200/600 | | | 1200/800 |
| | 7,5 | | 1000/400 | | | | 1000/600 | 1200/800 | | 1400/800 | |
| | 10,0 | | | | | 1200/600 | 1200/800 | 1400/800 | | | |
| | 12,5 | | 1000/600 | | | 1200/800 | | | | | |
| | 15,0 | | | | | 1400/800 | | | | | |
| | 17,5 | | | 1200/800 | | | | | | | |
| | 20,0 | | 1200/600 | | 1400/800 | | | | | | |
| | 25,0 | 1200/600 | 1200/800 | 1400/800 | | | | | | | |

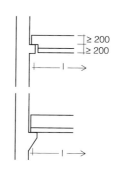

$b = b_0 + 400$

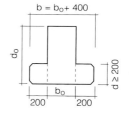

Dimensions [mm]

| $b_0\backslash d_0$ | 400 | 500 | 600 | 700 | 800 | 900 | 1000 | 1200 | 1400 |
|---|---|---|---|---|---|---|---|---|---|
| 300 | | | | | | | | | |
| 400 | | | | | | | | | |
| 500 | | | | | | | | | |
| 600 | | | | | | | | | |
| 800 | | | | | | | | | |

All dimensions are adequate for fire resistance class F 90 – A to DIN 4102.

Example:
span of beam     l = 7.50 m
pitch of beams   L = 10.0 m
imposed load     q = 5.0 kN/m²
read off from table: $d/b_0$ = 700/500 mm

Cross-sections below this line: only after consultation with precasting works

▨ Select a different beam

123

# Multistorey structures

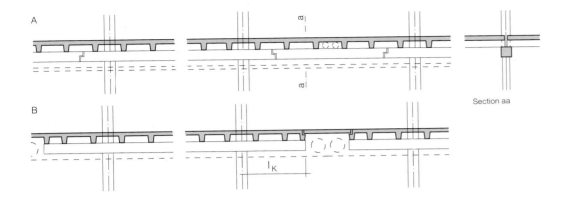

3.1.52  Support conditions for beams on columns

A   Beams supported on storey-height columns
B   Cantilever beams with gaps for building services

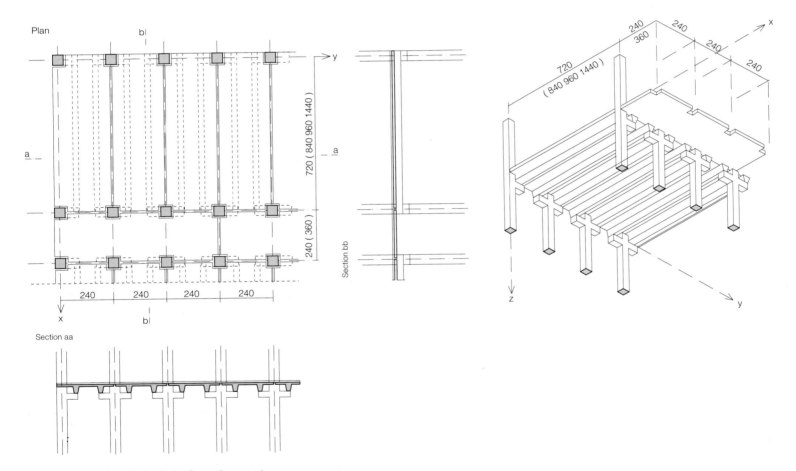

3.1.53  Construction with double-tee floor units supported on corbels on columns; joints between units positioned on column grid lines.

Floors and roofs

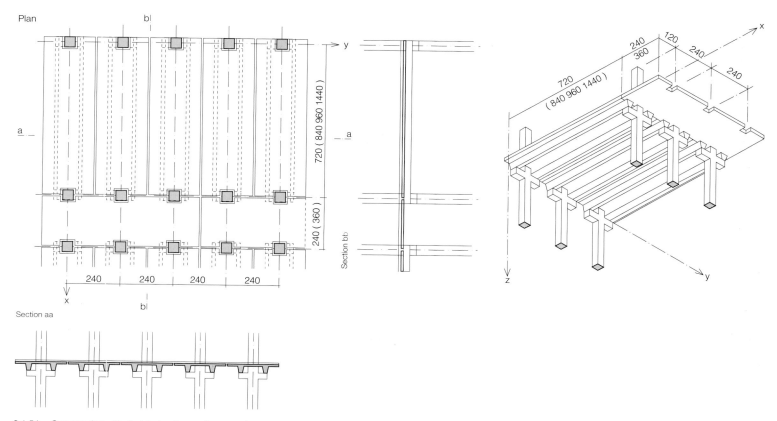

3.1.54 Construction with double-tee floor units supported on corbels on columns; joints between units positioned between column grid lines

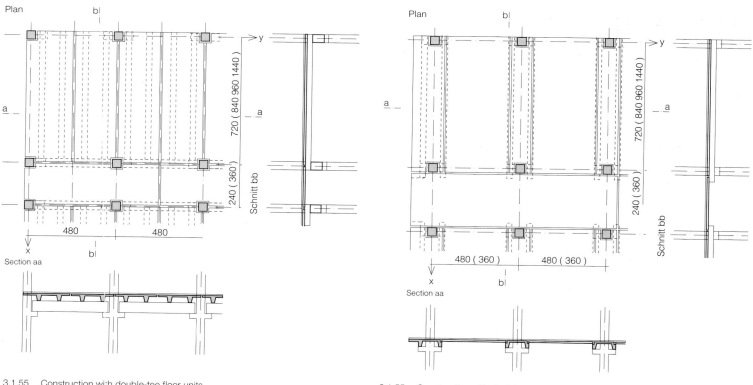

3.1.55 Construction with double-tee floor units supported on beams

3.1.55 Construction with double-tee floor units and intermediate single-span units

125

# Multistorey structures

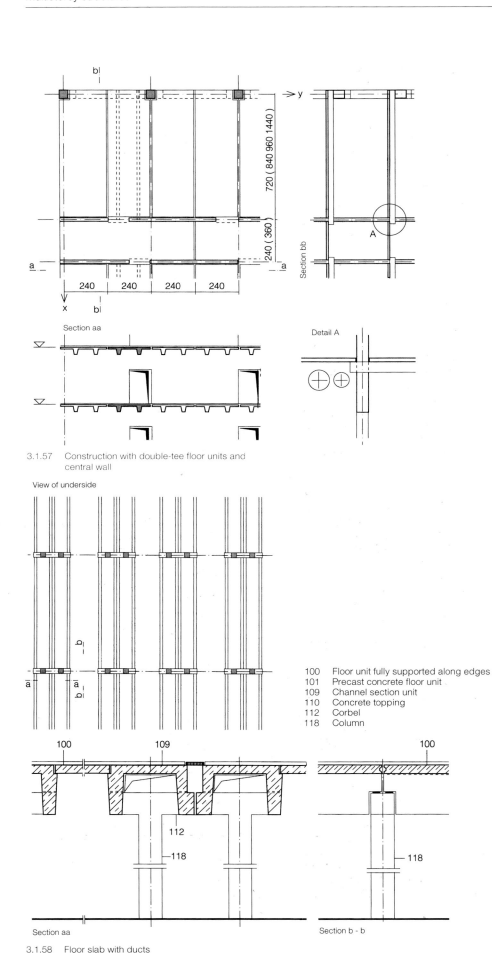

3.1.57 Construction with double-tee floor units and central wall

100 Floor unit fully supported along edges
101 Precast concrete floor unit
109 Channel section unit
110 Concrete topping
112 Corbel
118 Column

3.1.58 Floor slab with ducts

3.1.59 Floor slab with ducts

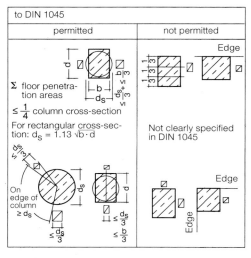
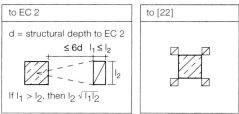

3.1.61 Restrictions on penetrations in floor slabs

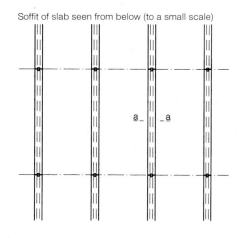
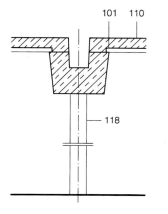

Omitting the cantilevering section and forming a notch in the top of the web as shown in figure 3.1.46 provides a bearing for a solid unit to be laid between the double-tee slabs. Double-tee units are available in widths up to 3 m, depths up to 800 mm and lengths up to 16 m. The webs are positioned at a spacing of 1.2 m.

Another variation on the double-tee floor slab is the channel section floor unit (figure 3.1.47). This is employed with heavier point loads or with systems in which the width of the unit matches the column grid. The width of the web can be 100 mm when the channel section units are erected adjacent to each other and the joint is no wider than 10 mm (otherwise the F 90 requirement calls for a minimum dimension of 200 mm). Channel section units can be combined with single-span solid units to produce an economic solution. Compared with the exclusive use of double-tee units, this employs only about half the number of webs. Apart from that, inserting the single-span units compensates for production and erection inaccuracies.

Floors formed with tee units can be manufactured with a thinner flange and provided with a concrete topping after erection. This achieves a good transverse distribution of the loads and, with the addition of a small amount of additional reinforcement, a plate effect. This type of construction is particularly suitable for parking decks without any further floor finishes; in this case the concrete topping is given vacuum dewatering treatment (see p. 114).

Beams
As a rule, the floor units are supported on beams which transmit the loads to the columns. Table 3.1.51 gives the standard dimensions of beams; deviations from these do not usually result in significantly higher costs because the formwork is very straightforward. The most common forms of support are shown in figure 3.1.49.

Single-span beams are supported on corbels on continuous columns (or in the case of a roof directly on the columns) (figure 3.1.50). If formed as cantilevers with a central beam, they are supported on the storey-height columns (figure 3.1.52). In terms of erection, the continuous column has advantages. However, the cantilevers and central beam arrangement is better for greater spans between columns because the span moment can be reduced by the negative moments. The position of the hinge is a matter of optimisation, where the goal is either to minimise the moment or the deflection. In any case, the hinge should not lie beneath a web.

With appropriate span ratios (small span for beam, large span for double-tee slab), the beam can be interrupted between the webs of the double-tee units (figure 3.1.52 B). Figures 3.1.53 to 3.1.57 illustrate structural systems for various column grids and for central wall arrangements which benefit the layout of the services.

Floors with ducts
The floors of exhibition halls require ducts for water and electrical services. If there is a basement below the hall, the floor slabs have to include ducts. Channel section units are suitable here; precast concrete floor units or planks are laid on these (figures 3.1.58 and 3.1.59).

Composite floors [11]
Steel beams can also be used. Headed studs are used to create a shear-resistant connection between the beams and the floor slab (figure 3.1.60 A), turning the beam into a composite member. The spaces between the flanges of steel beams can be filled with concrete to serve as fire protection. In this case reinforcement is placed in the concrete as a substitute for the bottom flange of the steel beam. This reinforcement is calculated according to the load on the beam. The concrete is secured by stirrups welded to the steel section.

Trapezoidal profile steel sheeting is often used for floors and as permanent formwork. Concrete is cast on the sheeting to distribute the load and prevent the sheeting from buckling. Additional profiling of the inclined webs of the sheeting can help to develop a bond between the metal and the concrete. Floors employing trapezoidal profile steel sheeting require additional fire protection to achieve class F 90. F 90-coatings are permitted, but they are very costly. Fire resistance F 90 can also be provided by F 60 coating if the steel section is dimensioned accordingly.

Holorib sheeting (figure 3.1.60 B) provides a good bond under certain conditions. This type of floor does not require additional measures to achieve class F 90 fire protection. Holorib sheeting must be supported at mid-span, with a camber, until the concrete is able to carry the load. To avoid this operation, the floor beams are often positioned at closer centres than is actually required structurally.

The floor slab can also be constructed using precast concrete planks with a concrete topping (figure 3.1.60 C), or with precast concrete units in which the composite effect is achieved by means of grouted headed studs (figure 3.1.60 D) or bolted connection (figure 3.1.60 E).

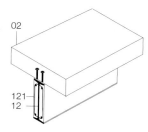

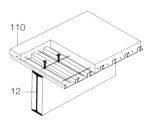

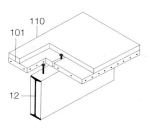

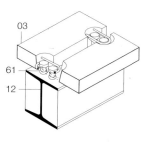

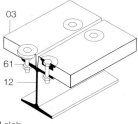

| | |
|---|---|
| A | Solid slab |
| B | Holorib metal decking |
| C | Precast concrete floor plank |
| D | Precast concrete units with grouted studs |
| E | Precast concrete units with bolted connection |

| | | | |
|---|---|---|---|
| 02 | Reinforced in situ concrete | 101 | Precast floor unit |
| 03 | Reinforced precast concrete | 110 | Concrete topping |
| | | 121 | Reinforcement |
| 12 | Steel | | |
| 61 | Steel stud/bolt | | |

3.1.60  Composite floor slabs

# Multistorey structures

3.1.62 Guide values for permissible load [kN/m] of axially loaded in situ concrete walls to DIN 1045 restrained by the floor slab (buckling length = half storey height):

| | μ [%] | d [cm] h [m] | 10 | 12 | 15 | 18 | 20 | 25 | 30 |
|---|---|---|---|---|---|---|---|---|---|
| B 25 | 0.5 | 2.85 | 732 | 934 | 1245 | 1564 | 1779 | 2333 | 2800 |
| | 0.5 | 3.35 | 689 | 885 | 1189 | 1502 | 1715 | 2253 | 2800 |
| | 0.5 | 3.60 | 669 | 862 | 1163 | 1473 | 1684 | 2219 | 2763 |
| | 4.0 | 2.85 | 1431 | 1811 | 2390 | 2907 | 3252 | 4083 | 4900 |
| | 4.0 | 3.35 | 1357 | 1728 | 2301 | 2873 | 3217 | 4080 | 4900 |
| | 4.0 | 3.60 | 1310 | 1673 | 2235 | 2810 | 3166 | 019 | 4846 |
| B 35 | 0.5 | 2.85 | 926 | 1181 | 1575 | 1978 | 2250 | 2988 | 3586 |
| | 0.5 | 3.35 | 871 | 1119 | 1504 | 1900 | 2168 | 2850 | 3586 |
| | 0.5 | 3.60 | 847 | 1091 | 1471 | 1863 | 2130 | 2807 | 3494 |
| | 4.0 | 2.85 | 1656 | 2095 | 2740 | 3333 | 3728 | 4738 | 5686 |
| | 4.0 | 3.35 | 1570 | 2000 | 2662 | 3294 | 3689 | 4677 | 5686 |
| | 4.0 | 3.60 | 1531 | 1955 | 2611 | 3275 | 3669 | 4657 | 5645 |

3.1.63 Guide values for permissible load [kN/m] of axially loaded in situ concrete walls to EC 2 restrained by the floor slab (buckling length = half storey height):

| | μ [%] | d [cm] h [m] | 10 | 12 | 15 | 18 | 20 | 25 | 30 |
|---|---|---|---|---|---|---|---|---|---|
| C 20/25 | 0.4 | 2.85 | 894 | 1181 | 1586 | 1986 | 2586 | 3233 | 3879 |
| | 0.4 | 3.35 | 779 | 1110 | 1520 | 1925 | 2192 | 3233 | 3879 |
| | 0.4 | 3.60 | 728 | 1047 | 1486 | 1891 | 2161 | 3233 | 3879 |
| | 4.0 | 2.85 | 1993 | 2591 | 3488 | 4382 | 5466 | 6832 | 8199 |
| | 4.0 | 3.35 | 1841 | 2436 | 3337 | 4237 | 4835 | 6832 | 8199 |
| | 4.0 | 3.60 | 1764 | 2359 | 3260 | 4163 | 4762 | 6832 | 8199 |
| C 30/37 | 0.4 | 2.85 | 1198 | 1615 | 2188 | 2758 | 3720 | 4650 | 5580 |
| | 0.4 | 3.35 | 1064 | 1479 | 2084 | 2660 | 3040 | 4650 | 5580 |
| | 0.4 | 3.60 | 995 | 1408 | 2032 | 2609 | 2991 | 4650 | 5580 |
| | 4.0 | 2.85 | 2486 | 3232 | 4352 | 5465 | 6600 | 8250 | 9900 |
| | 4.0 | 3.35 | 2294 | 3038 | 4163 | 5286 | 6029 | 8250 | 9900 |
| | 4.0 | 3.60 | 2202 | 2941 | 4065 | 5192 | 5940 | 8250 | 9900 |

3.1.64 Guide values for permissible load [kN/m] of axially loaded precast concrete walls to DIN 1045:

| | μ [%] | d [cm] h [m] | 8 | 10 | 152 | 15 | 18 | 20 | 25 |
|---|---|---|---|---|---|---|---|---|---|
| B 35 | 0.5 | 2.65 | 535 | 807 | 1117 | 1281 | 1648 | 1899 | 2546 |
| | 0.5 | 3.15 | 449 | 709 | 978 | 1397 | 1542 | 1786 | 2414 |
| | 0.5 | 3.40 | 417 | 649 | 932 | 1347 | 1495 | 1734 | 2354 |
| | 4.0 | 2.65 | 1079 | 1547 | 2045 | 2312 | 2951 | 3385 | 4497 |
| | 4.0 | 3.15 | 967 | 1388 | 1859 | 2565 | 2785 | 3207 | 4294 |
| | 4.0 | 3.40 | 916 | 1329 | 1776 | 2483 | 2700 | 3126 | 4200 |
| B 45 | 0.5 | 2.65 | 603 | 937 | 1233 | 1488 | 1915 | 2208 | 2959 |
| | 0.5 | 3.15 | 497 | 786 | 1112 | 1617 | 1793 | 2075 | 2806 |
| | 0.5 | 3.40 | 448 | 730 | 1063 | 1554 | 1737 | 2016 | 2736 |
| | 4.0 | 2.65 | 1172 | 1678 | 2187 | 2560 | 3267 | 3749 | 4980 |
| | 4.0 | 3.15 | 1043 | 1516 | 2021 | 2794 | 3084 | 3552 | 4755 |
| | 4.0 | 3.40 | 987 | 1439 | 1937 | 2711 | 2999 | 3461 | 4651 |
| B 55 | 0.5 | 2.65 | 665 | 1064 | 1371 | 1637 | 2107 | 2429 | 3256 |
| | 0.5 | 3.15 | 548 | 863 | 1223 | 1806 | 1972 | 2283 | 3087 |
| | 0.5 | 3.40 | 502 | 812 | 1161 | 1743 | 1911 | 2217 | 3009 |
| | 4.0 | 2.65 | 1238 | 1771 | 2173 | 2708 | 3456 | 3965 | 5256 |
| | 4.0 | 3.15 | 1105 | 1603 | 2147 | 2964 | 3262 | 3757 | 5030 |
| | 4.0 | 3.40 | 1045 | 1523 | 2048 | 2853 | 3172 | 3661 | 4919 |

(Buckling length = clear storey height)

3.1.65 Guide values for permissible load [kN/m] of axially loaded precast concrete walls to EC 2:

| | μ [%] | d [cm] h [m] | 8 | 10 | 12 | 15 | 18 | 20 | 25 |
|---|---|---|---|---|---|---|---|---|---|
| C 30/37 | 0.4 | 2.65 | 159 | 495 | 928 | 1570 | 2199 | 2639 | 3607 |
| | 0.4 | 3.15 | 85 | 259 | 620 | 1281 | 1920 | 2340 | 3396 |
| | 0.4 | 3.40 | λ > 140 | 184 | 522 | 1121 | 1782 | 2202 | 3288 |
| | 4.0 | 2.65 | 988 | 1607 | 2317 | 3406 | 4530 | 5287 | 7182 |
| | 4.0 | 3.15 | 789 | 1327 | 1967 | 3039 | 4141 | 4893 | 6792 |
| | 4.0 | 3.40 | λ > 140 | 1608 | 1813 | 2861 | 3952 | 4698 | 6594 |
| C 35/45 | 0.4 | 2.65 | 161 | 571 | 1069 | 1831 | 2566 | 3078 | 4208 |
| | 0.4 | 3.15 | 86 | 273 | 715 | 1493 | 2241 | 2731 | 3962 |
| | 0.4 | 3.40 | λ > 140 | 186 | 602 | 1293 | 2079 | 2576 | 3836 |
| | 4.0 | 2.65 | 1072 | 1743 | 2490 | 3659 | 4869 | 5682 | 7718 |
| | 4.0 | 3.15 | 872 | 1439 | 2132 | 3266 | 4450 | 5258 | 7299 |
| | 4.0 | 3.40 | λ > 140 | 1310 | 1965 | 3081 | 4246 | 5048 | 7085 |
| C 45/55 | 0.4 | 2.65 | 175 | 726 | 1359 | 2312 | 3241 | 3871 | 5292 |
| | 0.4 | 3.15 | 93 | 284 | 908 | 1899 | 2829 | 3446 | 4982 |
| | 0.4 | 3.40 | λ > 140 | 203 | 688 | 1645 | 2643 | 3242 | 4827 |
| | 4.0 | 2.65 | 1169 | 1877 | 2654 | 3900 | 5188 | 6055 | 8232 |
| | 4.0 | 3.15 | 968 | 1569 | 2291 | 3480 | 4741 | 5602 | 7778 |
| | 4.0 | 3.40 | λ > 140 | 1429 | 2129 | 3283 | 4524 | 5378 | 7551 |

(Buckling length = clear storey height)

In buildings, headed studs are employed to transfer the shear forces. DIN 18806 specifies two values for the loading capacity:

- headed studs with helix
- headed studs without helix

The helix is intended to accommodate the splitting tensile force. The forces specified are less for headed studs without helix. But it is more economic to attach a larger number of headed studs without helix.

### Openings in floor slabs

Openings in floors for internal stars or service shafts should be located parallel to the primary loadbearing direction to ensure that the load-bearing action is not disrupted more than is necessary.

In the case of flat slabs on columns, openings around the columns are subjected to a number of restrictions (DIN 1045, section 22.6; figure 3.1.61, p. 126). The distance between the opening in the slab and the centre of the column depends on the theoretical shear stress. The details given in the German standard, as well as those in the Eurocode, are based on incorrect theories. The openings must be positioned where the stresses are highest, i.e. in the corners, so that the primary loadbearing effect is displaced to where the loadbearing strip is not disrupted, i.e. near the column [19]. (This applies to both plastic and elastic theory.) In this case the analysis should be carried out using two intersecting strips of floor slab.

An excessive number of openings should be avoided. For example, in commercial kitchens it is advisable to position the reinforced concrete floor slab at a lower level and run the services on top in a layer of concrete with pumice aggregate. This is also useful when the equipment is changed, as the reinforced concrete floor slab does not need to be altered.

### Walls

Like with columns, imposed loads may be reduced according to the number of storeys (see p. 129).

*In situ concrete*
The minimum wall thickness below continuous floor slabs is 100 mm, and 120 mm elsewhere (tables 3.1.62 and 3.1.63).

*Walls with precast panels*
The sandwich panel represents an economic combination of precast concrete panels and in situ concrete. It is constructed of two 40-60 mm thick precast concrete panels connected at the works by lattice beams. The cavity between these is filled with in situ concrete on site

(figure 3.1.71). The walls are assumed to act as a uniform total cross-section. They may be constructed with or without reinforcement, according to DIN 1045, section 25.5 (figure 3.1.72).

The total cross-section can carry the vertical and horizontal loads (e.g. bracing, earth pressure). Such walls can also be used in conjunction with external tanking provided impermeable concrete is used and the joints are sealed appropriately.

In contrast to partly prefabricated panels with a structural layer of in situ concrete, the sandwich panel is not covered by a standard. It requires approval by the German Building Technology Institute. The filling of the cavity between the precast panels cannot be checked directly during construction. Therefore, care should be taken with the consistency of the fresh concrete and the compaction.

*Precast walls*
The minimum wall thickness below continuous floor slabs is 80 mm, and 100 mm elsewhere (tables 3.1.64 and 3.1.65).

Precast concrete wall panels are produced on the ground on one set of forms using concreting trains, on tilting tables or in battery forms. Production with concreting trains requires plenty of space because the panels must remain horizontal until they have reached sufficient strength to be transported. Tilting tables call for a higher capital outlay but the forms can be struck sooner. In the battery forms the panels are cast vertically, adjacent to each other and separated by steel formwork, and are heated to accelerate the hardening. The concrete does not require finishing; both sides of the wall have an excellent finish. However, sandwich panels cannot be manufactured in battery forms (see p. 136ff.).

Please refer to [3] for masonry walls.

## Columns

The task of the columns is to transfer predominately vertical loads from the floors to the foundation. Wherever possible, horizontal forces should be resisted by walls and/or service cores.

*Loading*
The imposed load on the floors may be reduced by the factor $\alpha_N$ when designing the columns:

$$\alpha_N = (2 + [n - 2] \cdot 0.7)/n$$

where n = number of storeys.
However, $\alpha_N = 1$ for multistorey car parks and garages.

*In situ concrete*
Figure 3.1.66 shows the minimum cross-section dimensions for linear in situ concrete compression members (DIN 1045, section 25.2.1). A column is classed as a wall when the maximum value of 5b is exceeded.

*Precast columns*
Precast concrete columns are usually cast in the horizontal position. The exception is columns with flared heads, which are cast vertically. In the case of columns cast horizontally, it is helpful for production if two opposing sides are flat. One of these sides lies on the bottom of the form, while the top side is finished smooth. Columns that will later stand exposed in a room and whose surface finish is important can have a formwork element pressed onto the surface after striking the formwork. Figure 3.1.67 specifies the minimum cross-section dimensions for linear precast concrete compression members (DIN 1045, section 25.2.1). A column is classed as a wall when the maximum value of 5b is exceeded.

Columns with a circular cross-section, as well as cross-sections symmetrical about two axes, can be produced using the spun concrete method. The reinforcement and the concrete are placed in a horizontal, open steel mould, which is then closed. The mould is mounted loose on wheels driven by electric motors; these cause the mould to rotate. The centrifugal force compacts the concrete at a speed of n = 300-400 r.p.m., reaching a radial acceleration of 25-27 g while doing so. This method of manufacture guarantees that the outer surface of the column is free from blowholes. A compacting factor of 0.89 can be achieved. If the mould is filled completely, the concrete is compacted such that a circular void is created in the middle. The diameter of this void is equal to about 1/3 of the outside diameter and can be used for downpipes, cables etc. Concrete grades of B 85 and even higher can be achieved using this method.

As this type of compaction was not considered when compiling DIN 1045, grade B 55 is also taken to be the maximum limit for spun concrete columns, and the maximum reinforcement content 9%. DIN 4228 also includes concrete grade B 65 for precast concrete masts. In Switzerland, columns with 30% reinforcement content have been used (see p. 206) [67].

Tables 3.1.73 to 3.1.76 show the permissible loads for axially loaded, internal columns held in position.

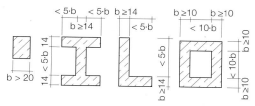
3.1.66 Minimum dimensions of in situ concrete columns [mm]

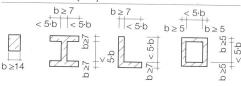
3.1.67 Minimum dimensions of precast concrete columns [mm]

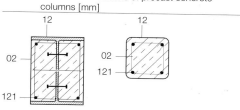

02 Reinforced concrete
12 Steel
121 Reinforcement

3.1.68 Sections through composite columns

3.1.69 Guide values for axially loaded columns of multistorey buildings for:

$s_k$ storey height / d column width < 15
$s_k$ storey height / ∅ column diameter < 12
column load in kN = column cross-section in cm² because B 25 perm $\sigma_1^*$ = 10.5 N/mm²
(B 35 perm $\sigma_1^*$ = 13.8 N/mm²)

\* with 5% reinforcement

3.1.70 Detail of junction between flat slab and multistorey spun concrete column

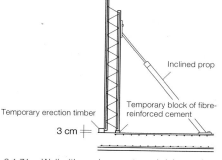
3.1.71 Wall with semi-precast sandwich panels

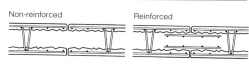
3.1.72 Horizontal sections through joint between semi-precast sandwich panels

# Multistorey structures

3.1.73 Guide values for permissible load [kN] for axially loaded columns with **circular cross-section** and both ends held in position to DIN 1045

| d [cm] / $s_k$ [m] | | $\lambda < 70$ 20 | 25 | 30 | $\lambda < 45$ 35 | 40 | 45 | 50 | 55 | $\lambda < 20$ 60 | 65 | 70 |
|---|---|---|---|---|---|---|---|---|---|---|---|---|
| 2,50 | B35 | 585 | 1129 | 1626 | 2213 | 2890 | 3658 | 4508 | 5454 | 6491 | 7618 | 8835 |
|  | B55 | 829 | 1590 | 2290 | 3118 | 4072 | 5153 | 6325 | 7653 | 9108 | 10689 | 12397 |
| 3,00 | B35 | 550 | 948 | 1626 | 2213 | 2890 | 3658 | 4508 | 5454 | 6491 | 7618 | 8835 |
|  | B55 | 754 | 1325 | 2290 | 3118 | 4072 | 5153 | 6325 | 7653 | 9108 | 10689 | 12397 |
| 3,50 | B35 | 506 | 891 | 1382 | 2213 | 2890 | 3658 | 4508 | 5454 | 6491 | 7618 | 8835 |
|  | B55 | 688 | 1222 | 1930 | 3118 | 4072 | 5153 | 6325 | 7653 | 9108 | 10689 | 12397 |
| 4,00 | B35 |  | 824 | 1284 | 1881 | 2890 | 3658 | 4508 | 5454 | 6491 | 7618 | 8835 |
|  | B55 |  | 1148 | 1824 | 2655 | 4072 | 5153 | 6325 | 7653 | 9108 | 10689 | 12397 |
| 4,50 | B35 |  |  | 1219 | 1814 | 2890 | 3658 | 4508 | 5454 | 6491 | 7618 | 8835 |
|  | B55 |  |  | 1696 | 2540 | 4072 | 5153 | 6325 | 7653 | 9108 | 10689 | 12397 |
| 5,00 | B35 |  |  | 1171 | 1726 | 2370 | 3658 | 4508 | 5454 | 6491 | 7618 | 8835 |
|  | B55 |  |  | 1612 | 2367 | 3355 | 5153 | 6325 | 7653 | 9108 | 10689 | 12397 |
| 5,50 | B35 |  |  |  | 1615 | 2283 | 3073 | 4508 | 5454 | 6491 | 7618 | 8835 |
|  | B55 |  |  |  | 2251 | 3204 | 4294 | 6325 | 7653 | 9108 | 10689 | 12397 |
| 6,00 | B35 |  |  |  | 1553 | 2176 | 2926 | 3793 | 5454 | 6491 | 7618 | 8835 |
|  | B55 |  |  |  | 2165 | 3054 | 4103 | 5301 | 7653 | 9108 | 10689 | 12397 |
| 6,50 | B35 |  |  |  |  | 2081 | 2853 | 3658 | 4590 | 6491 | 7618 | 8835 |
|  | B55 |  |  |  |  | 2941 | 3912 | 5125 | 6486 | 9108 | 10689 | 12397 |
| 7,00 | B35 |  |  |  |  | 2023 | 2706 | 3568 | 4481 | 5528 | 7618 | 8835 |
|  | B55 |  |  |  |  | 2752 | 3817 | 4948 | 6272 | 7719 | 10689 | 12397 |
| 7,50 | B35 |  |  |  |  |  | 2634 | 3432 | 4372 | 5398 | 6487 | 8835 |
|  | B55 |  |  |  |  |  | 3674 | 4771 | 6058 | 7549 | 9159 | 12397 |
| 8,00 | B35 |  |  |  |  |  |  | 3297 | 4208 | 5202 | 6335 | 7612 |
|  | B55 |  |  |  |  |  |  | 4595 | 5845 | 7295 | 8860 | 10621 |
| 8,50 | B35 |  |  |  |  |  |  | 3206 | 4098 | 5072 | 6182 | 7435 |
|  | B55 |  |  |  |  |  |  | 4477 | 5702 | 7040 | 8660 | 10391 |
| 9,00 | B35 |  |  |  |  |  |  |  | 3934 | 4942 | 6029 | 7258 |
|  | B55 |  |  |  |  |  |  |  | 5559 | 6871 | 8462 | 10160 |
| 9,50 | B35 |  |  |  | $\lambda > 70$ |  |  |  | 3825 | 4812 | 5877 | 7081 |
|  | B55 |  |  |  |  |  |  |  | 5488 | 6616 | 8163 | 9929 |
| 10,00 | B35 |  |  |  |  |  |  |  |  | 4682 | 5648 | 6904 |
|  | B55 |  |  |  |  |  |  |  |  | 6531 | 7964 | 9583 |
| 10,50 | B35 |  |  |  |  |  |  |  |  | 4552 | 5571 | 6727 |
|  | B55 |  |  |  |  |  |  |  |  | 6192 | 7765 | 9467 |
| 11,00 | B35 |  |  |  |  |  |  |  |  |  | 5419 | 6462 |
|  | B55 |  |  |  |  |  |  |  |  |  | 7566 | 9120 |
| 11,50 | B35 |  |  |  |  |  |  |  |  |  |  | 6373 |
|  | B55 |  |  |  |  |  |  |  |  |  |  | 8890 |
| 12,00 | B35 |  |  |  |  |  |  |  |  |  |  | 6196 |
|  | B55 |  |  |  |  |  |  |  |  |  |  | 8659 |

The tables of guide values for internal columns with both ends held in position are based on the method of analysis given in DIN 1045, section 25.
The values determined are serviceability loads and are valid for concrete grades:
- B 35 – with 6% reinforcement (upper figure)
- B 55 – with 9% reinforcement (lower figure)

The load specified represents a guide value. An appropriate structural analysis must be carried out to determine the final size.

**Square columns**
In order to use the table with square columns, take into account the following factors:
a) $\lambda < 20$
The size of the given square column (a x a) is multiplied by the factor 1.13 and hence corresponds to the area of a circular column. The guide value can then be read off the table.
b) $20 < \lambda < 70$
The size of the given square column (a x a) is multiplied by the factor 1.13 and hence corresponds to the area of a circular column. The guide value read off the table can then be increased by 10%.

Example:
Given parameters:
- column: a x a = 400 x 400 mm
- storey height: h = 7.00 m
- concrete: B 35

Conversion to circular column d = 1.13 x 400 = 450 mm
Read off: perm N = 2706 x 1.1 = 2976 kN

Columns

3.1.73 Guide values for permissible load [kN] for axially loaded columns with **circular cross-section** and both ends held in position to DIN 1045 ($d_i = 1/3d$)

| d [cm] / $s_k$ [m] | | λ < 70 20 | 25 | 30 | λ < 45 35 | 40 | 45 | 50 | 55 | λ < 20 60 | 65 | 70 |
|---|---|---|---|---|---|---|---|---|---|---|---|---|
| 2,50 | B35 | 519 | 1006 | 1449 | 1972 | 2576 | 3260 | 4000 | 4840 | 5760 | 6760 | 7840 |
|      | B55 | 726 | 1412 | 2034 | 2768 | 3616 | 4576 | 5625 | 6806 | 8100 | 9506 | 11025 |
| 3,00 | B35 | 477 | 826 | 1449 | 1972 | 2576 | 3260 | 4000 | 4840 | 5760 | 6760 | 7840 |
|      | B55 | 666 | 1156 | 2034 | 2768 | 3616 | 4576 | 5625 | 6806 | 8100 | 9506 | 11025 |
| 3,50 | B35 |     | 770 | 1202 | 1972 | 2576 | 3260 | 4000 | 4840 | 5760 | 6760 | 7840 |
|      | B55 |     | 1077 | 1685 | 2768 | 3616 | 4576 | 5625 | 6806 | 8100 | 9506 | 11025 |
| 4,00 | B35 |     | 721 | 1135 | 1648 | 2576 | 3260 | 4000 | 4840 | 5760 | 6760 | 7840 |
|      | B55 |     | 1008 | 1587 | 2313 | 3616 | 4576 | 5625 | 6806 | 8100 | 9506 | 11025 |
| 4,50 | B35 |     |     | 1073 | 1571 | 2164 | 3260 | 4000 | 4840 | 5760 | 6760 | 7840 |
|      | B55 |     |     | 1499 | 2196 | 3038 | 4576 | 5625 | 6806 | 8100 | 9506 | 11025 |
| 5,00 | B35 |     |     | 1017 | 1496 | 2079 | 2751 | 4000 | 4840 | 5760 | 6760 | 7840 |
|      | B55 |     |     | 1421 | 2090 | 2906 | 3862 | 5625 | 6806 | 8100 | 9506 | 11025 |
| 5,50 | B35 |     |     |     | 1427 | 1990 | 2658 | 3408 | 4840 | 5760 | 6760 | 7840 |
|      | B55 |     |     |     | 1994 | 2781 | 3715 | 4784 | 6806 | 8100 | 9506 | 11025 |
| 6,00 | B35 |     |     |     |     | 1908 | 2555 | 3306 | 4136 | 5760 | 6760 | 7840 |
|      | B55 |     |     |     |     | 2666 | 3571 | 4624 | 5805 | 8100 | 9506 | 11025 |
| 6,50 | B35 |     |     |     |     | 1832 | 2459 | 3191 | 4022 | 4933 | 6760 | 7840 |
|      | B55 |     |     |     |     | 2560 | 3437 | 4460 | 5632 | 6925 | 9506 | 11025 |
| 7,00 | B35 |     |     |     |     |     | 2371 | 3083 | 3899 | 4809 | 5802 | 7840 |
|      | B55 |     |     |     |     |     | 3314 | 4308 | 5450 | 6740 | 8144 | 11025 |
| 7,50 | B35 |     |     |     |     |     | 2289 | 2981 | 3777 | 4679 | 5666 | 7840 |
|      | B55 |     |     |     |     |     | 3198 | 4166 | 5278 | 6538 | 7949 | 11025 |
| 8,00 | B35 |     |     |     |     |     |     | 2886 | 3662 | 4542 | 5529 | 6593 |
|      | B55 |     |     |     |     |     |     | 4033 | 5118 | 6348 | 7727 | 9255 |
| 8,50 | B35 |     |     |     |     |     |     |      | 3554 | 4414 | 5379 | 6451 |
|      | B55 |     |     |     |     |     |     |      | 4966 | 6169 | 7518 | 9016 |
| 9,00 | B35 |     |     |     |     |     |     |      | 3452 | 4293 | 5237 | 6287 |
|      | B55 |     |     |     |     |     |     |      | 4824 | 5999 | 7319 | 8787 |
| 9,50 | B35 |     |     |     |     |     |     |      |      | 4178 | 5103 | 6132 |
|      | B55 |     |     |     |     |     |     |      |      | 5839 | 7131 | 8570 |
| 10,00 | B35 |     |     |     |     |     |     |      |      | 4069 | 4975 | 5984 |
|       | B55 |     |     |     |     |     |     |      |      | 5687 | 6953 | 8363 |
| 10,50 | B35 |     |     |     |     |     |     |      |      |      | 4853 | 5843 |
|       | B55 |     |     |     |     |     |     |      |      |      | 6783 | 8166 |
| 11,00 | B35 |     |     |     |     |     |     |      |      |      |      | 5708 |
|       | B55 |     |     |     |     |     |     |      |      |      |      | 7978 |
| 11,50 | B35 |     |     |     |     |     |     |      |      |      |      | 5580 |
|       | B55 |     |     |     |     |     |     |      |      |      |      | 7798 |
| 12,00 | B35 |     |     |     |     |     |     |      |      |      |      |      |
|       | B55 |     |     |     |     |     |     |      |      |      |      |      |

λ > 70

3.1.75 Guide values for ultimate load [kN] for axially loaded columns with **circular cross-section** and both ends held in position to EC 2

| d [cm] / $s_k$ [m] | | λ < 140 | | | | λ < 25 | | | | | | |
|---|---|---|---|---|---|---|---|---|---|---|---|---|
| | | 20 | 25 | 30 | 35 | 40 | 45 | 50 | 55 | 60 | 65 | 70 |
| 2,50 | C30/37 | 964 | 1652 | 2515 | 3551 | 5152 | 6519 | 8050 | 9739 | 11592 | 13603 | 15778 |
| | C45/55 | 1359 | 2328 | 3544 | 5005 | 7225 | 9144 | 11289 | 13659 | 16256 | 19079 | 22128 |
| 3,00 | C30/37 | 880 | 1547 | 2394 | 3417 | 4614 | 5982 | 8050 | 9739 | 11592 | 13603 | 15778 |
| | C45/55 | 1240 | 2181 | 3374 | 4814 | 6500 | 8426 | 11289 | 13659 | 16256 | 19079 | 22128 |
| 3,50 | C30/37 | 792 | 1440 | 2269 | 3276 | 4459 | 5815 | 7343 | 9042 | 11592 | 13603 | 15778 |
| | C45/55 | 1125 | 2030 | 3196 | 4614 | 6281 | 8192 | 10344 | 12737 | 16256 | 19079 | 22128 |
| 4,00 | C30/37 | 716 | 1337 | 2140 | 3129 | 4295 | 5639 | 7155 | 8843 | 10704 | 13603 | 15778 |
| | C45/55 | 995 | 1884 | 3015 | 4407 | 6050 | 7943 | 10080 | 12460 | 15079 | 19079 | 22128 |
| 4,50 | C30/37 | 631 | 1226 | 2013 | 2979 | 4127 | 5454 | 6958 | 8635 | 10484 | 12506 | 14697 |
| | C45/55 | 897 | 1740 | 2836 | 4197 | 5814 | 7684 | 9801 | 12164 | 14770 | 17617 | 20704 |
| 5,00 | C30/37 | 564 | 1113 | 1888 | 2829 | 3956 | 5265 | 6752 | 8416 | 10255 | 12265 | 14448 |
| | C45/55 | 802 | 1575 | 2662 | 3987 | 5573 | 7418 | 9513 | 11856 | 14446 | 17278 | 20353 |
| 5,50 | C30/37 | 504 | 1022 | 1754 | 2682 | 3782 | 5072 | 6542 | 8191 | 10015 | 12014 | 14186 |
| | C45/55 | 718 | 1450 | 2486 | 3778 | 5332 | 7147 | 9217 | 11538 | 14109 | 16924 | 19983 |
| 6,00 | C30/37 | 452 | 933 | 1612 | 2535 | 3613 | 4877 | 6329 | 7960 | 9769 | 11753 | 13913 |
| | C45/55 | 642 | 1325 | 2254 | 3577 | 5090 | 6874 | 8915 | 11212 | 13762 | 16557 | 19600 |
| 6,50 | C30/37 | 406 | 854 | 1507 | 2375 | 3445 | 4685 | 6112 | 7723 | 9515 | 11485 | 13631 |
| | C45/55 | 579 | 1211 | 2137 | 3369 | 4856 | 6601 | 8611 | 10882 | 13406 | 16181 | 19203 |
| 7,00 | C30/37 | 365 | 781 | 1397 | 2192 | 3274 | 4495 | 5898 | 7485 | 9259 | 11212 | 13343 |
| | C45/55 | 522 | 1110 | 1982 | 3087 | 4625 | 6333 | 8309 | 10546 | 13043 | 15794 | 18797 |
| 7,50 | C30/37 | | 715 | 1296 | 2087 | 3094 | 4306 | 5682 | 7249 | 9000 | 10934 | 13048 |
| | C45/55 | | 1016 | 1839 | 2959 | 4386 | 6067 | 8008 | 10213 | 12679 | 15404 | 18382 |
| 8,00 | C30/37 | | 656 | 1202 | 1955 | 2844 | 4110 | 5471 | 7011 | 8741 | 10654 | 12749 |
| | C45/55 | | 935 | 1708 | 2773 | 4025 | 5808 | 7711 | 9878 | 12314 | 15009 | 17960 |
| 8,50 | C30/37 | | 604 | 1117 | 1834 | 2762 | 3907 | 5265 | 6777 | 8481 | 10369 | 12447 |
| | C45/55 | | 859 | 1586 | 2599 | 3915 | 5538 | 7417 | 9548 | 11947 | 14611 | 17535 |
| 9,00 | C30/37 | | | 1038 | 1719 | 2608 | 3714 | 5040 | 6546 | 8223 | 10089 | 12142 |
| | C45/55 | | | 1475 | 2437 | 3697 | 5265 | 7130 | 9221 | 11585 | 14214 | 17108 |
| 9,50 | C30/37 | | | 968 | 1614 | 2465 | 3530 | 4816 | 6318 | 7967 | 9808 | 11839 |
| | C45/55 | | | 1376 | 2289 | 3495 | 5006 | 6826 | 8901 | 11226 | 13820 | 16679 |
| 10,00 | C30/37 | | | 903 | 1514 | 2329 | 3357 | 4601 | 6066 | 7715 | 9531 | 11536 |
| | C45/55 | | | 1282 | 2150 | 3304 | 4758 | 6521 | 8589 | 10872 | 13426 | 16249 |
| 10,50 | C30/37 | | | 842 | 1423 | 2202 | 3192 | 4277 | 5819 | 7464 | 9255 | 11235 |
| | C45/55 | | | 1198 | 2020 | 3124 | 4523 | 6078 | 8247 | 10522 | 13039 | 15827 |
| 11,00 | C30/37 | | | | 1338 | 2083 | 3035 | 4200 | 5583 | 7187 | 8983 | 10934 |
| | C45/55 | | λ > 140 | | 1899 | 2955 | 4303 | 5954 | 7912 | 10182 | 12656 | 15405 |
| 11,50 | C30/37 | | | | 1261 | 1972 | 2885 | 4013 | 5224 | 6920 | 8702 | 10638 |
| | C45/55 | | | | 1790 | 2798 | 4092 | 5686 | 7355 | 9807 | 12279 | 14988 |
| 12,00 | C30/37 | | | | 1190 | 1869 | 2746 | 3836 | 5139 | 6661 | 8405 | 10346 |
| | C45/55 | | | | 1688 | 2648 | 3896 | 5436 | 7282 | 9440 | 11911 | 14576 |
| | | | | | | λ < 140 | | | | | | |

The tables of guide values for internal columns with both ends held in position are based on the method of analysis given in EC 2, i.e. the values are comparable with the effects determined with the partial safety factors.
The values determined are ultimate loads and are valid for concrete grades:
C 30/37 – with 6% reinforcement (upper figure)
C 45/55 – with 8% reinforcement (lower figure)
The load specified represents a guide value. An appropriate structural analysis must be carried out to determine the final size.

**Square columns**
In order to use the table with square columns, take into account the following factors:
a) λ < 25
The size of the given square column (a x a) is multiplied by the factor 1.13 and hence corresponds to the area of a circular column. The guide value can then be read off the table.
b) 20 < λ < 140
The size of the given square column (a x a) is multiplied by the factor 1.13 and hence corresponds to the area of a circular column. The guide value read off the table can then be increased by 10%.

3.1.76 Guide values for ultimate load [kN] for axially loaded columns with **annular cross-section** and both ends held in position to EC 2

| d [cm] $s_k$ [m] | | λ < 140 | | | | | λ < 25 | | | | | |
|---|---|---|---|---|---|---|---|---|---|---|---|---|
| | | 20 | 25 | 30 | 35 | 40 | 45 | 50 | 55 | 60 | 65 | 70 |
| 2,50 | C30/37 | 834 | 1439 | 2203 | 3120 | 4191 | 5790 | 7149 | 8650 | 10295 | 12082 | 14014 |
| | C45/55 | 1174 | 2027 | 3103 | 4397 | 5905 | 8120 | 10024 | 12129 | 14435 | 16941 | 19649 |
| 3,00 | C30/37 | 756 | 1339 | 2086 | 2991 | 4051 | 5264 | 6627 | 8650 | 10295 | 12082 | 14014 |
| | C45/55 | 1065 | 1888 | 2940 | 4214 | 5707 | 7415 | 9338 | 12129 | 14435 | 16941 | 19649 |
| 3,50 | C30/37 | 670 | 1240 | 1967 | 2854 | 3901 | 5103 | 6456 | 7964 | 10295 | 12082 | 14014 |
| | C45/55 | 952 | 1748 | 2772 | 4023 | 5496 | 7189 | 9097 | 11219 | 14435 | 16941 | 19649 |
| 4,00 | C30/37 | 593 | 1139 | 1845 | 2716 | 3745 | 4932 | 6276 | 7772 | 9422 | 11223 | 14014 |
| | C45/55 | 844 | 1611 | 2602 | 3826 | 5276 | 6948 | 8841 | 10949 | 13273 | 15811 | 19649 |
| 4,50 | C30/37 | 526 | 1036 | 1729 | 2573 | 3585 | 4757 | 6085 | 7571 | 9210 | 11002 | 12947 |
| | C45/55 | 749 | 1472 | 2436 | 3626 | 5051 | 6700 | 8572 | 10665 | 12973 | 15499 | 18239 |
| 5,00 | C30/37 | 467 | 940 | 1604 | 2433 | 3423 | 4576 | 5889 | 7359 | 8988 | 10768 | 12705 |
| | C45/55 | 665 | 1335 | 2273 | 3428 | 4821 | 6447 | 8296 | 10369 | 12661 | 15170 | 17897 |
| 5,50 | C30/37 | 415 | 854 | 1481 | 2297 | 3262 | 4393 | 5688 | 7144 | 8757 | 10526 | 12450 |
| | C45/55 | 592 | 1213 | 2100 | 3236 | 4596 | 6189 | 8013 | 10064 | 12336 | 14828 | 17539 |
| 6,00 | C30/37 | 371 | 775 | 1366 | 2149 | 3102 | 4211 | 5485 | 6923 | 8520 | 10276 | 12187 |
| | C45/55 | 529 | 1103 | 1939 | 3047 | 4372 | 5933 | 7728 | 9752 | 12003 | 14476 | 17168 |
| 6,50 | C30/37 | 331 | 707 | 1261 | 2006 | 2944 | 4032 | 5283 | 6699 | 8279 | 10019 | 11916 |
| | C45/55 | 473 | 1003 | 1789 | 2832 | 4153 | 5679 | 7442 | 9438 | 11663 | 14114 | 16787 |
| 7,00 | C30/37 | | 644 | 1164 | 1873 | 2772 | 3854 | 5079 | 6476 | 8036 | 9758 | 11639 |
| | C45/55 | | 915 | 1653 | 2630 | 3931 | 5429 | 7156 | 9123 | 11309 | 13746 | 16398 |
| 7,50 | C30/37 | | 588 | 1075 | 1747 | 2611 | 3668 | 4880 | 6252 | 7789 | 9493 | 11358 |
| | C45/55 | | 837 | 1527 | 2479 | 3701 | 5185 | 6875 | 8807 | 10976 | 13375 | 16000 |
| 8,00 | C30/37 | | 537 | 995 | 1631 | 2458 | 3479 | 4685 | 6031 | 7546 | 9227 | 11074 |
| | C45/55 | | 765 | 1412 | 2314 | 3486 | 4929 | 6602 | 8496 | 10631 | 13000 | 15601 |
| 8,50 | C30/37 | | | 921 | 1524 | 2314 | 3276 | 4474 | 5812 | 7303 | 8962 | 10787 |
| | C45/55 | | | 1309 | 2163 | 3281 | 4599 | 6330 | 8190 | 10290 | 12626 | 15198 |
| 9,00 | C30/37 | | | 855 | 1425 | 2178 | 3123 | 4263 | 5595 | 7061 | 8696 | 10501 |
| | C45/55 | | | 1213 | 2021 | 3091 | 4429 | 6041 | 7889 | 9951 | 12254 | 14793 |
| 9,50 | C30/37 | | | 793 | 1332 | 2052 | 2961 | 4018 | 5357 | 6827 | 8436 | 10214 |
| | C45/55 | | | 1128 | 1891 | 2912 | 4197 | 5670 | 7592 | 9618 | 11884 | 14390 |
| 10,00 | C30/37 | | | 737 | 1248 | 1934 | 2807 | 3869 | 5128 | 6581 | 8177 | 9931 |
| | C45/55 | | | 1050 | 1772 | 2744 | 3978 | 5485 | 7267 | 9290 | 11520 | 13991 |
| 10,50 | C30/37 | | | | 1170 | 1824 | 2660 | 3687 | 4865 | 6323 | 7921 | 9648 |
| | C45/55 | | | | 1663 | 2588 | 3773 | 5226 | 6833 | 8961 | 11160 | 13595 |
| 11,00 | C30/37 | | | | 1099 | 1720 | 2524 | 3514 | 4697 | 6074 | 7648 | 9371 |
| | C45/55 | | λ > 140 | | 1559 | 2441 | 3579 | 4981 | 6657 | 8608 | 10808 | 13204 |
| 11,50 | C30/37 | | | | 1033 | 1625 | 2395 | 3350 | 4494 | 5796 | 7371 | 9098 |
| | C45/55 | | | | 1465 | 2305 | 3396 | 4747 | 6371 | 8092 | 10442 | 12819 |
| 12,00 | C30/37 | | | | | 1536 | 2273 | 3192 | 4302 | 5604 | 7100 | 8794 |
| | C45/55 | | | | | 2181 | 3224 | 4527 | 6095 | 7637 | 10061 | 12441 |
| | | | | | | λ < 140 | | | | | | |

# Multistorey structures

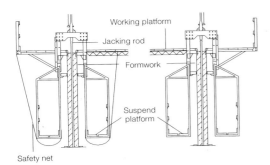

3.1.77  Slipforming platform

3.1.78  Slipforming (see figure 3.1.98)

Columns for special constructions, which are also subjected to bending moments, are produced in prestressed concrete in order to avoid cracking (Bonn Art Gallery, p. 206).

The prestressing tendons (strands) are anchored in the cap and base plates either with wedges or, if the column loading allows it, through the bond with the concrete. During the prestressing operation the jacks are supported against the formwork. To avoid cracks during transport, it is recommended to produce spun concrete columns in prestressed concrete.

From the erection point of view it is better to build multistorey columns (figure 3.1.70). Precast concrete columns are also used in combination with in situ concrete floor slabs, especially flat slabs. The concrete is omitted from the column at the junction with the floor slab so that the slab reinforcement can be threaded through.

*Composite columns* [11]
Columns can also be built using a steel-reinforced concrete composite system. The concrete lends the steel section the necessary fire resistance. Hollow sections and tubes are filled with concrete after inserting a reinforcing cage.

The spaces between the flanges of open sections (H-sections) can be filled with concrete. This concrete is reinforced similarly to that for beams (see p. 127).

## Service cores, stairs

Service cores contain stairs, lift shafts and, if necessary, shafts for building services. Their primary structural function is to stiffen the building, i.e. to carry the horizontal forces (e.g. wind) acting on the building and the horizontal load components of building components appropriately supported, or loadbearing components unintentionally out of plumb.

Shafts for ventilation ducts and building services should be arranged in such a way that they are accessible from the side after being integrated in the service core. This permits easier installation and maintenance.
The loadbearing construction of service cores must comply with the F 90 class of fire protection. Even with steel framed buildings, the service core is normally built in reinforced concrete.

### In situ concrete
Storey-by-storey construction
The service cores of in situ concrete structures with fewer than 8-10 storeys are built together with the individual storeys. Formwork panels are used for the walls. The stair flights and landings are also built of in situ concrete, but the stair flights are seldom joined to the walls. This avoids having to cut the formwork for the walls. The stair flights span from landing to landing. It is also economical to precast the stair flights and position these between the landings (see pp. 190-191).

In taller buildings, particularly those employing precast concrete or steel frames, it is advisable to build the service core first as this provides the necessary stability during construction. This can be achieved using climbing formwork or by slipforming.

### Climbing formwork
A climbing formwork is usually produced from storey-height formwork units. After the concrete has hardened, the form can be raised ready for casting the next storey. Climbing formwork is also used for producing concrete towers. It is preferred when the concrete surface is to have a textured or very smooth finish.

Stair flights and landings are usually precast components. Starter bars are left projecting from the walls ready to connect the floor slabs.

### Slipforming
This method has proved to be economic for producing service cores. It was originally used for building chimneys, silos and television towers (figure 3.1.77).

Threaded bars are fixed in the foundation and the form plus working platform are screwed upwards on them. The reinforcement is placed continuously, followed by placing and compaction of the low-slump concrete. The task of the scaffolding is taken over by the hardening concrete. The position of the formwork is checked continuously for accuracy. Starter bars are incorporated for attaching the floor slabs; these are initially folded flat against the concrete to allow the formwork to pass by and are folded out later as required.

Formwork for openings is fixed to the reinforcement or the threaded bars. It must be positioned with great care, and its position must be checked after compacting the concrete. A slipforming speed of 5 m/d can generally be expected. Concrete operations must continue uninterrupted, day and night. (Night operations are not permitted in residential areas.) Newly developed methods involving concrete additives to delay the setting process permit the work to be interrupted.

However, since the formwork is pulled past the concrete surface before it is fully hardened, it is not possible to attain the same standard of surface finish as with conventional methods. A textured surface is not possible with this method.

## Facades [10]

The facades dealt with in this book are those constructed from or clad in reinforced concrete. We shall first look at claddings to in situ concrete (or masonry) walls and then discuss precast concrete external walls. Please refer to p. 65 for details of surface finishes.

### Claddings to in situ concrete structures

The loadbearing walls require thermal insulation. The insulation is, however, normally a soft, porous material that has to be protected against rainfall and mechanical actions. Reinforced concrete facade panels are fitted to the loadbearing wall in such a way that an air space is created to allow any moisture diffusing through the wall to escape (figure 3.1.79 A). The minimum thickness of the panel, taking into account anchors and transport, should be $s = 80$ mm. A precast concrete panel must be able to expand and contract unhindered as the temperature fluctuates. The maximum dimensions are a result of the system of suspension. This is always by means of two anchors per panel positioned symmetrically regardless of the system or the manufacturer used. The loads that these anchors can carry are limited.

Apart from that, the maximum possible joint width between two precast concrete panels determines the height and width of the panels. Various approved anchor systems are available for facade panels up to 30 m². Tension anchors to secure the panels against wind suction may be necessary around the edges of panels. The anchor systems allow for adjustment in three directions to permit proper alignment of the cladding panels (see p. 189). The anchor fittings must be made of a rustproof material.

### Parapet panels

The cladding to a parapet may be provided in a similar manner to the general wall cladding. Parapet panels may hook over the top of the parapet and be fixed there with pins (figure 3.1.80). The anchor is given the necessary flexibility, for example, by including a permanent slip layer on the bearing plate or by embedding the top of the pin in an elastic material. The compression-resistant, and if necessary tension-resistant, anchor at the base of the panel is identical to those used for the facade panels.

---

Text passages from [10] have been incorporated here with the kind consent of the author; however, these have not been designated as such.

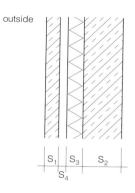
A  Double leaf

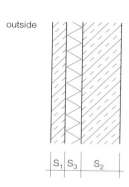
B  Sandwich

3.1.79  Section through facade

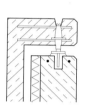  

3.1.80  Parapet panel fixings

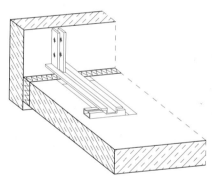

3.1.81  Spandrel panel fixing

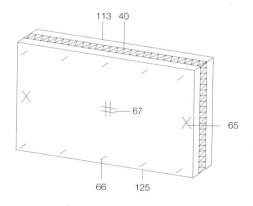

| 40 | Thermal insulation | 67 | Facade anchor (supporting anchor) |
| 65 | Torsion anchor | 113 | Loadbearing leaf |
| 66 | Retaining anchor | 125 | Facing leaf |

3.1.82  Fixing the facing leaf to the loadbearing construction

Spandrel panels (exposed)
The height is primarily determined by building regulations; it is normally between 900 and 1100 mm. The thickness should be s = 100 mm. The same criteria apply here as for facing leaves used as a surface finish.

The type of fixing determines the maximum width possible. A rigid connection between spandrel panel and floor slab calls for a fixing to the loadbearing structure by means of, for example, overlapping reinforcement cast into the concrete. In this case the maximum length should be 2.5 m because otherwise the stresses due to temperature influences can lead to damage (e.g. deformation, cracks). The use of special spandrel panel anchors permits much longer panels. These anchors include, for example, elastomeric bearings to compensate for movement due to temperature fluctuations (figure 3.1.81).

**Precast concrete external walls**
Precast concrete external walls are constructed in three layers (p. 182, figure 4.2 B). They consist of the loadbearing leaf, the layer of thermal insulation and the facing leaf (figure 3.1.79 B).

The loadbearing leaf is 80-150 mm thick according to the load in the main loadbearing direction; 80 mm is possible when the wall panel only performs an enclosing function. Thicker panels are encountered with tall walls and those carrying heavy in-plane loads (e.g. in industrial buildings). The thickness of the thermal insulation is governed by the regulations. Thicknesses of 80-100 mm are common in residential and office buildings. Due to the production process, the facing leaf cannot be cast directly on the layer of thermal insulation without an intervening air space. The need to provide adequate concrete cover to the reinforcement means that the minimum thickness of the facing leaf is 60 mm. This dimension should be increased when the exposed surface is to have a special finish (e.g. exposed aggregate) to allow for the removal of the surface layer of concrete. This minimum dimension applies to the thinnest part of profiled cross-sections. The layer of thermal insulation consists of rigid expanded foam with closed pores. It is usually applied in two layers with staggered joints. A single layer should employ insulating batts with a stepped joint in order to avoid thermal bridges.

The loadbearing leaf (inner leaf) acts as the loadbearing part and is analysed and reinforced accordingly. Production, transport and erection are the main factors determining the maximum dimensions. For economic reasons, it is recommended to use the largest feasible size of precast concrete panel. Typical sizes range from 4 to 10 m. The maximum dimensions of the facing leaf are limited by the ability to move freely to accommodate temperature effects. The area of the facing leaf should therefore not exceed 15 $m^2$, with a width of max. 5 m. This recommendation means that the facing leaf could have more joints than the loadbearing leaf, which is an effective way of preventing the facing leaf from distorting.

Experience has shown that the three-ply panel functions perfectly in building science terms even without an air space. If there is no air space behind the facing leaf, the temperature difference between the inside and outside surfaces, when exposed to solar radiation, is smaller, and therefore bending stresses due to temperature restraint are lower.

*Fixing the facing leaf to the loadbearing leaf*
stainless steel anchors (DIN 17440) are used to attach the facing leaf to the loadbearing leaf. We distinguish between supporting, retaining and torsion anchors (figure 3.1.82). While the horizontal loads, primarily wind pressure and suction, are carried by all anchors proportionately, the rigid supporting anchor carries the vertical loads. These are mainly the self-weight of the facing leaf and, if applicable, the weight of windows and other fittings. The supporting anchor is positioned above the centre of gravity of the facing leaf. The retaining anchors (needles) are placed around the edges, as well in other positions as in larger panels. Their shape allows them to follow the changes in length of the facing leaf almost without resistance.

Torsion anchors, rigid in one direction but movable perpendicular to the rigid direction, prevent the facing leaf from twisting. Such twisting can occur – even when the supporting anchor is positioned correctly in the centre – due to unavoidable manufacturing inaccuracies and the forces acting when the panel is lifted out of the form, or during erection. Approved systems are generally employed (figure 3.1.83).

*Fixing the sandwich panels to the structure*
The loadbearing leaf is attached to the loadbearing structure primarily by means of:

- overlapping reinforcement (figure 3.1.84)
- cast-in inserts, steel sections and bolting
- welding

# Facades

All steel parts that are not permanently protected against corrosion by being fully cast into the concrete or surrounded by grout, according to DIN 1045 section 6.7.1, must be given suitable protection against corrosion, e.g. corrosion protection systems to DIN 55928. Otherwise they must be fabricated from stainless steel.

Production
Facades have to satisfy high demands regarding aesthetics and serviceability. Only production facilities, production methods and good quality building materials can meet these demands. The production facilities include both the formwork in which the concrete hardens and also the equipment and facilities for storing and batching the concrete constituents; for mixing, transporting, placing and compacting the concrete; for lifting, transporting and storing the finished parts; and for working and subsequent treatment of the exposed surfaces.

One essential requirement is that the manufacturer should have his own laboratory for carrying out suitability and monitoring tests. And of course, production should only be carried out in suitable buildings protected from the weather.

Facade panels for vertical applications on the structure are always produced horizontally. This is important for achieving consistent quality (p. 136). We distinguish between the positive and negative methods.

In the positive method the form is arranged with the outside of the facing leaf on the top. This can be finished by brushing, by working in other materials or by exposing the aggregate.

In the negative method the side of the facing leaf that will be exposed in the final condition lies on the bottom of the form. The necessary finish is provided by the bottom surface of the form, which represents the "negative". The negative method is the most popular by far. Steel beam grids covered with negative steel are used as steel sheeting. Such casting tables (see p. 129) are appropriately sturdy to cope with the loads during compaction of the concrete and to guarantee a high re-usability and the necessary plane surface.

These are set up as circulating pallets or as stationary tables depending on the method of production. In some systems they can also be tilted up for stripping the forms. The dimensions of the tables are the result of the typical panel dimensions; they are max. 4 m wide and 10 m long. The negative of the element to be manufactured is attached to this table. The sheet steel covering is used as one formwork surface for planar facade elements. The profile of the facade panel is easily created by fixing further formwork elements to this surface. The texture of the facade is generally created using rubber or plastic dies attached to the table with adhesive. Special architectural requirements are catered for by fabricating special negatives from suitable materials. Plastic-coated plywood panels on suitable steel frames serve as the formwork to the sides; these can be folded or slid out of the way. Tightness of the joints is especially important. To avoid ridges being left on the concrete elements, the corners and edges are normally chamfered by including triangular fillets in the forms.

Just as important as providing a grout-tight, dimensionally accurate and stable form are the preparations for concreting, and cleaning after stripping the concrete element. Preparations include applying a release agent. The purpose of this is to prevent the concrete adhering to the form and ensure easy stripping without damaging the surface of the concrete or the formwork.

Providing adequate concrete cover to the reinforcement is crucial for the durability of a facade. A sufficient number of spacers should be used to secure the reinforcement in position during concreting (see DBV Memo). The type of spacer that can be used depends on the intended finish of the facing leaf. This issue should be clarified between designer and manufacturer.

It is particularly important to maintain a consistent water/cement ratio. To achieve adequate durability, this should not exceed 0.60. Facing leaves with special aggregates are placed in the forms in layers approx. 30-40 mm thick, depending on the size of the aggregate.

The method of production generally only lends itself to producing panels flat on one side only. When choosing the aggregate, compatibility with the cement and weathering resistance (e.g. frost resistance) must also be considered.

Generally, the concrete hardens without needing to apply heat to the form. It can usually be stripped and stored on the following day thanks to the use of early and high-strength cements. At this stage the newly placed concrete has still not reached its final strength and is very vulnerable to the effects of heat, cold, wind and precipitation as well as chemical and mechanical attack. Fair-face concrete surfaces must therefore be protected from the weather, and cured. This is carried out with sheeting or tarpaulins. In doing so, care should be taken to ensure that the sheeting surrounds the fair-face concrete surface but does not touch it. Another means of protection is to spray the surface with a special curing agent.

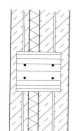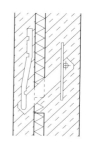

Supporting anchors (vertical sections)

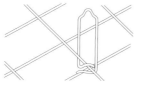

Retaining anchors

3.1.83 Anchor systems

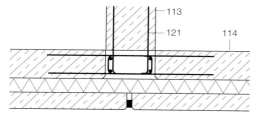

Horizontal section

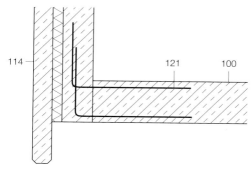

Vertical section

The precast concrete unit is placed in the formwork and the concrete of the floor slab cast up against it.

| 100 | Floor slab supported on beam or wall | 114 | Sandwich panel |
| 113 | Wall | 118 | Column |
| | | 121 | Reinforcement |

3.1.84 Fixing sandwich panels with overlapping reinforcement

# Multistorey structures

3.1.85  Lake Point Tower,
        Chicago, 1968
        A: G. Schipporeit
3.1.86  Messeturm, Frankfurt am Main, 1991
        A: H. Jahn
        E: F. Nötzold, E. Cantor
3.1.87  Colonia Tower,
        Cologne, 1969-71
        A: Busch-Berger
        E: W. Naumann, Strabag
3.1.88  High-rise office block, Dortmund, 1994
        A: E. Gerber
        E: S. Polónyi, Hochtief
3.1.89  National Bank Headquarters,
        Tampa, 1982-88
        A: Harry C. Wolf
3.1.90  Marina City,
        Chicago, 1964-67
        A: B. Goldberg
3.1.91  Torhaus Messe,
        Frankfurt am Main, 1980-83
        A: O. M. Ungers
        E: Hochtief
3.1.92  Ontario Center,
        Chicago, 1979-86
        A: Skidmore Oweings & Merrill

During storage of the elements, care must also be taken to ensure that no undesirable deformations take place and that fair-face concrete surfaces are not spoiled by the effects of support timbers or soiling. The ingress of rainwater into the void between the facing and loadbearing leaves of sandwich panels should be prevented by attaching a self-adhesive tape, for instance.

External walls of aerated concrete
The reader is referred to [27] for details of such walls.

**Anchor connections**
A wide range of fixing systems is available for attaching components to existing concrete, masonry or aerated concrete constructions. The principles for such anchors are regulated by the German Building Technology Institute. Only approved anchors may be used.
The anchor transfers the forces that occur into the base material. Many parameters are important when choosing a type of anchor (e.g. adhesive, expansion). Generally speaking, an anchor can carry only as much load as the building material in which it is anchored. The base material is important when selecting an anchor. Furthermore, the forces can be reliably transferred only when the appropriate edge distances are maintained (i.e. distance from edge of component to centre of anchor). The load-carrying capacity of the anchor is determined by the pitch of multiple anchors and the edge distances. Details and design parameters are specified in the approval documentation.

## High-rise buildings

### General
Up to now the loadbearing structures of high-rise buildings (federal state building code: finished floor level of an occupied room more than 22 m above ground level) have preferably been built in steel. Owing to the higher dead load and the low strength of the concrete, the dimensions of reinforced concrete columns were too large compared to steel. In recent years, however, tall buildings have been built in Chicago using reinforced concrete with a concrete strength exceeding 65 N/mm$^2$, and in Seattle with 131 N/mm$^2$. It is predicted that concretes with a strength of 159 N/mm$^2$ will soon be feasible. Reference [12] reports on the use of grade B 85 in individual cases with the approval of the highest building authority.

The design of high-rise buildings demands an intricate consideration of utilisation, structure, services, construction methods and economics (in terms of both construction and maintenance). With respect to utilisation, the ratio of facade area to gross floor area is especially important. The loadbearing structure and the building services must be harmonised. Large ducts are required for ventilation and air conditioning. These must be routed so as not to affect stability.

### Stability
The provision of stabilising and stiffening elements to accommodate wind and other horizontal loads is particularly important in high-rise structures. The wind load depends very much on the shape of the building. Admittedly, a circular, cylindrical shape (figures 3.1.89 and 3.1.90) is best. However, a circular plan often has disadvantages for the utilisation because the zone provided with daylight is too small related to the total area. George Schipporeit developed a plan shape for his Lake Point Tower apartment building in Chicago (figure 3.1.85) that represents an optimum between utilisation and reducing the wind load. The building is stiffened by three reinforced concrete shear walls.

The most popular plan shape is a modified square. The horizontal forces are thereby accommodated by the service core (figure 3.1.86) and, particularly in the case of tall buildings, by frames positioned in the facade (Vierendeel girder), a perforated facade (figure 3.1.91) or lattice girders (figure 3.1.92) forming a tube.

In residential buildings the party walls between apartments form the stiffening shear walls (figure 3.1.87).

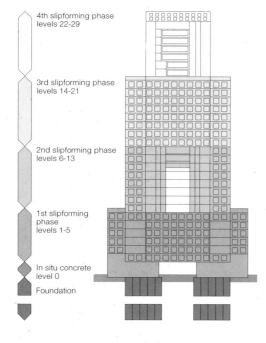

3.1.93  Slipforming phases (see figure 3.1.91)

# High-rise buildings

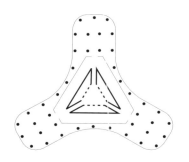
3.1.85

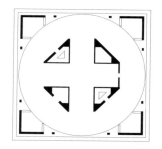
3.1.86

3.1.87

3.1.88

3.1.85

3.1.86

3.1.87

3.1.88

3.1.89

3.1.90

3.1.91

3.1.92

# Multistorey structures

3.1.94  Hotel Budapest, 1967
A: G. Szrog

In contrast to American office buildings, office workers in Germany should have adequate daylight for their work. This results in a typical plan layout with a building width of about 12 m. If the corridor is not to be disrupted by the service core, the depth of the core is about 5 m. This results in a moment of inertia that can accommodate horizontal forces in buildings of 20 storeys. Taller buildings require additional shear walls, e.g. in the gable, or overlapping cores (see p. 264). Larger columns are necessary if the gable wall is to act as a Vierendeel girder to carry the wind load. If constructed as a Vierendeel girder, the spandrel panels, coupled with the lintels, form the top and bottom chords (see p. 182).

The centre of gravity of the stiffening structure should be positioned to coincide with the centre of gravity of the total area wherever possible. This ensures that the load in the stiffening members due to eccentric loading is kept small. The core should have a large load transfer area so that tension loads from wind are precompressed.

It is very important that the stiffening concept is carried down to the subsoil without being weakened. The stiffening members are connected rigidly to the tube from which the walls and floors of the basement and the ground slab are formed. The walls should be arranged in the basement storeys such that the tube can transfer loads into the subsoil without any special type of foundation.

In order to simulate the loadbearing effect realistically, the entire loadbearing system should be analysed as one unit (by means of FEM). The analysis is very time-consuming because of the optimisation required.

Slipforming is usually employed for constructing cores (figure 3.1.78). Perforated facades can also be produced in the same way (figure 3.1.93). Hotel Budapest in the Hungarian capital was constructed as long ago as 1967 using the slipforming technique (figure 3.1.94). Here, the core was constructed with radial walls and cylindrical perforated facade. The floor bays were constructed from the top down. The formwork to each bay was suspended on four threaded bars and screwed down each time to the floor below, ready for fixing the reinforcement and pouring the concrete. The advantage of this method was that the formwork for the floor did not have to be dismantled and re-erected each time. One set of formwork was enough to cast all the floors. The disadvantage was that the reinforcement had to be passed in through the window openings. On the other hand, placing the concrete by way of concrete pumps did not present any problems. When constructing the floors starting at the bottom, the structure can be erected faster because it is not necessary to wait until the roof can carry the loads. However, additional sets of formwork and scaffolding are required.

### Joints in structures
Structures with an almost square plan shape (point-block buildings) require no expansion joints. Buildings rectangular in plan can generally only be stiffened without movement joints (see p. 10).

### Floors
In designing the floors, particularly for the upper storeys, vertical displacements near the core due to horizontal loads must also be taken into account.

## Suspended high-rise buildings

These high-rise structures consist of a central reinforced concrete core and a supporting structure at the very top, possibly with further intermediate supporting structures (figures 3.1.95 and 3.1.96). The suspension members are fixed to the ends of the supporting structure in the facade. The floors span between the core on the inside and the suspension members on the outside. This arrangement of the loadbearing structure means that the core carries the maximum possible axial loading but is subjected to large bending moments in an asymmetric imposed load. Suspended buildings with two cores and bridge beams between or above are subjected to this disadvantageous loading only transverse to the bridge beams (figure 3.1.97). The bridge beams are costly and complicated.

When designing suspended buildings, special attention must be given to the way in which the floors are supported on the core and the suspension members, as these undergo different vertical movements. The relative vertical displacement of the support at the suspended member is the sum of the compression of the core, including creep, and the expansion of the suspended member. To ensure that this displacement does not have an adverse effect, it should be kept small. At the Standard Bank Centre in Johannesburg, South Africa, the suspended construction is divided into three segments (figure 3.1.100). Each group of 10 storeys is suspended below its own supporting structure (figure 3.1.99). At BMW's headquarters in Munich (figure 3.1.98) the top seven storeys are carried on stilts on the primary supporting storey in the middle of the building. The other storeys are suspended below it.

Suspended high-rise buildings

3.1.95   Marl Town Hall, 1958-66
         A: J. H. van den Broek & J. B. Bakema
         E: Franz Vaessen (Hochtief) & H. J. Ehrhardt
3.1.96   Deutschlandfunk offices, Cologne
         A: Planungsgruppe Stildorf
         E: Leventon, Werner, Schwarz
3.1.97   Hospital ward block, Cologne University Clinic, 1969-72
         Supporting structure and core
         A: Heinle Wischer & Partner
         E: Bole & Partner
3.1.98   BMW headquarters, Munich
         A: K. Schwanzer
         E: H. Bomhard (Dyckerhoff & Widmann)
3.1.99   Standard Bank Johannesburg, 1965-70
         A: Hentrich-Petschnigg + Partner
         E: Ove Arup & Partners
3.1.100  Standard Bank Johannesburg, 1965-70
         A: Hentrich-Petschnigg + Partner
         E: Ove Arup & Partners

3.1.95

3.1.96

3.1.97

3.1.98

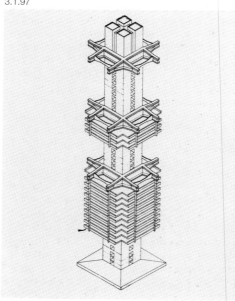
3.1.99

3.1.100

# Single-storey sheds

3.2.1 Influences on the shape and design of single-storey sheds

Function
  homogeneous/changing
  linear (railway station)
  dominant (church)

Plan shape
  rectangular
  circular

Means of transport
  conveyor (store for tipped goods)

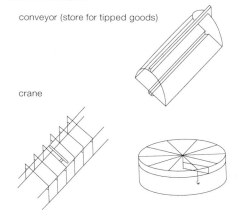

  crane

  fork-lift (impact load)
  automatic conveyor system (high-level racking)

Light (rooflights, northlight roof)

Weather protection

Sound insulation

Fire protection

Costs

These are buildings one storey high with a large span that covers an extensive open area. Of course, a single-storey shed could also be built on top of a multi-storey building.

The utilisation has a major influence on the architecture of the building and the load-bearing structure. This aspect is summarised in table 3.2.1.

The plan dimensions are a consequence of the building plot and the utilisation. The goods stored and the method of storage or production determine the height of the building. For tipped or bulk goods, the method of tipping defines the shape of the building (e.g. arch construction). In the case of high-level racking, the racking itself forms the primary loadbearing structure to which the external envelope is attached. The treatment of warehouses with such racking is not dealt with any further here.

The method of transporting goods within the building has a major influence on the design. If fork-lifts are in use, an impact load on the columns must be taken into account. Table 3.2.2 lists the magnitude of impact loads. When cranes are used, the type and size of crane governs the dimensions of the building and the loadbearing construction.

*Cranes*
The types of crane found in single-storey sheds are travelling overhead cranes, roof-slung gantry cranes and column-mounted cranes. Travelling overhead cranes with various types of control are illustrated in figure 3.2.4. Clearance below the crane hook, the type of crane, its lifting capacity and type of control determine the height of the building. The width of the building is the result of the width of the floor area to be served by the crane, but there are also other restraints (e.g. shape of plot).

Travelling overhead cranes run on crane gantry girders, supported on corbels on the columns. The crane gantry girders are made from rolled steel sections or combinations thereof. The crane rail itself is generally welded to the crane gantry girder. Precast concrete crane gantry girders have not really been very successful because of difficulties in attaching the rail to the concrete.

3.2.2 Standard fork-lift vehicles

| 1 | 2 | 3 | 4 | 5 | 6 | 7 | |
|---|---|---|---|---|---|---|---|
| Permissible total weight | Rated carrying capacity | Statical axle load (standard load) | Average track width | Total width | Total length | Uniformly distributed imposed load (standard load) | |
| t | t | kN | a m | b m | l m | [kN/m²] | |
| 2.5 | 0.6 | 20 | 0.8 | 1 | 2.4 | 10 | |
| 3.5 | 1 | 30 | 0.8 | 1 | 2.8 | 12.5 | |
| 7 | 2,5 | 65 | 1 | 1.2 | 3.4 | 15 | |
| 13 | 5 | 120 | 1.2 | 1.5 | 3.6 | 25 | |

Impact load: 5 x permissible total weight applied 0.75 m above floor level

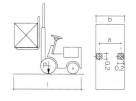

Where smaller loads have to be lifted, a roof-slung gantry crane can be used. The crab travels on the bottom flange of a crane beam, which is fixed to the underside of the roof structure. Roof-slung gantry cranes can be split into sections and coupled together, and also joined via a transfer station so that the crab can travel from one bay to another.

*Light*
Lighting requirements affect the type of single-storey shed chosen and the loadbearing construction. If lighting requirements are low, but the window areas along the longitudinal walls are not sufficient, lantern lights are positioned on the roof (monitor roof). These can also be used for ventilating the building. When the production demands light but not direct sunlight, a northlight roof is the answer. Placing the windows at an angle on a northlight roof improves lighting efficiency. Horizontal rooflight strips are increasingly favoured. They admit the greatest amount of daylight but do need to be provided with some form of sunshading.

Single-storey shed structures are divided into two groups. First, buildings assembled from individual frames comprising columns and trusses or arches – primarily preferred for industrial production and warehousing; and second, plate and shell structures. Although industrial buildings can be erected economically using stressed skin structures, as we shall see later, this structural form is mainly used for exhibition halls, large restaurants and churches.

## Single-storey sheds of trusses and frames

Single-storey sheds assembled from individual frames are erected exclusively in precast components owing to the high cost of scaffolding and formwork for in situ construction.

**Roof**
*Roof covering [31]*
Trapezoidal profile sheeting

Trapezoidal profile sheeting is employed as the roof covering for nearly all single-storey sheds. This material is characterised by its low price, low self-weight, easy erection and high flexural strength.

The trapezoidal profile sheeting is given the task of bracing individual members (purlins, trusses etc.). It can also be designed to act as a shear diaphragm and hence transfer horizontal loads to the columns and other bracing members. The depth of the trapezoidal profile sheeting is chosen according to the span; the maximum deflection of the roof covering should not exceed l/300.

For single-storey sheds without thermal insulation, the roof covering of trapezoidal profile sheeting can be laid in the negative position, provided this follows the fall of the roof. Purlins are required. Therefore, all roofs without thermal insulation inevitably include purlins, in other words a secondary loadbearing construction (figure 3.2.6). Roofs with thermal insulation require additional waterproofing. This consists of either synthetic or bituminous roofing felts.

Adding gravel chippings to the roof is not possible owing to the unfavourable relationship of the weight to the imposed load (snow); this is especially so on buildings with a large span. The trapezoidal profile sheeting may span in either direction on roofs with additional waterproofing. Once the direction of the sheeting is no longer tied to its waterproofing function, it may be laid directly on the primary roof structure (roof without purlins) (figure 3.2.8) or on purlins.

One of the most important analyses required for a single-storey shed is to determine the optimum spacing of the frames. This also governs whether a roof with or without purlins is economical [49].

*Aerated concrete roofing panels [59]*
The roofs of insulated buildings may use reinforced aerated concrete roofing panels. These introduce more weight into the construction but have advantages in terms of thermal and sound insulation. The use of appropriate details means that the roof can be formed as a plate to withstand horizontal loads.

*Purlins*
With a primary loadbearing construction of reinforced or prestressed concrete, trapezoidal cross-sections are used for the purlins. The dimensions of the purlins are based on the recommendations of the "Fachvereinigung Deutscher Betonfertigteilbau e.V." (German Trade Association for Precast Concrete Components) (table 3.2.5).

*Loadbearing roof structure*
Roof members pin-jointed on fixed-base columns are preferred in reinforced and prestressed concrete structures. This structural system has become firmly established because a rigid corner to the frame is very costly in precast concrete (see table 3.1.3, p. 109). The dimensions of the roof members and columns are based on the recommendations of the German Trade Association for Precast Concrete Components. These cross-sections have been optimised on the basis of many years' experience and forms for these sizes are kept in stock by the precasting industry. Inserts are placed in the basic forms to obtain the various dimensions. Sizes other than those proposed

3.2.3 Single-storey sheds with cranes

Type of crane
   travelling overhead crane
   roof-slung gantry crane
   column- or wall-mounted crane
   Scotch derrick crane

Crane control
   from the floor
   from a cabin

Design parameters
   clearance below hook
   lifting capacity
   type of control
   (transfer station)
   type of crane (travelling overhead, roof-slung gantry)

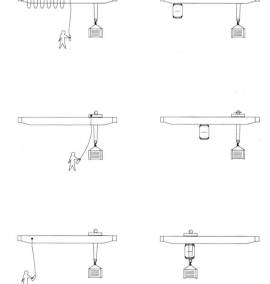

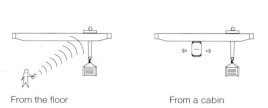

From the floor      From a cabin

3.2.4 Forms of crane control [44]

# Single-storey sheds

## 3.2.5 Roof purlins [29]

| | Dimensions [mm] | | | Span | Fire resistance class to DIN 4102 | |
|---|---|---|---|---|---|---|
| | d | $b_u$ | $d_o$ | max. l [m] | Reinforced concrete | Prestressed concrete |
| | 400 | 150 | 190 | 7.5 | F 90 – A | F 30 – A |
| | | 190 | 230 | 10.00 | | F 90 – A |
| | 500 | 150 | 200 | 10.00 | | F 30 – A |
| | | 190 | 240 | 12.50 | | F 90 – A |
| | 600 | 150 | 210 | 11.00 | | F 30 – A |
| | | 190 | 270 | 17.50 | | F 90 – A |
| | d | $b_u$ | $d_0$ | max. l [m] | | |
| | 850 | 190 | 250 | 20.00 | F 90 – A | F 90 – A |
| | 950 | 190 | 270 | 20.00 | | |

Take account of deflection or camber

Bottom edges chamfered 10 mm

Loads:
- included in table: purlin self-weight $g_1$
- user-defined: roof load q

| Span l (m) | Spacing a (m) | Depth of purlin d [mm] for roof load q [kN/mm²] |
|---|---|---|
| | | 1,0  1,5  2,0  2,5  3,0  3,5  4,0  4,5  5,0 |
| 7,5 | 3,0 | |
| | 4,0 | 400 |
| | 5,0 | 500 |
| | 6,0 | 600 |
| 10,0 | 3,0 | |
| | 4,0 | 400  500  600  800 |
| | 5,0 | |
| | 6,0 | |
| 12,5 | 3,0 | 500 |
| | 4,0 | 600  800 |
| | 5,0 | 850 |
| | 6,0 | 950 |
| 15,0 | 3,0 | 600 |
| | 4,0 | 800 |
| | 5,0 | 850  950 |
| | 6,0 | |
| 17,5 | 3,0 | |
| | 4,0 | 800  850 |
| | 5,0 | 950 |
| | 6,0 | |
| 20,0 | 3,0 | |
| | 4,0 | |
| | 5,0 | 850  950 |
| | 6,0 | |

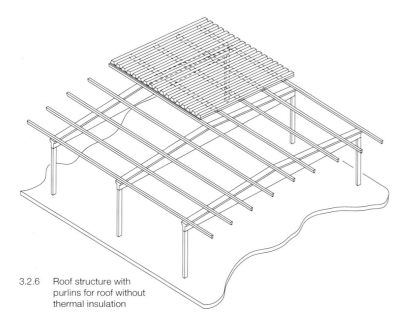

3.2.6 Roof structure with purlins for roof without thermal insulation

Example 1:
lightweight roof covering    $g_2 = 0.50$ kN/m²
snow    $s = 0.75$ kN/m²
roof load    $q = 1.25$ kN/m²
   $\approx 1.50$ kN/m²
span    l = 10.00 m
spacing    a = 5.0 m
read off    d x $b_u$ = 500 x 150 mm
(F 90 – A) (reinforced concrete)

Example 2:
aerated concrete roof
waterproofing and snow    $g_1 = 1.00$ kN/m²
self-weight of aerated concrete
(d = 150 mm)    $g_2 = 1.10$ kN/m²
roof load    $q = 2.10$ kN/m²
   $\approx 2.00$ kN/m²
span    l = 12.00 m
   $\approx 12.50$ m
spacing    a = 4.0 m
read off    d x $b_u$ = 600 x 190 mm
(F 90 – A) (prestressed concrete)

▨ select a different roof beam

# Single-storey sheds of trusses and frames

### 3.2.7 Reinforced concrete roof beams (primary beams) [29]

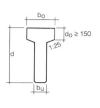

| Dimensions [mm] | | | Span |
|---|---|---|---|
| d | $b_o$ | b | max. l [m] |
| 600 | 400 | 190 | 15.00 |
| 800 | 400 | 190 | 20.00 |
| 1000 | 400 | 190 | 25.00 |
| 1200 | 500 | 190 | 25.00 |
| 1400 | 600 | 190 | 30.00 |
| 1600 | 700 | 250 | 35.00 |
| 1800 | 800 | 250 | 35.00 |
| 2000 | 800 | 250 | 40.00 |

All dimensions are adequate for fire resistance class F 90 – A to DIN 4102

$d = l/20 - l/16$
$b_o = l/50 - l/40$
$d_U = 150 - 350$
$b_U = 300 - 400$

| Pitch [%] | Section for design $x_o$ | Ridge height $d_e$ |
|---|---|---|
| up to 5.0 | 0.40 l | 1.05 d |
| 5.0-10.0 | 0.33 l | 1.10 d |
| 10.0-15.0 | 0.25 l | 1.25 d |

Pitched beam

Parallel beam

Loads:
If additional purlins are required due to the roof construction (e.g. for trapezoidal profile steel sheeting with spans ≥ 7.5 m, for aerated concrete units with spans ≥ 6.0 m), they must be included in the roof load using a value of approx. 0.75 kN/m².
- included in table: beam self-weight $g_1$
- user-defined: roof load q

| Span l | Spacing a | Depth of beam d [mm] for roof load q [kN/mm²] | | | | | | | | |
|---|---|---|---|---|---|---|---|---|---|---|
| | | 1,0 | 1,5 | 2,0 | 2,5 | 3,0 | 3,5 | 4,0 | 4,5 | 5,0 |
| 15,0 | 5,0 | 600 | | | | 800 | | | | |
| | 6,0 | | | | | | | 1000 | | |
| | 7,5 | 800 | | | 1000 | | 1200 | | | |
| | 10,0 | | | 1000 | | 1200 | | 1400 | | 1600 |
| 20,0 | 5,0 | 800 | | 1000 | | | | | | |
| | 6,0 | | | | | | | | 1400 | |
| | 7,5 | | | 1200 | | 1400 | | 1600 | | |
| | 10,0 | | 1200 | 1400 | | 1600 | | 1800 | 2000 | |
| 25,0 | 5,0 | | | 1200 | | | | 1600 | | |
| | 6,0 | | | | 1400 | | 1600 | 1800 | | |
| | 7,5 | | | 1400 | 1600 | 1800 | | | | |
| | 10,0 | | 1400 | 1600 | 1800 | 2000 | | | | |
| 30,0 | 5,0 | 1400 | | | | | | 1800 | | 2000 |
| | 6,0 | | | 1600 | | 1800 | 2000 | | | |
| | 7,5 | 1600 | | 1800 | 2000 | | | | | |
| | 10,0 | | | 1800 | 2000 | | | | | |
| 35,0 | 5,0 | | | | 1800 | 2000 | | | | |
| | 6,0 | | | 1800 | 2000 | | | | | |
| | 7,5 | | 1800 | 2000 | | | | | | |
| | 10,0 | 1800 | 2000 | | | | | | | |
| 40,0 | 5,0 | 2000 | | | | | | | | |
| | 6,0 | | | | | | | | | |
| | 7,5 | | | | | | | | | |
| | 10,0 | | | | | | | | | |

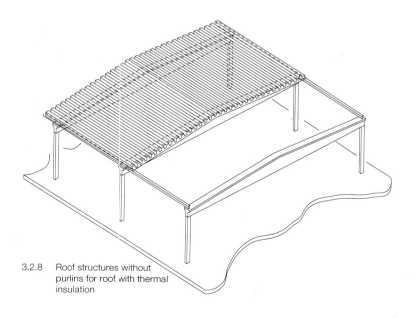

3.2.8 Roof structures without purlins for roof with thermal insulation

Example 1:
lightweight roof covering    $g_2$ = 0.50 kN/m²
snow    s = 0.75 kN/m²
roof load    q = 1.25 kN/m²
   ≈ 1.50 kN/m²
span    l = 20.00 m
spacing    a = 6.0 m
read off    d × $b_o$ = 800 × 400 mm
(F 90 – A) (reinforced concrete)

Example 2:
aerated concrete roof    $g_2$ = 2.00 kN/m²
snow    $s_2$ = 0.75 kN/m²
services etc.    = 0.25 kN/m²
roof load    q = 3.00 kN/m²
span    l = 25.00 m
spacing    a = 5.0 m
read off    d × $b_o$ = 1200 × 500 mm
(F 90 – A) (prestressed concrete)

▨ select a I-beam

# Single-storey sheds

3.2.9 Prestressed concrete roof beams (primary beams) [29]

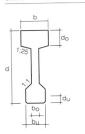

| Dimensions [mm] | | | | Span |
|---|---|---|---|---|
| d | $b_o$ | b | $d_u$ | max. l [m] |
| 800 | 400 | 120 | 150 | 20.00 |
| 1000 | 400 | 120 | 150 | 25.00 |
| 1200 | 500 | 120 | 160 | 30.00 |
| 1400 | 600 | 120 | 250 | 35.00 |
| 1600 | 700 | 120 | 250 | 40.00 |
| 1800 | 800 | 150 | 250 | 40.00 |
| 2000 | 800 | 150 | 350 | 40.00 |
| 2200 | 800 | 150 | 350 | 40.00 |
| 2400 | 800 | 150 | 350 | 40.00 |

All dimensions are adequate for fire resistance class F 90 – A to DIN 4102.

$d = l/20 - l/16$
$b_o = l/50 - l/40$
$d_u = 150 - 350$
$b_u = 300 - 400$

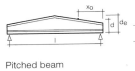

Pitched beam

| Pitch [%] | Section for design $x_o$ | Ridge height $d_e$ |
|---|---|---|
| up to 5.0 | 0.40 l | 1.05 d |
| 5.0–10.0 | 0.33 l | 1.10 d |
| 10.0–15.0 | 0.25 l | 1.20 d |

Parallel beam

Loads:
If additional purlins are required due to the roof structures (e.g. for trapezoidal profile steel sheeting with spans ≥ 7.5 m, for aerated concrete units with spans ≥ 6.0 m), they must be included in the roof load using a value of approx. 0.75 kN/m².
- included in table: beam self-weight $g_1$
- user-defined: roof load q

| Span l (m) | Spacing a (m) | Depth of beam d [mm] for roof load q [kN/mm²] | | | | | | | | |
|---|---|---|---|---|---|---|---|---|---|---|
| | | 1,0 | 1,5 | 2,0 | 2,5 | 3,0 | 3,5 | 4,0 | 4,5 | 5,0 |
| 20,0 | 5,0 | | 800 | | | 1000 | | | 1200 | |
| | 6,0 | | 800 | | | | | | | 1400 |
| | 7,5 | | | | | | | | | |
| | 10,0 | 1000 | | | | 1200 | 1400 | | 1600 | |
| 25,0 | 5,0 | | | 1000 | | 1200 | | 1400 | | |
| | 6,0 | | | | | | | | | 1600 |
| | 7,5 | | | | | | 1400 | 1600 | | |
| | 10,0 | | | 1200 | 1400 | 1600 | | | 1800 | |
| 30,0 | 5,0 | | | | | 1400 | | | 1600 | |
| | 6,0 | | | | | | | 1600 | | 1800 |
| | 7,5 | | 1200 | | | | | | 1800 | 2000 |
| | 10,0 | | | 1400 | 1600 | 1800 | | 2000 | | 2200 |
| 35,0 | 5,0 | | | | | | 1600 | 1800 | | |
| | 6,0 | | 1400 | | | | | | 2000 | 2200 |
| | 7,5 | | | | | | 1800 | | | |
| | 10,0 | | | 1600 | 1800 | 2000 | 2200 | | 2400 | |
| 40,0 | 5,0 | | | | | | | | | 2200 |
| | 6,0 | | 1600 | | 1800 | | 2000 | | | |
| | 7,5 | | | | | | 2000 | 2200 | 2400 | |
| | 10,0 | | | | | 2000 | 2200 | 2400 | | |

Example 1:
lightweight roof covering   $g_2 = 0.50$ kN/m²
snow   $s = 1.50$ kN/m²
roof load   $q = 2.00$ kN/m²
span   $l = 25.00$ m

spacing   $a = 6.0$ m
read off   $d \times b_o = 1000 \times 400$ mm
(F 90 – A) (reinforced concrete)

Example 2:
aerated concrete roof   $g_2 = 2.00$ kN/m²
snow   $s = 0.75$ kN/m²
services etc.   $= 0.25$ kN/m²
roof load   $q = 3.00$ kN/m²
span   $l = 30.00$ m

spacing   $a = 6.0$ m
read off   $d \times b_o = 1400 \times 600$ mm
(F 90 – A) (prestressed concrete)

▨ select a different type of construction

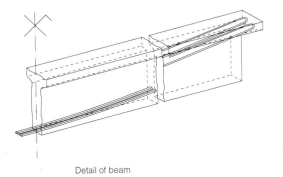

Detail of beam

3.2.10 Bundled reinforcement according to the new reinforced concrete concept [47, 49]

require the construction of expensive, special forms and are therefore avoided.

Reinforced concrete roof beams
See table 3.2.7 for details of the dimensions of such roof beams. Roof beams with conventional reinforcement are employed with typical frames up to lengths of about l = 24 m. The spacing of the frames varies depending on the type of roof selected, from a = 4.8 m to a = 10.8 m.

Reinforced concrete roof beams can be built with bundled reinforcing bars according to the new reinforcing concept (figure 3.2.10) [45]. In doing so, the soffit of the roof beam can also be formed as a parabola, thus saving concrete and weight for transportation. Such beams require a fork support (figure 3.2.11 A) or another form of retaining detail at the supports which provides lateral restraint. A single-storey shed must always be planned considering the complete construction. It is therefore possible – when using reinforced concrete wall panels – to utilise the wall panels to achieve such a fork support without additional measures (figures 3.2.14 and 3.2.16). The beam can be notched at the supports so that the desired fork support arrangement for the wide top flange is provided automatically above the position of the centre of gravity of the beams (figure 3.2.15). It is advantageous to support the beam on the columns with self-adjusting plug-in connectors.

Prestressed concrete roof beams
Precast prestressed beams can be fabricated and transported in one piece up to a length of max. l = 40 m. Larger beams are usually made from several segments which are then stressed together on site. However, such beams are employed only in exceptional cases.

Table 3.2.9 gives the dimensions of precast prestressed components as recommended by the German Trade Association for Precast Concrete Components. Prestressed beams make use of the I-form because the widened bottom flange has to withstand the force due to beam self-weight and prestress. Above all, accurate workmanship is vital during production of these beams. They must be produced and stored straight so that there are no undesirable deformations, which could reduce the lateral buckling resistance of the beam.

Lateral guys may be necessary during the erection of large spans in order to secure the beam against lateral buckling. This temporary support must be provided until the beam is secured by the purlins or the trapezoidal profile sheeting. The fork supports for the beam can be accomplished by way of recesses in the tops of the columns (figure 3.2.11 B) or by means of wall panels or precast parapet elements (figures 3.2.17 and 3.2.18). Bolting the bottom flange is an alternative.

3-pin truss with tie
Roof members with a low depth at the eaves are very economical because the area of the facade to the single-storey shed, which constitutes a major part of the cost, is thus kept to a minimum. It seems appropriate, especially for greater spans for which prestressed concrete beams can no longer be produced or transported in one piece, to develop a reinforced concrete truss with a tie. This is similar to the Polonceau truss common in structural steelwork (figure 3.2.19).

The truss consists of two shallow, triangular reinforced concrete truss sections propped against each other at the ridge and with their bases connected by a tie (figure 3.2.19 A). The sections of the truss are produced in a form with a wedge-shaped cross-section. This is designed so that one form can be used to cast various spans by using appropriate elements to limit the size (figure 3.2.19 B). The hardened elements are lifted out of the form without having to open it. Erection is carried out using a mobile trestle (figure 3.2.19 C). Threaded steel bars are preferred for the ties. These can be connected both at the supports and also in the bottom corners of the triangular elements.

Arches
The shape, and hence the loadbearing structure, of a single-storey shed used for storing tipped goods matches the tipped form. Even today, the solution to this task is still demonstrated admirably by the salt store designed in 1953 by Miklós Gnädig (figure 3.2.12). The three-pin arched ribs were cast in the horizontal position, turned through 90° into the vertical position and then lifted into their final position. A simple steel plate, slid in from the side, forms the pin joint at the ridge. The waffle panels mounted on the arches have grouted joints to create shear-resistant connections. These guarantee the longitudinal stability of the building. Admittedly, such a building would now be erected with large mobile cranes, and trapezoidal profile sheeting would be used for the roof covering – provided it is able to withstand the aggressiveness of the stored product.

Northlight roofs
The loadbearing construction benefits from a larger column spacing in the east-west direction. Girders – containing the windows – can then be used over the large span. The roof elements, e.g. reinforced concrete waffle panels, are supported at one end on the bottom chord and at the other end on the top chord. This roof surface, in the form of a rigid plate, provides lateral restraint to the top chord of the girder. The erection is therefore carried out starting in the south and proceeding northwards (figure 3.2.13).

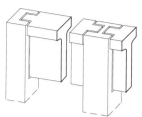

A   Reinforced concrete

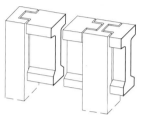

B   Prestressed concrete

3.2.11   Lateral restraint

3.2.12   Salt store

3.2.13   Northlight roof

# Single-storey sheds

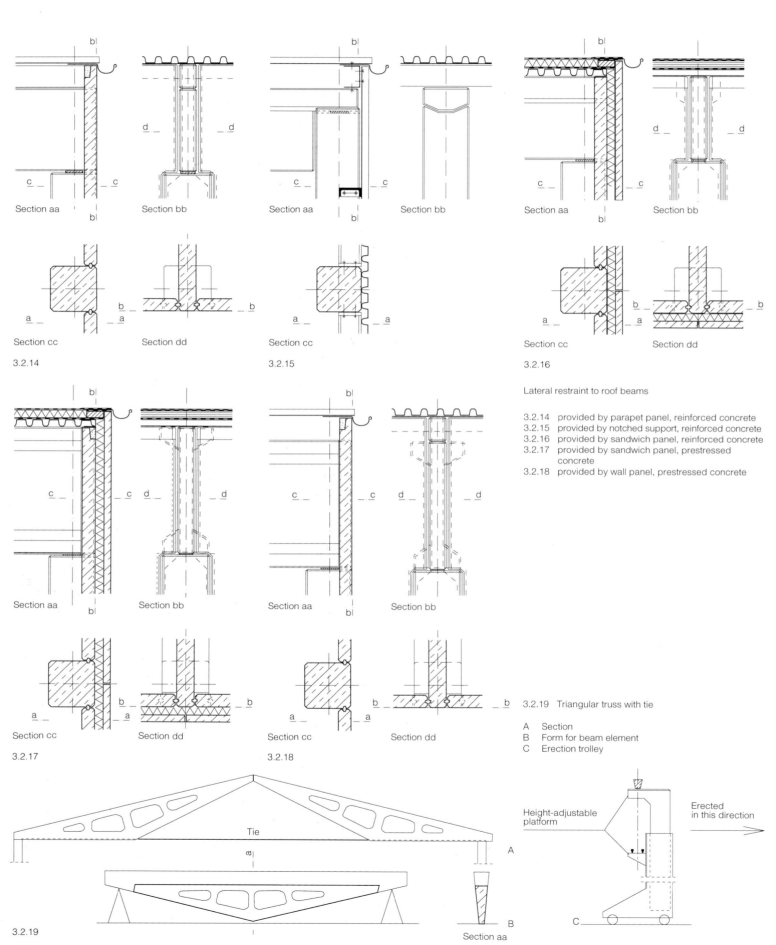

Lateral restraint to roof beams

3.2.14 provided by parapet panel, reinforced concrete
3.2.15 provided by notched support, reinforced concrete
3.2.16 provided by sandwich panel, reinforced concrete
3.2.17 provided by sandwich panel, prestressed concrete
3.2.18 provided by wall panel, prestressed concrete

3.2.19 Triangular truss with tie

A Section
B Form for beam element
C Erection trolley

# Single-storey sheds of trusses and frames

The reinforced concrete girders are precast in the horizontal position and transported in bundles. Special attention must be given to anchoring the reinforcing bars of the end diagonals. The bars are either welded to steel anchor plates or provided with loops transverse to the loadbearing plane. To ensure that the loops fit within the concrete cross-section, bundles of small reinforcing bars are employed.

Northlight roofs with smaller spans are built using (prestressed concrete) valley beams. In designing these roofs the aim should be to keep the depth of construction below the valley as small as possible so that the area of the external wall is minimised. Other solutions for northlight roofs are shown in figure 3.2.25 with beam-type stressed skin structures.

If the larger column spacing has to be in the north-south direction, prestressed beams for example are used for the large span and the actual northlight roof structure it supports.

**Columns**
The columns are fixed at the base to provide the necessary transverse stability, and longitudinal stability as well if necessary. The detail at the base usually involves pockets in the foundations. However, it has been shown that despite the extra work involved, and despite storage and transport problems, a foundation cast monolithically with the column is usually more economical because the complicated pocket is then no longer required.

An impact analysis of a column shows that reinforced concrete columns with typical dimensions, even without a protective plinth, are able to withstand an impact load. Appropriate additional reinforcement is generally sufficient. However, it is recommended to provide columns with some form of protection against impact.

Buildings with travelling overhead cranes require corbels on the columns to support the crane gantry girders (figure 3.2.20).

**Walls**
Trapezoidal profile sheeting or aerated concrete panels are the most inexpensive walls for single-storey sheds. If thermal insulation is required, the trapezoidal profile sheeting is provided with a layer of rigid expanded foam on the inside, protected by a suitable inner lining. Aerated concrete or precast concrete panels are alternatives when the use of trapezoidal profile sheeting is ruled out owing to its vulnerability to mechanical damage, or for safety reasons.

*Trapezoidal profile sheeting*
This is attached vertically. Halfway up the wall a channel section rail is bolted to steel plates cast into the reinforced concrete columns.

The gable walls of low-rise single-storey sheds do not require a framework, but merely connecting pieces. A Z-section rail transmits the support reactions of the purlins into the trapezoidal profile sheeting of the gable wall. An angle section is provided in each corner of the wall for attaching the horizontal cladding rails. The vertical members framing the door opening continue up to the roof. These form the supports for the cladding rails.

*Aerated concrete [59]*
Aerated concrete walls are used primarily for thermally insulated single-storey sheds. Panels placed horizontally enable walls of any height to be built. However, it may be necessary to include intermediate supports according to the type of fixing and height of wall. The spacing of the frames is 7.50 m to match the maximum length of panel available. Aerated concrete walls of panels placed horizontally can be built up to a height of 12 m without any intermediate support.

The slenderness ratio may be critical for wall panels placed vertically.

It may be necessary to include steel door framing in gable walls, depending on the type of door. Such framing also serves as the support necessary for the gable wall.

Owing to their hygroscopic properties, aerated concrete panels need to be protected at the base of the wall by means of a suitable coating (e.g. epoxy), or by placing them on a reinforced concrete plinth (figure 3.2.21).

If a layer of gravel of sufficient depth is not provided beneath the ground slab, a frost apron of precast (normal-weight) concrete is required. This is supported on the foundations.

*Reinforced concrete*
The maximum width of a precast concrete wall panel is assumed to be 4.20 m. This width can still be transported tilted up at an angle on a

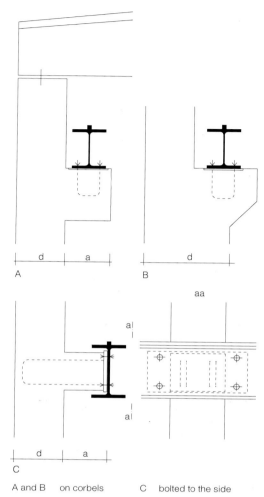

A and B on corbels    C bolted to the side

3.2.20    Supporting the crane gantry girder

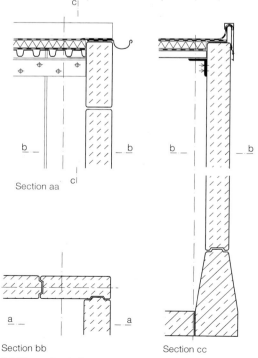

3.2.21    Aerated concrete external wall

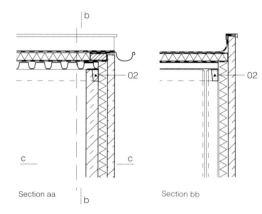

Section aa    Section bb

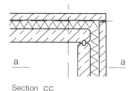

Section cc

02  In situ concrete ring beam
3.2.22  Gable wall constructed with sandwich panels

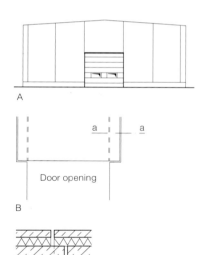

Section aa
A  Elevation on gable
B  Part-elevation from inside

3.2.23  Door lintel with wall panels

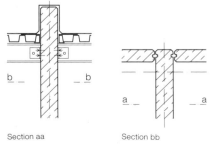

Section aa    Section bb

3.2.24  Fire wall

low-loader. It is advisable to incorporate such wall panels horizontally in longitudinal walls, and vertically in gable walls.

Where escape doors are positioned in the longitudinal walls, any wall panels placed horizontally should be taller than the door openings to ensure that they are not completely severed by the doors. Such panels include a continuous tie, which is concealed by the subsequent door threshold. It is also feasible to separate the panel into two pieces, i.e. erected with a gap between them corresponding to the width of the door. These are then held by the wall panel above. The panels are positioned between the columns and held in place by grout injected into the groove formed between end of panel and column.

Mortar should not be injected if there is a risk of frost. If freezing weather is expected during the period of erection, the connections should use cast-in steel parts, usually connected by welding. Such connections are best made with a few narrow weld seams so that not too much heat is introduced, which could lead to cracks in the concrete. The cast-in steel parts must be protected against corrosion.

Different companies have different views on the issue of injected grout versus cast-in steel parts. There is, however, no noticeable difference in the cost.

As figure 3.2.6 shows, a non-insulated roof requires secondary framing. The wall panels continue to the underside of the trapezoidal profile sheeting and simultaneously support the edge (figure 3.2.22). As the wall panels provide lateral restraint for roof members supported on their bottom edge, they must be erected first. This calls for a filler piece between the wall panels above the roof member. The wall panels do not have to provide a fork support in the case of notched supports; the area above the column can be closed off with cantilevering sections of wall panel. A ring beam is formed by fixing reinforcement in a longitudinal groove. After erection, a formwork board is used to close off the open side and the beam cast in situ.

Temporary struts must be provided to hold the gable wall panels during erection. It is advisable to proceed with the roof construction without delay because the final support to the walls is guaranteed by the purlins in conjunction with the trapezoidal profile sheeting, which acts as a shear diaphragm.

Door openings are formed with the help of the wall panels. The self-supporting wall panels above the door are supported in notches in the two adjoining elements and held in place by injecting grout into the joints (figure 3.2.23).

Fire walls
Reinforced concrete wall panels 120 mm thick are provided for reinforced and prestressed concrete primary loadbearing structures. These are held on both sides by the roof construction so that the side not exposed to the fire always guarantees stability (figure 3.2.24).

**Bracing**
The following effects must be considered when designing the bracing:
· Wind
· Out-of-plumb effects of vertical bracing components
· Out-of-plumb effects (intended and unintended) of loadbearing, bracing components
· Crane, braking and lateral surges
· Temperature changes
· Shrinkage and creep

The trapezoidal profile sheeting forming the roof covering provides lateral restraint to the purlins or roof members. Furthermore, they can be constructed as a shear diaphragm; a wind girder in the roof is then unnecessary. Studies have shown that despite the extra work required for the roof decking (fixings in every rib, edge trimmers), the shear diaphragm arrangement is somewhat less costly than the wind girder. However, structural analysis is essential.

As fixed column bases accommodate the horizontal loads in the transverse direction, the longitudinal forces can also be resisted by this means. In the case of shorter single-storey sheds the horizontal force can be transferred to the gable walls, with bracing provided by shear walls. It is also possible to brace a single-storey shed by way of girders. Threaded bars are especially useful here. They are inserted through sleeves embedded in the concrete and anchored at the rear with nuts. Fixed-base columns are not necessary in buildings in which stability is guaranteed by wall panels or girders. However, additional measures must be taken during erection.

**Economics**
The results of an extensive study of the economic aspects have been published in [49]. The most important dimensions, quantities of materials and cost relationships for many types of construction are given in tabular form in this publication.

As the external walls represent the lion's share of the costs for a single-storey shed, the aim should be to minimise them. This is carried out at the preliminary design stage at a scale of 1:500. When selecting the loadbearing construction, a low eaves height should be considered wherever possible.

# Stressed skin structures

## General

*Definition of the task*
The task is to create an envelope to enclose a space. The question is therefore: how should the envelope be shaped to provide the loadbearing function? Reinforced concrete stressed skin structures are able to fulfill this task.

*Definitions* (see pp. 107, 152-153)

Classification
- according to geometrical shape
  folded plate
    prismatic
    tapered
    arch-type
  shell
    single curvature
    double curvature
      synclastic
      anticlastic
- according to the loadbearing behaviour or loading
  beam-type, frame-type, arch-type stressed skin
  beam-type, frame-type, arch-type shell
  membrane shell (in tension, in compression, membrane-type)
  shell in bending
- according to the support conditions
  continuous (air-supported structure)
  on walls or beams
  on columns
- according to the type of surface
  continuous
  discontinuous (open 2-dimensional structures: lattice, net)

*Stressed skin structures for roofs*
In this book we shall deal with reinforced concrete stressed skin structures for roofs. Tanks and cooling towers will not be considered in any detail. Stressed skin structures are classified according to their loadbearing behaviour or their geometrical shape, which in turn defines the loadbearing behaviour. The loadbearing behaviour takes precedence in this context. The terms rotational shell, hyperbolic paraboloid shell and membrane-type shell also characterise a certain type of loadbearing behaviour. The following classification serves for the discussion in this book:

- beam-type, frame-type, arch-type stressed skins
- rotational shells
- hyperbolic paraboloid shells
- membrane-type shells
- free-form shells
- general folded plates
- suspended roofs

All these stressed skin structures are supported on beams, walls or columns.

**Beam-, frame- and arch-type stressed skin structures**
The primary loadbearing effect of the stressed skin structures dealt with here corresponds to a plane frame of individual members. As they have thin-walled, open cross-sections, a force develops in the transverse direction. This must be taken into account when designing these stressed skin structures.

*Beam-type stressed skin structures*
These are particularly suitable for the roofs of single-storey sheds rectangular in plan. We refer to
- beam-type folded plates (figure 3.2.25) or
- beam-type shells (figure 3.2.26)
according to whether flat or curved surfaces are involved.

There are no restrictions on width when using in situ concrete. However, the width of precast concrete components should not exceed 3 m for reasons of transport. This can lead to aesthetic problems when the division into smaller sections is not in proportion to the dimensions of the building.

Folded plates and shells should be designed in such a way that the thickness does not drop much below 100 mm.

Figure 3.2.25 shows the customary cross-sections of beam-type folded plates, figure 3.2.26 those of beam-type shells. A wider compression zone is desirable near the span moments. For continuous beams, a larger cross-sectional area in compression is also required around the internal supports. When using precast components it is advisable to position the joints between elements at the high points as movements occur at the joints, which can also put a strain on the sealing. It is therefore better for the joint not to be located in the valley.

Casting the inclined surfaces without formwork to the top surface limits the angle and hence the height of the stressed skin structure. Beam-type folded plates and shells covering longer spans therefore often include webs. If larger column grids in both directions are called for in multi-bay single-storey sheds, beam-type folded plates and shells can be combined with, for example, "Vierendeel-type" girders to form northlight roofs (figures 3.2.27 and 3.2.28).

Stiffening transverse plates are required for transferring large concentrated forces, i.e. at the supports. The transverse plates can also be formed as upstand or downstand beams (figure 3.2.33). The effect of the transverse plate can be achieved by a suitably thick slab or shell itself in certain instances.

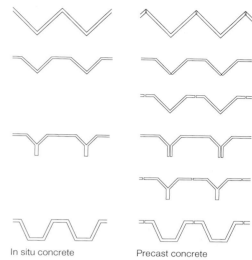

In situ concrete     Precast concrete

3.2.25 Beam-type folded plate structures

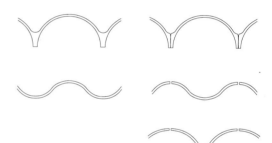

In situ concrete     Precast concrete

3.2.26 Beam-type shells

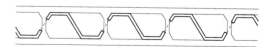

3.2.27 Vierendeel girder with folded plate

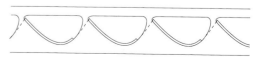

3.2.28 Vierendeel girder with shell

# Single-storey sheds

### 3.2.29 Concrete stressed skin roof structures

| Loadbearing behaviour | Folded plate structures | | | | Shells |
|---|---|---|---|---|---|
| **Defined geometrical form** | | | | | |
| beam-type | prismatic | | tapered | | parabolic cylinder |
| frame-type | | | | | |
| arch-type | | | folded plate arch assembled from ruled surfaces (Orly, p. 23) | | barrel vault |
| rotationally symmetric | | | | | single curvature (cone) |
| centrally symmetric | pyramidal folded structures | | | | spherical (ellipsoidal) dome |
| hyperbolic paraboloid (p. 159) | | | | | |
| conoid | | | | | straight edges |
| **Form determined from ideal stress state** | | | | | |
| membrane-type shells (prestressed shells) (p. 161) | | | | | |

Free form (p. 164)

Church of St Paul, Neuss-Weckhofen (figure 3.2.79)

TWA terminal, Kennedy Airport, New York (p. 33)

Stressed skin structures

3.2.29  Concrete stressed skin roof structures
Shells

segments of a
rotational hyperboloid

torus segments,
elliptical surface

groined vault

cloistered vault

cone

elliptical surface

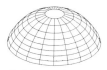
double curvature,
same direction (p. 23)
(sphere, ellipsoid)

opposite directions

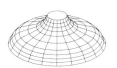
changing curvature

assembled from
elliptical surfaces

multiple

curved edges

multiple

conoid

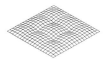

rotational surface

Keramion (p. 35)

Sydney Opera House
(p. 33)

Naples railway station
(figure 3.2.77)

153

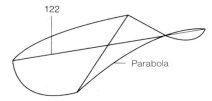

122 Prestressing tendon

3.2.30 Beam-type hyperbolic paraboloid shell

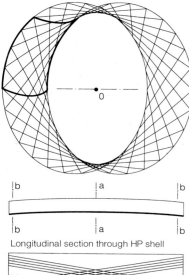

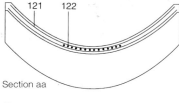

Section aa

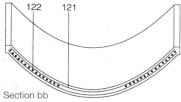

Section bb

121 Reinforcement    122 Prestressing tendon

3.2.31 The "HP shell"

Prestressed concrete is very useful for long-span beam-type stressed skin structures. Post-tensioning is used with in situ concrete, i.e. after the concrete has hardened, which allows the tendons to be laid to approximate a curve. The tendons are lifted at the supports of single-span beams. This reduces the compressive stresses on the underside and the primary stresses in the web-type sections. The tendons are positioned at the top over the supports of continuous members in order to accommodate the negative moments. The tendons must be positioned so that the forces due to the change in direction exhibit no, or only very small, components perpendicular to the surface. The cross-section must be thickened locally to accommodate the anchorages.

Precast prestressed concrete components are produced on a stressing bed. The tendons are stressed prior to pouring the concrete. Deflecting the tendons to approximate a curve is extremely complicated in this case and is therefore not usually carried out.

Beam-type stressed skin structures are cast without transverse plates so that several elements of different lengths can be produced in the same form in one prestressing operation. These must be of such a size, or lifted, stored and transported in such a way, that the self-weight (including dynamic effects) can be accommodated without transverse plates. The means of support on the adjoining component is to be arranged so that the effect of the transverse plate is available for the full load. If the transverse plates are to be cast monolithically, the formwork must be designed so as not to hinder the prestressing operation.

One of the most systematic beam-type shell structures is the "Silberkuhl shell" with curvature in the longitudinal direction as developed by W.J. Silberkuhl and E. Häussler. The task was to develop a beam-type prefabricated shell with the following conditions:
- 2.5 m wide, 12-24 m span (shells 150 mm thick and 3 m wide spanning 31.2 m have also been built)
- prestressed concrete without deflected strands
- cast without formwork to top surface

The shape can be clearly deduced from these conditions:
- A ruled surface, e.g. hyperbolic paraboloid, is suitable for straight prestressing strands (see p. 159).
- The depth of the cross-section results from the angle at which the concrete can be placed without formwork to the top surface; this angle was determined experimentally.
- The longitudinal increase was established so that the prestressing strands arrive at the right point in the end cross-sections for the maximum span.

As it was planned to cast the shell with a concreting trolley with a trailing tamping vibrator, a constant radius, i.e. a circle, was chosen for the curve in the longitudinal direction. Therefore, what was originally a hyperbolic paraboloid became a segment of a rotational paraboloid (see "General, geometry", p. 159). (The original name "HP shell" had already been introduced and was not altered again; figures 3.2.30 and 3.2.31.) The end plates are awkward to produce and transport and are only required in the final condition. The support beams are therefore designed accordingly (figure 3.2.36).

The gable columns can then transfer the wind load into the shell only when the end shells are suitably braced. Otherwise, the gable columns could only be restrained at the base. And owing to the lack of vertical load, this would lead to large foundations.

Northlight roofs can be formed by supporting the HP shells at an angle (figure 3.2.36).

Shells that follow the bending moment diagram
The beam-type shell forming the canopy over the inspection area of the "TÜV" facilities (German technical survey institute) in Darmstadt follows the shape of the bending moment diagram of a single-span beam with two cantilevers. Here, the tension chord is straight and boards could be used for the formwork to the shell. The surfaces to the cantilevers are hyperbolic paraboloids and conoids in the main span (figure 3.2.37).

Segmental prestressed beam-type prismatic and cylindrical stressed skin structures

If the dimensions of a beam-type stressed skin structure exceed the maximum permissible transport dimensions, it is advisable to divide it, transverse to the loadbearing direction, into transportable elements and prestress them together in their final, erected position. To do this, conduits with a slightly larger diameter than normal (to facilitate easier threading of the tendons later) are cast into the elements. The prestressing jacks are supported against the hardened concrete. The force in the tendons is transferred via the anchorages into the concrete. The contact faces can be formed using one of the following options:
- Position the elements 50-100 mm apart and fill these spaces with in situ concrete. Disadvantage: long waiting time until the prestressing force can be applied; the setting process can be shortened by using synthetic resin mortar.
- Create an accurate contact face with help of synthetic resin filler.
- Cast the elements against one another (match-casting) with a separating sheet at the mating face. Disadvantages: two con-

creting operations are necessary and the elements must be numbered and erected in the corresponding order.
- Provide accurate formwork, for example, with plate glass, although positioning the formwork face at 90° to the loadbearing axis is difficult.

It can be advantageous to provide interlocking mating faces to accommodate shear forces. Erection takes place on a scaffold or on the ground. Heavy lifting equipment is required to position the assembled loadbearing structure.

Figures 3.2.34 and 3.2.35 show single-storey sheds with northlight roofs made from segmental prestressed precast components acting together as a beam-type shell.

### Economics

When considering the economic aspects, in addition to the cost of the stressed skin structure itself, the roof covering (developed surface, extra cost of sloping and curved surfaces, forming valleys), supporting beams, gable rails and columns (especially gable columns and foundations) must be taken into account.

Prefabricated forms that can be used for elements with different spans are available only for HP shells. This means that other elements incur additional formwork costs. Comprehensive studies show that beam-type stressed skin structures are worth considering primarily for special architectural concepts and northlight roofs. Single- and double-tee units, and in particular roof structures with trapezoidal profile sheeting, are more economical owing to the less expensive roof covering and the ready availability of the forms.

### Folded plate structures

3.2.32 Petrol station near Milan, Favini (span: cylindrical surface, cantilever: conoid)
3.2.33 Cement works, Wiesbaden, E. Neufert, Dyckerhoff
3.2.34 Central warehouse for VSK, Switzerland, Heinz Hossdorf, Element AG, Tafers
3.2.35 Weaving shop at textile plant, Poland, J. Glowczewski, S. Sikorski, Walclaw & Zenon Zalewski, contractor: Industriebau Posen
3.2.36 Sports hall, Herne, Denzinger, Vestakon AG
3.2.37 Inspection bays, TÜV Darmstadt, H. Tuch, W. Fuchssteiner
3.2.38 Grandstand for 1st FC Cologne football club, H. Schulten, S. Polónyi
3.2.39 Church of St Hedwig, Oberursel, H. Günther, S. Polónyi
3.2.40 Bus depot, Budapest, I. Menyhárd
3.2.41 Aircraft hangar, Marseille-Marignane, France, Auguste Perret & Nicolas Esquillan, contractor: Société des Enterprises Boussiron, Paris

3.2.32

3.2.33

3.2.34

3.2.35

3.2.36

3.2.37

3.2.38

3.2.39

3.2.40

3.2.41

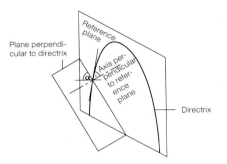

3.2.42 Definition of the ruled surface for the Church of St Hedwig, Oberursel (figure 3.2.39)

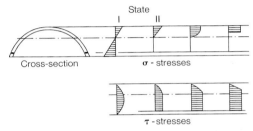

3.2.43 Distribution of longitudinal stress with associated shear stresses

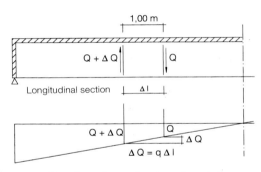

3.2.44 Shear force diagram for a beam-type shell Longitudinal section

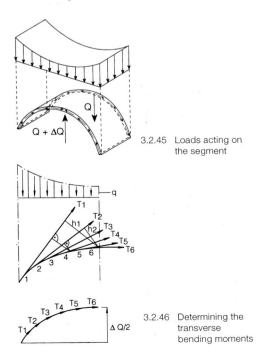

3.2.45 Loads acting on the segment

3.2.46 Determining the transverse bending moments

*Frame- and arch-type stressed skin structures*
Folded plates
Arch-type folded plates can be formed from planar triangular surfaces (see p. 152, column 1, row 3).

The grandstand at the stadium of the 1st FC Cologne football club is a three-pin folded plate with a large cantilever. The cross-section of the folded plate was chosen so that one strip is available for the compression zone. Transverse plates at the changes of direction of the folded plate resist the uplift forces. The transport conditions determined the shape of the element. The front plate originally envisaged was omitted so that the inevitable different creep deformations of the elements and the inaccuracies of erection would not be visible afterwards. Temporary support to the stepped terrace elements was necessary during erection (figure 3.2.38).

Ruled surfaces (one generated motion of a straight line with one degree of freedom)

In the first place these include thin-walled vaults formed as cylindrical or conical surfaces. These curved shells are stiffened by the end plates and stiffening ribs. Freyssinet's arch-type shell to the airship hangar at Paris-Orly (since demolished) (p. 23) achieved the necessary flexural stiffness through the folded cross-section. The surfaces themselves are not planar but ruled. In principle, the construction of the church in Oberursel is an arch-type shell (figure 3.2.39). The ruled surface is defined by the generators that lie in the plane perpendicular to a planar directrix, and form a constant angle with the plane of the directrix (figure 3.2.42).

Elliptical surfaces (surfaces in double curvature in one direction)

Examples of elliptical arch-type shells with a tie are the bus depot in Budapest (figure 3.2.40) and the aircraft hangar at Marseille-Marignane (figure 3.2.41). The latter was cast on the ground and raised into its final position by hydraulic jacks.

*Analysis*
The longitudinal stresses in a beam-type shell can be determined from the beam action corresponding to the stress states (figure 3.2.43). This results in the distribution of the shear stresses. The shear stresses cause the transverse bending in the shell. It is not wrong to assume that the shear stress distribution is constant throughout the cross-section. The transverse bending for a 1 m wide segment is determined from the shear force difference $\Delta Q$ (figure 3.2.44); $\Delta Q$ is distributed over the cross-section segments according to the assumed constant stress distribution (figure 3.2.45). The shear force components of cross-section segments T and the load q generate the moments at the individual positions (figure 3.2.46).

This simplified method deviates only slightly from an exact analysis and can be used for stressed skin structures with either conventional or prestressed reinforcement. It provides sufficiently accurate results for preliminary calculations, and in many cases for the final structural analysis too. The directrix can be optimised to minimise the transverse bending moments. This is the case when the positive and negative moments are equal.

This method can also be used for analysing beam-type folded stressed skin structures. It is suitable for both frame- and arch-type stressed skin structures.

Special attention should be paid to calculating the forces due to the changes of direction. Transverse plates or connecting members should be included here. Appropriate stirrups must be incorporated to hold the cranked reinforcement.

The edges of adjacent beam-type stressed skin structures provide mutual horizontal support. The end folds or shells require horizontal restraint in the form of a valley beam or girders.

**Rotational shells**
*Definition*
A rotational surface is formed by a curve rotated about an axis.

*Loadbearing behaviour*
Critical load:
- for shallow shells primarily rotationally symmetric: self-weight and snow
- for high shells not rotationally symmetric: wind

*Analysis*
Membrane loadbearing effect
Rotational shells can carry their distributed rotationally symmetric load only through membrane forces when the meridian curve – with the exception of the crown – does not have a horizontal tangent. The following generally applies: a shell carrying a parallel load may not have a tangential plane perpendicular to the loading direction along a closed contour line perpendicular to the loading direction because the force transfer would be through shear forces, i.e. through bending (figure 3.2.47).

The shells also exhibit edge disturbances in a rotationally symmetric support. These disturbances can be eliminated only with elaborate support systems. A circumferential prestress can reduce the edge disturbances in certain cases. Flexural disturbances are accommodated with bending reinforcement.

Internal forces: meridional and circumferential forces are rotationally symmetric for a rotationally symmetric support and loading. The majority of rotationally symmetric plates and shells are given in tabular form with ready-made equations in [45].

A graphical method is useful for determining the membrane forces of rotational shells with any meridian curve that are loaded parallel to the axis of rotation (figure 3.2.48) [30]. Divide the shell into segments by means of planes perpendicular to the axis of rotation (a). Apply the loads on these segments $V_I$, $V_{II}$,... as shown in (b). The load acting at the interfaces must be carried by a force direction acting tangentially to the interface 1, 2,... of the meridian curve. This is shown by $T_I$, $T_{II}$,... in the force diagram (b), which gives $n_{\varphi I} = T_I / 2\pi r_1$ etc. The circumferential forces result from the difference ($\Delta H_I$, $\Delta H_{II}$,...) between the forces acting on the upper and lower edges of the segment; $\Delta H_I$, $\Delta H_{II}$,... are the summations of the forces acting radially on the circumference of the circle. The radial force acting on the ring is $\Delta h_{III} = \Delta H_{III}/2\pi r_{III}$. The resultant internal force of a segment is $N_{\psi III} = -(\Delta H_{III}/2\pi r_{III}) r_{III}$. This resultant internal force is carried over length $s_{III}$, where the circumferential internal force $n_{\psi III} = -\Delta H_{III}/2\pi s_{III}$. If the H-force acting on the lower edge is greater than the H-force acting on the upper edge, then the ny-force is compressive, in the opposite case tensile. In our example the circumferential force above segment III is compressive, and below that, tensile. This method determines the internal forces from the equilibrium condition, which supplies the appropriate distribution of internal forces with corresponding support only near the supports.

DIN 1045 section 4.1 states: "The deformation and internal forces should be determined according to elastic loadbearing behaviour." Calculations based on plastic theory would be more favourable; for example, the amount of reinforcement can be reduced in cylindrical tanks. It can be shown that the minimum reinforcement required for the same load is given by heavier circumferential bars and less reinforcement in the plane of the shell than is the case with elastic theory.

### Discrete supports

In order to avoid bending stresses, concentrated forces and line loads should be applied to the shell tangentially whenever possible (figure 3.2.49). If the non-continuous support is not tangential, the support reactions can be introduced into the shell only by way of bending (slab action) (figure 3.2.57). The method of resolution is suitable for the preliminary design of such shells. In doing so, the resultants in the symmetry segments are determined from the equilibrium conditions (see p. 163).

### Buckling

The critical load for a spherical half-dome $p_k = \alpha E d^2 / R_1 / R_2$, where $\alpha$ = buckling factor, E = modulus of elasticity, d = shell thickness, $R_1$ and $R_2$ = primary curvature radii. Partly based on investigations of collapsed shells, Csonka recommends taking $a = 0.05$, which allows for a safety factor and inaccuracies during production [51]. It would be better to employ separate safety factors instead of the global safety factor hidden in the buckling factor. The inaccuracies during production could be taken into account by incorporating d and R in the analysis with their most unfavourable tolerances.

### Production
In situ concrete
Formwork:
- Pneumatic formwork
  As local loads can change the shape, coated textiles filled with air can be used only when sprayed concrete is used. Other methods of placing and compacting the concrete cause movement of the formwork.
- Timber formwork
  Boards: planks placed on edge and cut to match the shape of the meridian curve and boards laid in the radial direction or square timbers laid in a ring with meridional board formwork. One problem of design may be that the view through the falsework structure leads to the crown of the dome.
- Fibre-reinforced boards
- Permanent formwork
  Nervi employed diamond-shaped ferrocement waffle units as permanent formwork in his shells (figures 3.2.51 and 3.2.58). The ferrocement waffle units were produced from fine-aggregate concrete with wire reinforcement. They were laid on a scaffold. The reinforcement to the ribs was placed in the intermediate spaces. The actual loadbearing structure was formed by casting the ribs and a layer over the waffle units – the ribbed shell, a very impressive, decorative form of construction.

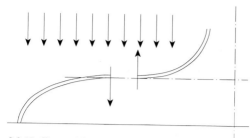

3.2.47 The meridian curve of a rotational shell may have a horizontal tangential plane only at the crown

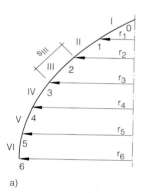

a)

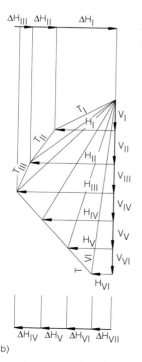

b)

3.2.48 Graphical determination of the circumferntial internal force of a rotational shell [30]

Single-storey sheds

3.2.49

3.2.50

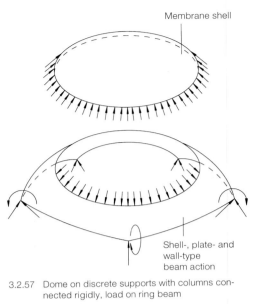
3.2.57 Dome on discrete supports with columns connected rigidly, load on ring beam

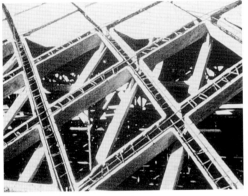
3.2.51

3.2.52

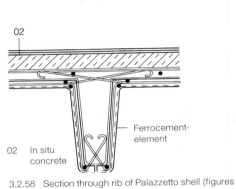
3.2.58 Section through rib of Palazzetto shell (figures 3.2.50-3.2.52)

3.2.53

3.2.54

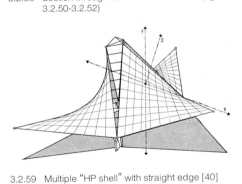
3.2.59 Multiple "HP shell" with straight edge [40]

3.2.60 Multiple "HP shell" with curved edge

3.2.55

3.2.56

**Shells**

3.2.49  Spherical dome on discrete supports, Barcelona trade fair
3.2.50 to 3.2.52
   Palazzetto, Rome: internal view, erection of ferrocement elements and rib reinforcement, shell supported on raking Y-columns, P. L. Nervi
3.2.53  Railway station, Düsseldorf
3.2.54  Department store, Mexico, F. Candela
3.2.55 and 3.2.56
   Church of St Suitbert, Essen: reinforcement, scaffold and formwork, J. Lehmbrock, S. Polónyi

In the Palazzetto the horizontal thrust of the dome is directed down to the foundation ring through raking columns (figures 1.69, 1.70 and 3.252). A tension ring at the eaves would have made the raking columns – dominant in architectural terms – unnecessary [37]. Admittedly though, this would have produced less interesting architecture.

Precast concrete
The aim of precast construction is to save scaffolding and formwork, and to transfer the fixing of the reinforcement and casting of the concrete into a factory. In this sense the design of Nervi's Palazzetto cannot be regarded as precast concrete construction. His method corresponds to production of partly prefabricated floors and roofs.

Domes can be assembled from radial precast elements temporarily supported on a trestle below the crown. Examples: St Hedwig's Cathedral, Berlin; Tonhalle, Düsseldorf.

**Hyperbolic paraboloid shells**
General, geometry
Hyperbolic paraboloid (HP) shells are very popular. The reasons for this are:
- Their aesthetically attractive form, especially when several HP shells are combined.
- Ease of production; there are two systems of straight generators; in reinforced concrete shells the formwork boards are laid in one direction, the square timbers of the supporting scaffold in the other.
- Favourable, clear loadbearing behaviour with a reasonably even distribution of internal forces for a corresponding support and relatively favourable stability behaviour.

The hyperbolic paraboloid is a non-developed surface in double curvature whose primary curvatures are in opposite directions (see p. 152).

The hyperbolic paraboloid can be defined as a ruled surface (surface with straight generators) and as a translation surface (figure 3.2.64). All straight lines $e_1$ (generators) which intersect two skewed straight lines $g_1$, $g_2$ (directrices) and are parallel with a reference plane ($L_1$) form a hyperbolic paraboloid (figure 3.2.61). (None of the directrices may be parallel to the reference plane.) A second system of generators $g_1$ results from the reference plane $L_2$ as well as the directrices $e_1$ and $e_2$. The same surface can be generated by means of translation (parallel displacement). A generating parabola slides on another generating parabola parallel to a reference plane (figure 3.2.62). Which description is used depends on the particular task. If a reinforced concrete shell is to be constructed and the formwork for the surface is to be formed with boards, then the definition as a ruled surface is better. But if the surface is to be formed by parallelograms as polyhedra (glazing), the definition of the translation surface should be used (figure 3.2.64).

Equation for the surface
According to the recommendations of Candela [32, 36], the ruled surface is useful for determining the internal forces, whereby the z-axis of the system of coordinates is parallel to the two reference planes, the x- and y-axes are parallel to reference planes $L_1$ and $L_2$ respectively and are at 90° to the z-axis. The angle $\alpha$ between the reference planes may be any size. The projection of the generators in the z-direction onto the xy plane produces a parallelogram (figure 3.2.63). If the hyperbolic paraboloid is defined by a three-dimensional quadrilateral OACB (straight generators), the origin O of the system of coordinates is placed in one corner of the quadrilateral. The surface is thus defined by the projection of the sides $l_x$ and $l_y$ as well as the z-coordinates a, b and c, which designate the distance of the corners A, B and C from the xy plane. The z-coordinates of any point P on the surface with the coordinates x, y can be described by the height relationships:
$z = z_b + x(z_c - z_b)/l_x$
where:
$z_b = b\,y/l_y$ and $z_c = a + (c - a)y/l_y$
From this we get:
$z = (b/l_y)y + [a + (c - a)y/l_y - b\,y/l_y]x/l_x$
using the designations:
$r = (c - a - b)/l_x l_y$, $r_x = a/l_x$, $r_y = b/l_y$,
$z = r\,x\,y + r_x\,x + r_y\,y$
r is a characteristic value for the curvature of the surface; if r is large, the surface is highly curved, while r = 0 represents a flat plane.

Determining the internal forces
First, we investigate the equilibrium condition of an HP shell with a load distributed uniformly over the plan area. During this investigation the shell should be in such a position that the generators form a parallelogram on plan. This means that the axis of the hyperbolic paraboloid is vertical and the loading direction parallel to it (figure 3.2.65).

The equilibrium condition for a "finite" element defined by four generators is easy to obtain. Only the equilibrium condition of the projection of the forces in the direction of the z-axis remains because all other equilibrium conditions are already fulfilled. In doing so, it is sensible to calculate using the projections of the internal forces in the xy plane. These are designated with N. The projection equation on the z-axis is:
$Z\,\Delta x\,\Delta y = N_{xy} \cdot \Delta y [(z_4 - z_2) - (z_3 - z_1)]/\Delta y$
$+ N_{yx} \cdot \Delta x [(z_4 - z_3) - (z_2 - z_1)]/\Delta x$
From the moment equation about z we get:
$N_{xy} = N_{yx}$
and hence:
$Z\,\Delta x\,\Delta y = N_{xy}(2z_4 + 2z_1 - 2z_3 - 2z_2)$
$= 2N_{xy}(z_4 + z_1 - z_3 - z_2)$
$Z/2N_{xy} = (z_4 + z_1 - z_3 - z_2)/(\Delta x\,\Delta y)$

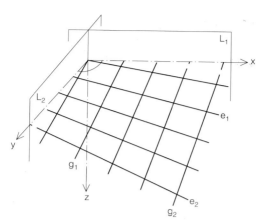

3.2.61 Definition of "HP shell" as ruled surface

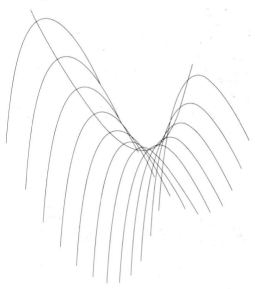

3.2.62 Definition of hyperbolic paraboloid as translation surface

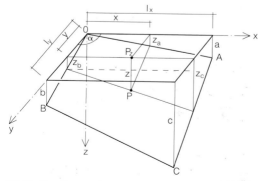

3.2.63 Coordinates of a random point on the surface

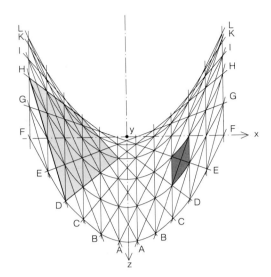
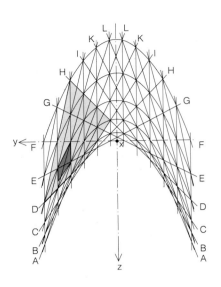

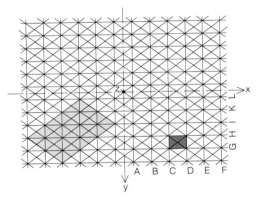

3.2.64  Hyperbolic paraboloid as ruled and translation surface [40]

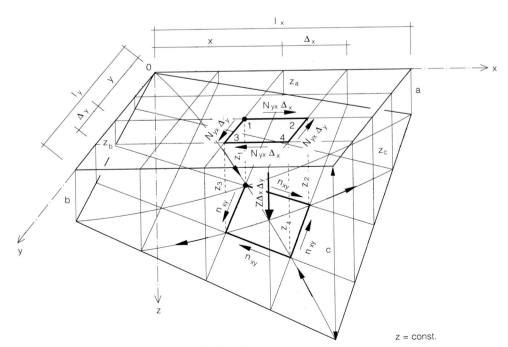

3.2.65  Internal forces of HP shell due to a distributed load parallel to the z-axis

$= (c - b - a)/(l_x l_y)$
from which we get:
$N_{xy} = Z l_x l_y /[2(c - b - a)] = Z/2r$

We can see from the equilibrium that with a uniformly distributed load parallel to the axis of the hyperbolic paraboloid:
- only shear forces occur along the generators,
- the straight edges of the shell must be provided with edge members in order to withstand the thrusts,
- the edge members for this loading case are subjected merely to longitudinal forces,
- the primary tensile forces in the shell run in the direction of the suspended primary curvature, the primary compressive forces in the direction of the vault curvature, and
- the shallower the shell, the greater the internal forces are.

The above deduction assumes that the edge members are infinitely stiff and that they undergo the same longitudinal deformation as the shell. This is, however, never the case; the effective stress state therefore deviates from that calculated according to membrane theory. The edge members are subjected to transverse loading, especially in shallow shells and/or hyperbolic paraboloids whose axis deviates substantially from the vertical. This is why they deflect outwards in the tangential plane of the surface. This gives rise to a redistribution of forces, where the forces diminish towards the corners and the edge members are relieved at midspan.

The method described for determining internal forces provides adequate information for a preliminary design. A finite element (FE) analysis must be used for the final structural analysis. In doing so, the stiffness of the edge members, and the adjoining components (e.g. walls), too, if applicable, is taken into account.

A shell spanning between edge members will not tend to buckle as long as the edge members do not yield. If the vault action fails, the suspension effect is activated.

The shell may be designed with an unsupported edge along a parabolic or hyperbolic edge because it does not need to transfer shear forces (figure 3.2.55). The buckling of the unsupported edge must be investigated carefully.

*Edge members*
The shell and its edge members form a unit. Although the engineer considers these separately in an analysis using discrete solutions, this is no reason to emphasise the edge members as an architectural element "merely for the sake of structural honesty". It is often regretted that the edge members hide the "thinness" of the shell. In a number of structures, in order to reveal how thin the shell really is, the shell cantilevers beyond the edge member. Although this shows the thickness of the shell, it does not indicate its loadbearing effect.

The shell transfers longitudinal forces to the edge member. This should be arranged in such a way that the force from the shell loads the edge member centrally, or the longitudinal force counteracts the bending moment generated in the edge member, e.g. due to self-weight. Figure 3.2.66 illustrates typical edge member arrangements. The edge members of reinforced concrete shells should be designed to be cast in one operation without additional formwork.

*Multiple hyperbolic paraboloid shells*
Joining together several hyperbolic paraboloid shells can result in interesting roof layouts. The shells can be joined either along the
- straight generators (p. 153), or the
- curved segments (p. 153).

The shell arrangement generated in this way can gain stability through the shells or the edge members providing mutual support for each other. It is advisable to look for combinations in which the internal forces in the shell cancel each other out at the common edge member. The cross-sections of the common edge members result from the shells meeting at that point. The edge members must carry the incoming internal forces in the shells and, if necessary, their self-weight.

Hyperbolic paraboloid shells can be employed to create individual or combined umbrella-type canopies. The forces in the edge members balance each other out at full load. The load is transferred through the valley beams into the columns (figure 3.2.53). The moment generated by an asymmetric snow load and wind is transferred into the column by the bending of the valley beams. Raking umbrella-type canopies can be employed for northlight roofs (figure 3.2.54).

*Construction*
Scaffolding, formwork
The square timbers of the supporting falsework are positioned in one generator direction, the boards of the formwork in the other generator direction (figure 3.2.56).

Reinforcement
Steel mesh reinforcement can be used to reinforce shells with a gentle curvature. Standard rectangular meshes are used, with the main reinforcement laid in the direction of the primary tensile forces, and the laps offset (figure 3.2.68). The offset overlap of mesh reinforcement with a bar diameter of 4 mm adds up to 20 mm when the bars nest together. With an appropriate roof covering, concrete cover of 10 mm is required on each side. This gives us our minimum shell thickness of 40 mm. However, it is very difficult to maintain the concrete cover on curved surfaces. It is therefore recommended to increase the concrete cover to 15 mm, which results in a minimum shell thickness of 50 mm. Shells with a severe curvature require reinforcement assembled from individual bars as the meshes cannot be bent to suit. The shell should be gradually strengthened at the transition to the edge member to accommodate flexural disturbances (figure 3.2.69). At the junction of two shells it is usually sufficient to strengthen the resulting cross-section slightly and provide appropriate reinforcement (3.2.70).

See p. 167 for details of concreting operations.

**Membrane-type shells** [51]
*Definition, analysis, properties*
The shells discussed up to now possess a geometrically predefined shape. The internal forces result from the load depending on the shape. An "ideal" distribution of internal forces can also be formulated and the form determined from it. A constant distribution is regarded as the "ideal" internal force distribution. This means that the internal force is equal at every point of the shell and in every direction. A constant distribution of the internal forces means that there are no shear forces in the shell. This loadbearing behaviour corresponds to that of the soap bubble. As the soap bubble cannot accommodate any shear forces, it deforms at every change in the load (change in air pressure) so that the surface is perpendicular to the load. In our structures we can determine this "ideal shape" only for the dominant load – self-weight and full snow load in the case of roofs. But this load is parallel (direction of the pull of gravity). This means that the surface corresponding to the condition described above is a horizontal plane with an infinitely large internal force. The original formulation is therefore altered in that it is not the internal force but rather its horizontal projection that must remain constant. Shells corresponding to this condi-

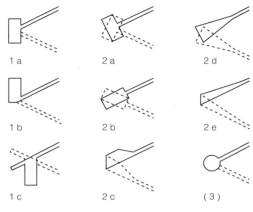

3.2.66 Principal design options for edge members

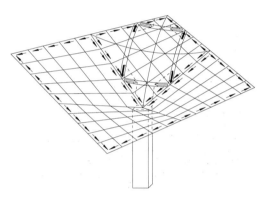

3.2.67 Loadbearing action of an HP umbrella-type roof

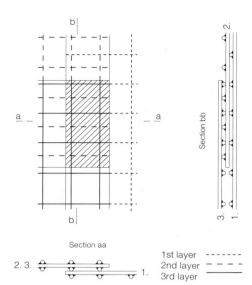

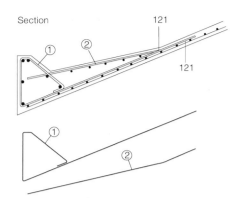

121   Reinforcement tied together with wire

3.2.69   Reinforcement at the transition to the edge member

3.2.68   Mesh reinforcement overlap with the mesh rotated

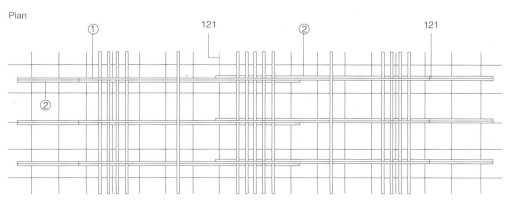

3.2.70   Reinforcement to ridge member

tion are known as membrane-type or prestressed shells. In the case of shallow shells this condition produces internal forces for the self-weight loading case that do not deviate significantly from the "soap bubble condition". In steeper shells or parts of shells the difference becomes significant. As the permissible stresses for shells cannot be fully exploited in most cases, these fluctuations in the internal forces do not make the shell uneconomic. If necessary, it is possible to achieve constant stresses by varying the thickness of the shell.

It is simple to determine the coordinates of the surface that fulfils the condition of "constant horizontal projection" of the internal forces. The continuum is replaced by a mesh with either an orthogonal or polar plan projection, according to the basic geometrical shape. By assuming the magnitude of the horizontal projection of the internal force n, or the projection of the force in the member determined from that, the height coordinate for every node of the mesh is determined from the equilibrium equation for the vertical force component as a result of the vertical force from the dominant load acting on the node. The equilibrium of the horizontal component is already fulfilled by assuming the horizontal projection of the internal force. The system of equations contains as many equations as there are nodes assumed in the mesh.

In the case of rotational shells, the meridian curve for the condition of the membrane-type shell can be determined directly by means of an equation.

From this basic condition it follows that:
- the shallower the shell, the larger the internal force n is.
- multiplying the heights of all points on a membrane-type shell by a random number changes only the magnitude of the internal force – the basic condition remains fulfilled. The heights of the supports must also be multiplied in the case of supports at different heights. If we wish to change the proportion of the shell while maintaining the heights of the supports, we need to determine the heights using the new n.
- edges shaped like an arc on plan are only subjected to constant tensile or compressive stresses when the heights of the individual points are determined according to the basic condition.
- edge members in tension on shells in compression are prestressed. The prestressing force is determined such that the edge member has the same compressive stress as the shell.
- when penetrating two membrane-type shells with equal horizontal internal force components, only vertical forces need to be carried by a valley beam/arch.
- in areas near edge or valley beams the

different strains in the edge member and in the shell lead to edge disturbances, but do not usually cause any special problems.
- flexural disturbances can occur due to unintended restraint in areas around edge or valley beams, but are usually small owing to the low flexural stiffness of the shell.
- small deviations from the exact form merely imply an internal force distribution different from the basic condition, but the shell remains in the membrane stress state.

Planning membrane-type shells
The planning of membrane-type shells involves the following steps:
- Establishing the edge conditions (supports and defining the surface).
- Establishing the horizontal component n of the shell forces (paying special attention to the supports).
- Selecting the mesh.
- Determining the loads.
- Setting up the system of equilibrium equations and its solution.
- If the height of the shell does not correspond to expectations, it can be changed by varying n: a small n is a steep shell, a large n a shallow one.
- Determining the internal forces and the stresses in the shell from the horizontal component n.
- Estimating the risk of buckling.
- Estimating the influences of the additional loading cases.
- Determining or estimating the deflection and establishing the camber. After deformation has taken place (including creep), the height coordinates of the central surface of the shell must agree with the calculated values because in the case of smaller heights larger internal forces would ensue than those calculated. This situation is particularly important in the case of shallow shells.

Estimating the risk of buckling
The critical load for membrane-type shells can be estimated by analysing domes. In the case of domes we use the equation $p_{KR} = \alpha E d^2 / R_1 , R_2$ (see p. 157). The Gaussian curvature $G = 1/R_1 R_2$ that occurs here with a square grid and with areas having a plane of symmetry at the crown can be simply determined by graphical or mathematical means using the z-coordinates of the neighbouring points. One primary curvature lies in the plane of symmetry, the other in the plane perpendicular to this and to the tangential plane of the surface. The critical load for single symmetry is:

$$q_{kr} = Z_{kr} = Ed^2/10a^4 \cdot (2z_{i,j} - z_{i-1,j} - z_{i+1,j}) \cdot (z_{i,j} - z_{i,j-1})$$

and for surfaces with two planes of symmetry:

$$q_{kr} = Z_{kr} = Ed^2/5a^4 \cdot (z_{i,j} - z_{i-1,j}) \cdot (z_{i,j} - z_{i,j-1})$$

where a = spacing of assumed mesh, d = shell thickness, z = vertical coordinates of nodes of lattice shell. A buckling factor $\alpha = 0.05$ has been included in both equations.

The risk of buckling can be reduced by:
- a thicker shell,
- increasing the rise of the shell,
- including stiffening ribs coupled with an increase in the curvature of the membrane-type shell between them.

The observations above apply generally to elliptical shells (with synclastic curvatures).

Estimating the influences of the other loading cases
Determining the shape of the shell according to the condition for membrane-type shells means that the stresses due to the dominant load, i.e. primarily self-weight and snow, are known. The effect of the wind also needs to be investigated. The stresses due to the wind in shallow shells over enclosed buildings are low compared to those due to the dominant load. As in shell structures the actual stresses are generally well below those permissible, it is usually sufficient to estimate the stresses caused by the wind load. This estimate can be carried out by:
- evaluating the proportion of the wind load compared to the dominant load. This method can only be used where the direction and distribution of the wind load do not deviate appreciably from the direction and distribution of the dominant load.
- using a comparative analysis of a similarly shaped but mathematically defined surface.
- using the method of resolution. In this case it is not the internal forces that are calculated but rather their resultants. The distribution of the internal forces is estimated with the resultants assumed to be in the extreme positions.

Just estimating the wind load is a problem on tall shells with large cantilevers because the shape factors for these surfaces cannot be covered by standards. A wind tunnel test may be necessary.

Finite element methods can be employed to obtain an accurate mathematical analysis. Such methods enable possible edge disturbances in the shell under dominant loads to be ascertained; however, these are not usually significant. In this case the method of calculation shown merely serves to determine a favourable shape for the shell. The buckling behaviour can also be ascertained with an FE analysis according to second order theory.

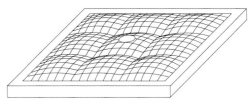

3.2.71  Membrane-type shell with straight edge member

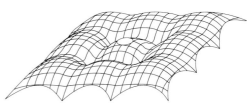

3.2.72  Membrane-type shell with curved edge member (circular arc on plan)

Single-storey sheds

**Shells by Heinz Isler**

3.2.73  Warehouse, Spreitenbach, Switzerland, 1970 & 1983
3.2.74  Garden centre, Solothurn, Switzerland, 1962
3.2.75  Garden centre, Camorino, Switzerland, 1971
3.2.76  Production plant and warehouse, Hasel-Rüegsau, Switzerland, 1956-91, scaffold for various sizes of shell

3.2.73

3.2.74

3.2.75

3.2.76

*Applications of membrane-type shells*
The above observations make it clear that through the appropriate choice of edge conditions and internal forces, the architect has unlimited design options at his disposal. A few basic forms are outlined below.

The shells of Heinz Isler over square plan layouts essentially correspond to the conditions for membrane-type shells. He determined the shape experimentally (figure 3.2.73) [34, 39]. As the shell is supported only at the corners, the surface was corrected. The membrane-type shell with horizontal, straight edge members has a horizontal tangential plane in the corners. So the shell itself contributes to resisting the shear force (support reaction), Isler altered the surface at the corners in such a way that the tangential plane of the surface is inclined. The edge members are of prestressed concrete. The prestressing force was determined under the condition that the stress in the edge member should be equal to the stress in the shell.

Figures 3.2.71 and 3.2.72 depict shells that are especially suitable for buildings with homogeneous or changing functions. In the first solution the roof of the perimeter aisle needs to be designed to accommodate the horizontal components of the support reactions to the shell. An arch is necessary at each junction between the shell bays. This is subjected to vertical loads only as a result of the dominant loading. However, it is possible to design the shell without ribs at the junctions and rigid edge beams. The edge beams are arcs on plan, which is why they are subjected exclusively to longitudinal forces due to the dominant loading.

The Keramion in Frechen (p. 35) is a membrane-type shell.

Isler's shells with unsupported edges and those with turned-up edges are designed in such a way that they attract only compressive stresses (figures 3.2.74 and 3.2.75) [34, 39].

*Construction of membrane-type shells*
Isler designed scaffolds of radial laminated timber ribs for his shells. The ribs can be moved to suit different sizes of shell (figure 3.2.76). The concrete is cast on thermal insulation panels, which serve as permanent formwork and form part of the final construction.

The formwork can itself constitute the reinforcement. In this case the reinforcement is suspended from a scaffold positioned above the shell. A wire mesh is stretched across the reinforcement and the concrete sprayed on (see p. 167).

**Free-form shells**
Shells can also be constructed with a completely free form, i.e. not according to any geometrical rules. Examples of this are the TWA terminal in New York (p. 33) and Sydney Opera House (p. 33), and the projects by Castiglioni for Naples railway station (figure 3.2.77) and the church in Syracuse (figure 3.2.78).

The loadbearing behaviour of the freely formed surfaces can be roughly estimated, at least section by section, by referring to one of the shell types outlined above.

In choosing a shape, care should be taken to ensure that the tangential plane is not horizontal at any closed contour line (p. 157). Gently curved surfaces result in large internal forces and have low buckling stability.

Great care should be exercised when designing the edge members, to ensure that there is a clean transfer of the shell forces into the edge members and their supports. The stability, previously determined with the help of model analysis, is now ascertained using FE techniques.

One of the methods described above is used as the basis for constructing such a shell.

**General folded plate structures**
Any large shape can be designed as a folded plate structure using triangular surfaces. This possibility is readily evident on the church at Neuss-Weckhofen (figure 3.2.79). The surfaces were reinforced by two layers of steel mesh reinforcement. This resulted in a thickness of $d = 70$ mm. The ridge and valley reinforcement was enclosed with hairpin bars along the edges. The ridge and valley reinforcement is three-dimensional; it can be bent only as instructed by the designer. The corners require heavy transverse reinforcement to accommodate the forces due to the change in direction (figure 3.2.80). Concreting was carried out similarly to the church in Oberursel (see p. 156). This folded plate structure was constructed using expanded metal as the formwork to the top surface. This was supported on square timbers and meant that it was possible to ensure that the concrete was placed correctly. Compaction was by way of external vibrators clamped to the formwork.

Although the loadbearing effect is provided by the folded surface with only a small quantity of concrete, the production of the formwork, fixing of the reinforcement and placing of the concrete was very involved and costly. It is therefore justifiable only for prestige structures.

## Suspended roofs

Frei Otto uses the term "suspended roof" in his book [46]. He defines it thus: "The suspended roof is a skin stretched between fixed points; it is the roof structure and the roof covering in one. The principal elements of the suspended roof basically lie in the roof covering, are essentially only subjected to tension and are curved in the negative direction (sagging) in at least one direction. The suspended roof is the reverse of the shell form subjected to compression. The buckling problem of the shell cannot occur with the suspended roof."

In this book we will consider only reinforced or prestressed suspended roofs. The options for suspending the roof always need to be investigated. Are there fixed points to which the suspension members can be anchored (e.g. rocks, neighbouring structures), or do suspension points (e.g. pylons, edge beams/arches) need to be built first?

The non-rigid member (cable) held at both ends changes its shape as the load varies. The utilisation and the construction itself permit only small deformations. The suspended roof therefore needs to be stabilised. This stabilisation can be achieved in the following ways:
- weight
- rigidity
- guying

When the dead load of the roof is considerably greater than the imposed load, movements due to varying imposed loads (snow, wind) are small.

### Free forms and suspended roofs

- 3.2.77 Naples railway station, design: E. Castiglioni
- 3.2.78 Pilgrims' Church, Syracuse, design: E. Castiglioni
- 3.2.79 Church, Neuss-Weckhofen, F. Schaller, S. Polónyi
- 3.2.80 Church, Neuss-Weckhofen, reinforcement to folded plate structure
- 3.2.81 Terminal, Dulles Airport, Washington DC, E. Saarinen, Ammann/Whitney
- 3.2.82 Schwarzwaldhalle, Karlsruhe, E. Schelling, U. Finsterwalder
- 3.2.83 and 3.2.84 Westfalen IV Halle, Dortmund, W. Höltje, view of prestressing tendons
- 3.2.85 Friedrich-Ebert-Halle, Ludwigshafen, R. Rainer, Dyckerhoff & Widmann
- 3.2.86 Casting a shell using the sprayed concrete method

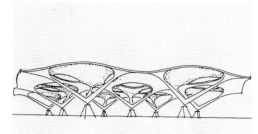

3.2.77

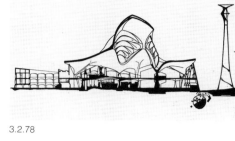

3.2.78

3.2.79

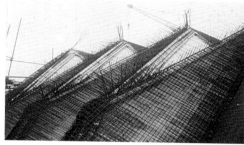

3.2.80

3.2.81

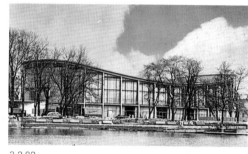

3.2.82

3.2.83

3.2.84

3.2.85

3.2.86

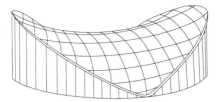

3.2.87 Stabilising a suspended roof by means of guys

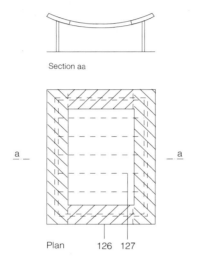

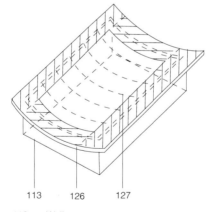

113 Wall
126 Compression member
127 Tie

3.2.88 Bracing the roof surface with a curved compression member

The movements of the roof can also be reduced by increasing the flexural rigidity of the suspension members of the shell. This is achieved by constructing the suspension members in prestressed concrete. The compressive stress in the concrete reduces the risk of cracking and so allows the stiffness of the full cross-section to be effective. The roof construction can be guyed both inside and outside the area covered. In order to use tension members as guys inside the area covered, the roof must be curved in opposite directions (figure 3.2.87). Reinforced and prestressed concrete suspended roofs curved in one direction need to be stabilised by a combination of weight and rigidity; all three options need to be combined for roofs of anticlastic curvature.

The most important example of a single curvature suspended roof is the terminal at Dulles Airport in Washington (figure 3.2.81). Interesting features are the edge of the shell turned up to form the edge beams, and the junction with the raking columns. The columns are also subjected to bending owing to the horizontal component of the support reactions for the suspended roof, and their cross-section therefore enlarges towards the base.

The edge strips carry the support reactions for the suspended roof. They act like beams with a large structural depth lying in the plane of the shell and they can transfer the load to the gable strips, provided the building is not too long (figure 3.2.88). The gable strips form a curved compression member that is able to transfer the support reactions of the edge beams because the member is stiffened by the gable walls. It is also possible to consider the edge of the roof as a closed frame that carries the forces from the suspension effect. In this situation the columns do not need to accommodate the horizontal components of the support reactions.

The closed ring beam carries the support reactions of the suspended roof to the Schwarzwaldhalle in Karlsruhe; the adjacent strips of the shell also contribute to this effect (figure 3.2.82). The rounded plan shape reduces the bending effect on the edge member and renders the loadbearing effect clearly visible. The suspended roof spans across the longer distance – evidently for architectural reasons. Temporary support to the entire prestressed concrete roof was necessary. The prestressing tendons were positioned at a pitch of a = 400 mm in the loadbearing direction and a = 5.0 m in the transverse direction. In the end zones, additional tendons carry the support reactions due to the suspension effect. After the concrete had hardened, the tendons were post-tensioned in sheaths cast into the concrete, subsequently anchored, and the sheaths filled with grout. The underside of the roof looks like a waffle slab – thus tracing the positions of the tendons.

As we can see from the loadbearing behaviour described above, the splendid gesture of the pylons to the Stadthalle in Bremen is unnecessary in structural terms (figure 3.2.89).

The edge members to the Westfalenhalle IV in Dortmund are supported on A-frames (figure 3.2.83). The roof was constructed on a 5.0 m wide mobile scaffold. The prefabricated rib elements were laid on the timber platform, the tendons threaded into the ducts and the concrete planks with pumice aggregate laid between the ribs (figure 3.2.84). Cement mortar was used to fill the joints between the elements. The longitudinal curvature of the suspended roof is very small. This does not contribute to stabilising the shell and serves merely to drain rainwater.

The Friedrich-Ebert-Halle in Ludwigshafen (figure 3.2.85) is a suspended roof over a diamond-shaped plan. The edge members are straight. The edge members had been built, the tendons were suspended and the precast concrete units positioned. The in situ concrete topping was cast on the preloaded suspended construction. A preload was suspended from the tendons; removing this load induced a prestress in the concrete shell. The load due to insulation and waterproofing, as well as snow and wind, is carried by the shell effect of a hyperbolic paraboloid. The arch-shaped edge member is advantageous for the loadbearing effect of the suspended roof because it means that bending stresses in the edge member can be avoided. The straight edge members of an HP shell with a parallelogram shape on plan are subjected only to longitudinal forces as a result of the dominant load. So the optimum form for an edge member in this method of construction is the curve between the arch of the pure suspended construction (circle) and the straight edge of the HP shell. Its shape depends on the proportion of the load components of the two loadbearing effects.

The prestressed concrete suspended roof to the Berlin Congress Hall is stabilised by the anticlastic curvature (figure 3.2.90). The architect Hugh A. Stubbins wanted a cantilever roof. So that the roof structure cannot rotate about the axis defined by the two supports, the following solution was chosen. The actual suspended roof spans merely between the curved edge beams above the outer wall. The perimeter arches were suspended from this ring beam. The structure chosen was the subject of a lively debate among architects and engineers from the time of its erection because of its inconsistency. The external appearance feigns a structural effect that is not present. The south arch collapsed 20 years after completion due to corrosion of the tendons, which had not been properly grouted. The repair work involved building an additional, completely independent suspended roof in lightweight prestressed concrete above the intact roof of the hall. The perimeter arches to the new roof are now restrained at the foundations to accommodate an asymmetric snow load as well as the wind load.

It should be mentioned at this point that in the original construction, the thin shell, which was in the shade, was joined to the massive arch exposed to the sunlight. The temperature difference led to cracks at the junction, into which moisture was able to penetrate. This accelerated the corrosion of the tendons. Generally, large differences in concrete cross-sections should be avoided in areas exposed to solar radiation because the warm concrete cools down faster in thinner areas. This leads to restraint stresses, which in turn can lead to cracks.

### The casting of stressed skin structures

The special problems of concreting the different types of stressed skin structures have been mentioned in the individual sections. Generally, shells can be cast only under suitable weather conditions and concreting operations must be carried out with great care. It must be ensured that the reinforcement is not pressed down under the weight of operatives and plant. Shallow shells can be cast in the conventional manner and compacted with vibratory tampers. The maximum size of aggregate should not exceed 7 mm. Steep, thin shells are produced in sprayed concrete (figure 3.2.86). Either the dry- or wet-mix method may be used. In the dry-mix method angular-shaped quartz sand is mixed with cement and forced by compressed air through a hose to the nozzle where the water is added from a second hose. In the wet-mix method, the concrete is mixed in the conventional way but using a small aggregate ($\leq 7$ mm), and forced by compressed air to the spray nozzle. The concrete is placed starting at the bottom and working upwards. Upon starting a new section, the edge faces are cleaned thoroughly (with a wire brush) to remove any loose material (rebound). The spray nozzle must also be panned back and forth so that the space between the reinforcement and the formwork is filled completely with concrete. If the concrete is sprayed onto a wire mesh, care must be taken to ensure that there are no overlaps in the mesh, because voids can form here. Stressed skin structures with a thickness $d \geq 70$ mm can be constructed with formwork to the top surface (see 'General folded plate structures", p. 164).

### Roof coverings to concrete stressed skin structures

Enclosed structures in Central Europe require thermal insulation because of the climate. The roof covering to steeper roof sections can present problems.

Roof covering with thermal insulation:
- Rigid polyurethane foam sprayed on; the adhesion, especially on steep sections, has not been sufficiently proved.
- Conventional roof covering

Galvanised wires are placed around the reinforcing bars and project from the concrete. They can be used to attach square timbers to the shell. The thermal insulation (mineral wool) is laid between the timbers. The formwork on which the concrete surface was cast is then removed from beneath the concrete and nailed onto the square timbers above (some boards will undoubtedly need to be renewed). Space must be guaranteed to allow the air to circulate between the thermal insulation and the timber boards; the square timbers are therefore interrupted at certain intervals. Roofing felt is then laid on top of the boards and covered with metal decking or a bonded plastic decking. Fibre-reinforced cement or slate coverings can only be used on surfaces at the suitable pitch.

Isler's shells are formed, or their edges prestressed, in such a way that the shell is subjected exclusively to compressive stresses and so is free from cracks. With a suitable concrete mix and careful workmanship, the 80 mm thick shell remains waterproof and weather-resistant. Under typical climatic conditions and typical uses (no operations with high air humidity), the internal thermal insulation complies with the regulations. The insulation does not become saturated with condensation because this can diffuse through the concrete shell.

3.2.89 Stadthalle, Bremen

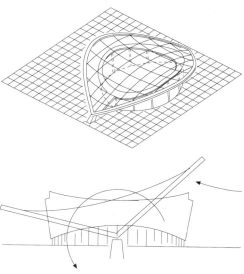

3.2.90 Berlin Congress Hall: plan (top) and sketch of system (bottom) [57]

# Foundations

02 In situ concrete

3.3.1 Foundations integrated in the ground slab

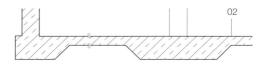

3.3.2 n-values for the spread of the load

| perm $s_b$ [kN/m²] | 100 | 200 | 300 | 400 | 500 |
|---|---|---|---|---|---|
| B 5 | 1.6 | 2.0 | 2.0 | not perm. | |
| B 10 | 1.1 | 1.6 | 2.0 | 2.0 | 2.0 |
| B 15 | 1.0 | 1.3 | 1.6 | 1.8 | 2.0 |
| B 25 | 1.0 | 1.0 | 1.2 | 1.4 | 1.6 |
| B 35 | 1.0 | 1.0 | 1.0 | 1.2 | 1.3 |

$h_{reqd} \geq nb$

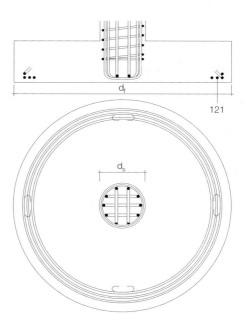

121 Reinforcement/bundle

3.3.3 Reinforcement for a circular pad foundation

Structures rest on a shallow or deep foundation, according to the depth of the subsoil with suitable bearing capacity.

## Shallow foundations

Strip footings under walls and pad foundations under columns are classed as shallow foundations. These are generally formed by increasing the thickness of the ground slab (figure 3.3.1). For heavy loads and/or a subsoil with a low bearing capacity, the structure is supported by a raft foundation.

Subsoil susceptible to frost heave brings with it the risk that the foundation could be lifted by the formation of ice in the ground. To prevent this, the underside of the perimeter foundations is founded below the depth affected by frost or, if necessary, a layer of concrete fill is placed below them.

### Strip footings

Reinforcement is not required in foundations with suitable proportions (table 3.3.2). Starter bars are often left projecting from the foundation for connecting to a reinforced concrete wall. These are generally superfluous. They may be necessary when the wall is subjected to earth pressure or when the foundation is intended to act as part of the plate effect.

The walls of single-storey buildings without basements can be placed directly on the ground slab. A frost apron is necessary on the perimeter of the ground slab when the capillarity of the subsoil and the ingress of moisture cannot be prevented by any other means, e.g. a layer of gravel. In this situation the strip footings to the external walls are essentially frost aprons.

### Pad foundations

Isolated foundations carrying a central load represent a problem of rotational symmetry. Accordingly, they are constructed as circular pad foundations with circumferential reinforcement. In doing so, it is easiest to provide the circumferential reinforcement in the form of a bundle (figure 3.3.3). DIN 1045 section 18.11.1 defines a bundle of bars as "two or three individual bars". This is true for longitudinal reinforcement transferring the force to the concrete via the bond. Here, the force from the bundle is transferred to the concrete as a single bar. For this reason, bundles consisting of more bars are also possible. Bundles are best formed using 12 mm dia. bars because these are supplied on reels. EC2 also recognises 16 mm dia. bars delivered in coils. These bars do not need to be bent – they are merely fixed at the necessary diameter. Care should be taken to ensure that the reinforcement is fixed correctly [61]. To prevent punching shear, the reinforcement should be placed outside a critical circle.

Formwork for foundations not integrated in the ground slab can be produced from a strip of sheet metal. This can be simply held together with packing tape.

The amount of steel reinforcement required in a circular foundation is only half that required in a square foundation. Foundations reinforced with a ring of bars are also more robust than square foundations because no cracks occur along the reinforcement.

Circular foundations are also suitable for eccentric loads. The required diameter for various permissible soil pressures can be determined using diagram 3.3.4 according to the eccentricity. The cross-section of the associated circumferential reinforcement can be found using diagram 3.3.5 [66].

Precast concrete columns are founded on pad foundations. Provided the maximum transport dimensions and weights are not exceeded, it is economical to deliver the precast concrete columns to site, with the foundations already cast monolithically with the columns (see "Columns", p. 129). Circular foundations are advisable. The columns with their attached foundations are erected on the carefully prepared subgrade. They are subsequently adjusted and, to guarantee a solid bearing, provided with a bed of grout underneath.

If foundations and columns are cast separately, pocket foundations are employed. The pockets in which the columns are seated, forming a fixed base to the column, are shaped like a truncated pyramid tapering towards the bottom of the pocket (figures 3.3.7 and 3.3.8). This enables the formwork for the pocket to be lifted out in one piece. The dimensions of the pocket are such that inaccuracies, both vertical and horizontal, can be compensated for.

# Raft foundations

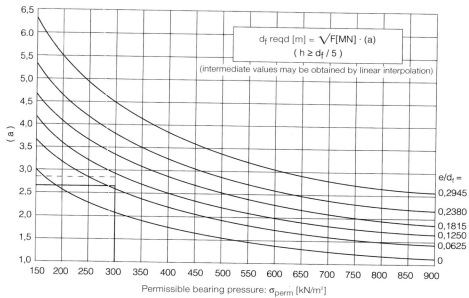

3.3.4　Determining the diameter of foundation required [66]

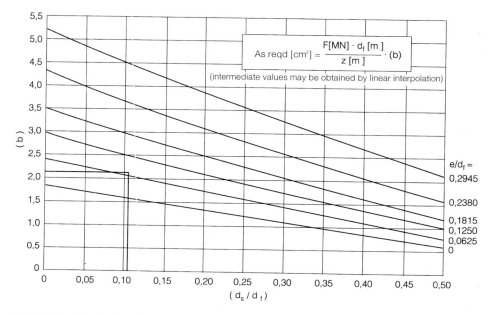

3.3.5　Determining the circumferential reinforcement required [66]

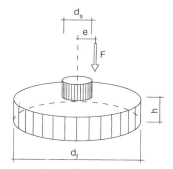

Preliminary values:
column dia.　　　　　　　　$d_s$ = 0.4 m
vertical load　　　　　　　　F = 1.74 MN
bending moment　　　　　　M = 0.609 MN
permissible bearing pressure　$\sigma_{perm}$ = 300 kN/m²
required:　　　　　　　　　$d_f$ = diameter of foundation
　　　　　　　　　　　　　h = depth of foundation
　　　　　　　　　　　　　$A_s$ reqd = reinforcement required

1st iteration (- - - - -)
Assume eccentricity with proposed foundation diameter:
$d_f$ = 3.0 m; h ≥ $d_f$/5 = 3.0/5 = 0.60 m;
$G_f$ = π 3.0²/4 x 0.60 x 25.0 = 106 kN
ΣF = 1.74 + 01.06 = 1.846 MN
e/$d_f$ = M/ΣF x $d_f$ = 0.609/1.846 x 3.0 = 0.11
read off 3.3.4: (a) = 2.8
$d_f$ reqd = 2.8 √1.846 = 3.8 > 3.0 m

2nd iteration (———)
$d_f$ = 3.8 m; h = 3.8/5 = 0.76: select: h = 0.8 m
$G_f$ = π 3.8²/4 x 0.80 x 25.0 = 227 kN
ΣF = 1.74 + 0.227 = 1.967 MN
e/$d_f$ = 0.609/1.967 x 3.8 = 0.0814
read off 3.3.4: (a) = 2.65
$d_f$ reqd = 2.65 √1.967 = 3.72 ≈ 3.8 m

h = 0.8 m; d = 0.75 m
z = 0.85d 0.85 x 0.75 = 0.64
$d_s$/$d_f$ = 0.4/3.8 = 0.105; e = 0.0814 $d_f$
read off 3.3.5: (b) = 2.15
$A_s$ reqd = 1.967x 3.8/0.64 x 2,15 = 25.1 cm²

3.3.6　Example of the design of a circular pad foundation

Foundations

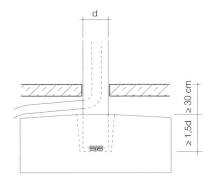

3.3.7  Pocket foundation

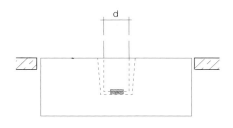

3.3.8  Pocket foundation with adjacent ground floor slab

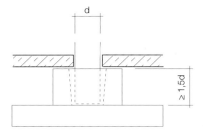

3.3.9  Pocket foundation with plinth

3.3.10  "Swedish foot" precast concrete foundation

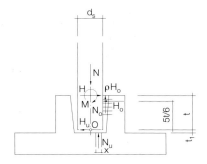

3.3.11  Forces acting on a pocket foundation

As a rule, the columns have a centring screw. The sleeve for this is a steel plate with a corresponding hole, which is placed on the base of the pocket – set out accurately on a bed of grout – before the column arrives. To prevent columns with symmetrical dimensions but asymmetrical reinforcement being erected incorrectly, the centring screw is installed off-centre. Wedges are placed between the wall of the pocket and the column to line up the column approximately vertically. The crane hook can now be detached. The column is moved into position by adjusting the wedges. After the main structure of the building has been erected, the pockets are filled with grout. The timber wedges are removed after the grout has hardened.

The depth of the pocket is usually set at 1.5 times the largest dimension of the column cross-section. As pocket foundations are very easy to handle on site, they are also used for pin-jointed column bases. In this case the pocket does not need to be as deep.

On larger foundations the pocket is placed in a plinth (figure 3.3.9). Such foundations are constructed in two operations; however, they do save material. Considering the materials-labour relationship, the pocket in a plinth on a larger base is only economical for foundation diameters, or side lengths, > 3.0 m. Of course, such a base can be built circular. A considerable amount of reinforcement is saved in this way. Apart from that, the outer formwork is simple.

Pocket foundations can also be constructed in precast concrete. The "Swedish foot" (a Swedish patent, figure 3.3.10) is the most popular type. The elaborate shape and complicated reinforcement reduces the weight of the foundation to a minimum.

In dealing with the structural analysis of pocket foundations, most publications (including EC2, section 5.4.10) place the axial force centrally in the base of the pocket. But this ignores the fact that the fixed part of the column twists due to the H-force or the bending moment, and therefore causes the axial force to act eccentrically (figure 3.3.11).

This eccentricity reduces the H-force, i.e. the load on the pocket, and also the primary tensile stress in the base of the column. It is also correct to take into account the friction due to H. Providing serrations to both faces in contact with the grout ensures that the moment is transferred to the walls of the pocket by way of the couple.

**Raft foundations**
The ground slab is supported elastically on the subsoil. It must be analysed as a slab on an elastic bed because this comes close to simulating the effective loadbearing behaviour and leads to economical reinforcement. It often proves beneficial to increase the thickness of the raft below columns and walls.

**Cellular rafts**
Basements below the water table can be constructed in concrete with external or internal tanking (waterproofing). The basement floor acts as a raft. Rooms requiring absolutely dry conditions, e.g. bank vaults, archives, computer rooms and other sensitive uses, generally make use of internal tanking. External tanking is preferred for basement car parks and other rooms, e.g. storage for damp and non-sensitive goods, or basements for residential premises.

Internal tanking consists of a layer of concrete beneath the ground slab, plus walls of concrete or masonry waterproofed to DIN 18195 parts 1-6. The concrete construction within the waterproofing can also be built as for external tanking.

External tanking consists of reinforced concrete with a minimum slab and wall thickness of 250 mm. By means of a suitable mix – low-heat cement, w/c ratio max. 0.60 or 0.55 and waterproofing additive – as well as appropriate placing, compacting and curing, the concrete can be made impermeable. "Impermeable" is understood to mean "that even when subjected to water for a long period, such water does not penetrate the concrete of the component and does not leak through to the side not directly exposed to the water, and this side does not exhibit any damp patches" [Beton-Kalender 1995, part 1, p. 79].

This can be guaranteed only when no cracks penetrate the component and the width of other cracks is limited by the reinforcement. In this respect it is important to limit the restraint stresses in the concrete. The majority of restraint stresses are caused by the heat of hydration as the concrete cools down. Care should be taken to ensure that the concrete cools down slowly so that the build-up of heat during hydration remains small (see "Movement joints", p. 110).

## 3.3.12 Piles [60]

A  Reinforced concrete driven pile
B  Bored pile with helical reinforcement
C  Column on pile cap

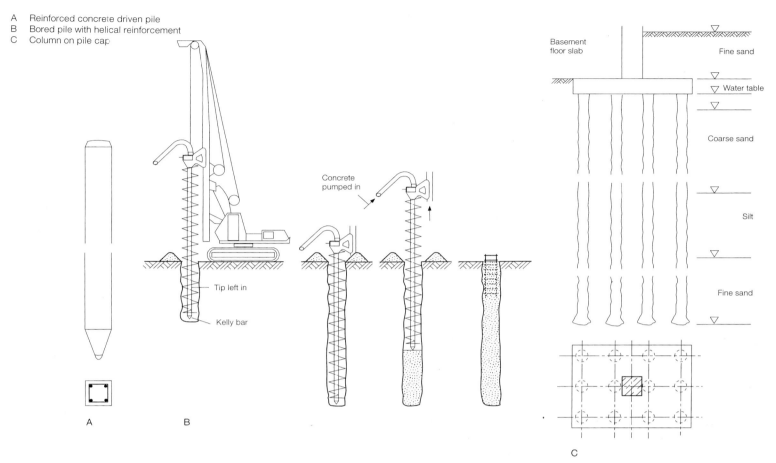

3.3.13  Stages in forming a bored and cast-in-place pile

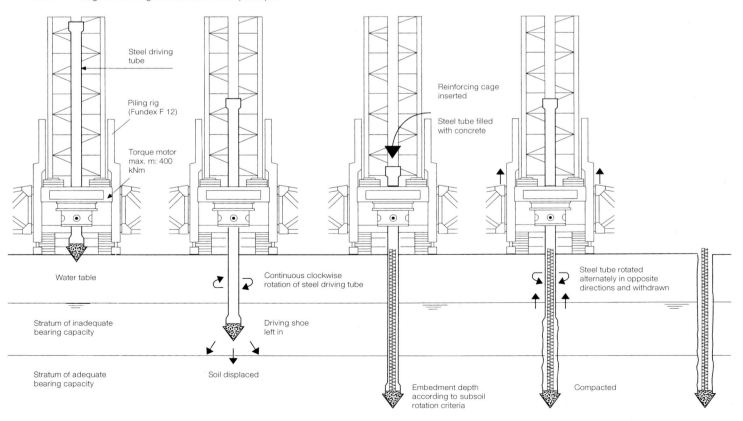

# Foundations

3.3.14 Skin friction values for micro injection piles

| Type of soil | Compression pile MN/m$^2$ | Tension pile MN/m$^2$ |
|---|---|---|
| Moderate and coarse gravel | 0.20 | 0.10 |
| Sand and coarse gravel | 0.15 | 0.08 |
| Cohesive soil | 0.10 | 0.05 |

3.3.15 Safety factors h for micro injection piles

| Micro injection pile as: | | η for loading case to DIN 1054 | | |
|---|---|---|---|---|
| | | 1 | 2 | 3 |
| Compression pile | | 2.0 | 1.75 | 1.5 |
| Tension pile with deviation from the vertical | 0-45° | 2.0 | 1.75 | 1.5 |
| | 80° | 3.0 | 2.5 | 2.0 |

The values between 45° and 80° may be interpolated for tension piles.

3.3.16 Max. angle of inclination for raking piles

| Type of pile | max. inclination to the vertical |
|---|---|
| driven pile | ≤ 45° |
| bored pile | ≤ 14° |
| large-diameter bored pile | ≤ 11° |
| micro injection pile | ≤ 80° |

## Deep foundations, piled foundations [60]

As soon as subsoil strata with a suitable bearing capacity occur at greater depths, the load of the structure must be carried at these deeper levels. This is usually achieved by way of piles. Other special forms of deep foundations are not considered in this book.

Piles transfer the load by way of end bearing and/or skin friction. Concrete foundation piles are classified as in situ or precast according to their method of production. In situ reinforced concrete piles are cast in a void in the ground created beforehand by removing or displacing the soil. Piles can be classified as driven, driven and cast-in-place, jacked, bored and cast-in-place, or composite (figure 3.3.12). A recent development is precast concrete piles that are "screwed" into the ground. Precast concrete piles can be produced on site or in a precasting plant (spun concrete method). They are driven, vibrated, water-jetted or screwed into the ground or inserted into a hole formed beforehand. In the following we distinguish between displacement and bored (or replacement) piles depending on the way in which the construction of the pile affects the subsoil. The shaft for a bored pile is produced without appreciable displacement of the soil.

### Displacement piles
The classical displacement pile is the driven pile of timber, steel or concrete. Concrete piles have a square or circular cross-section (figure 3.3.12 A). They are cast horizontally, circular ones using the spun concrete method. The piles are driven into the ground by a "slow-acting pile driver" with a weight to suit the pile and the driving conditions. Today, fast-acting pile drivers or vibration techniques are frequently used instead in suitable soils. This type of pile is also known as a displacement pile because an amount of soil corresponding to the volume of the pile shaft has to be displaced in all cases. The displacement piles also include the screw pile with a circular cross-section and thread-type outer surfaces at the toe. These are screwed into the ground under the action of a vertical load.

Driving is not usually permitted in densely populated areas. Furthermore, obstacles to driving above the loadbearing stratum make progress difficult or even impossible. Bored piles are preferred in such cases.

### Bored piles (DIN 4014)
Bored piles are characterised by the fact that a hole is formed in the subsoil and the pile shaft cast in place against the surrounding soil (figure 3.3.12 B, C). Generally, the side of the hole is supported before concreting by means of a tube or bentonite slurry in order to prevent – as far as possible – the surrounding soil from collapsing or loosening. The pressure of the wet concrete creates a good interlock with the subsoil. The load transfer can be further improved by subsequently compressing the toe and skin of the pile. Figure 3.3.13 illustrates the procedure for constructing a bored and cast-in-place pile with the steel tube subsequently withdrawn.

The toe of certain bored piles can be enlarged (underreamed) for heavier loads (Solétanche system). When planning such piled foundations, the minimum spacing of the piles must be adhered to.

It is feasible to carry part of the load on piles and part on the ground slab of the structure (shallow foundation).

### Small-diameter micro injection piles (DIN 4128)
Micro injection piles with a small diameter (< 300 mm) have proved to be an economical form of pile foundation. They were developed from injection ground anchors.

Micro injection piles have continuous longitudinal bar reinforcement (the loadbearing member), preferably made from threaded bars to allow easy connections. Bars with a diameter of 32, 40 and 50 mm are employed, in bundles of three bars as well. Concrete cover is 30 mm, in aggressive groundwater conditions up to 45 mm. The minimum diameter of a micro injection pile is 100 mm. The loadbearing member is inserted into a (bored) hole and grouted over its full length. It is also possible to fill the hole with grout first and then insert the loadbearing member. The surrounding soil provides this compression member with lateral restraint against buckling. The minimum shear strength of the soil required to guarantee this is 10 kN/m$^2$. Lower shear strengths require a buckling analysis of the bar to be carried out. Small-diameter micro injection piles cannot accommodate bending moments or shear stresses. They are classed as having pin-jointed ends. Therefore, appropriate arrangements of three are required to fix a point, six to fix a three-dimensional body (tables 3.3.14 and 3.3.15).

### Raking piles
Raking piles are constructed to accommodate horizontal force components. Their maximum inclination is limited by the characteristics of the pile-driving plant (table 3.3.16).

## Securing excavations

Where there is no room for an embankment, the sides of an excavation must be provided with temporary support [60]. In this book we will look at diaphragm walls and contiguous piling because these can usually form part of the final external wall to a structure.

### Diaphragm walls

The diaphragm wall is often a beneficial means of securing an excavation in urban civil engineering works, and for deep excavations in close proximity to existing structures where the water table is high. The diaphragm wall method is a low-noise and low-vibration technique that does not contaminate the groundwater and permits the construction of reinforced concrete walls directly within the pressure bulb of existing foundations. The majority of diaphragm walls are 600-1000 mm thick, but in special cases walls up to 1500 mm thick are also feasible. Their depth is theoretically unlimited, but at depths exceeding 40 m gaps can occur owing to deviations in vertical alignment.

To construct a diaphragm wall, a row of individual excavation and concreting segments, the "diaphragm wall panels", is required. First, a guide wall approx. 1 m high is built to guide the grab used for excavating the soil. Bentonite slurry is pumped into the excavation as the work proceeds to support the sides. Steel tubes are incorporated against the soil to define the ends of the excavation. After lowering in the cage of reinforcement, the concrete is placed according to the techniques of underwater concreting using a tremie pipe. The bottom end of the pipe remains immersed in the fresh concrete already placed. This ensures that only a very small quantity of the fresh concrete mixes with the bentonite slurry. The slurry displaced by the concrete is pumped out ready for treatment prior to being reused. After the concrete has hardened, the steel tubes at the ends are extracted. The wall panels are usually constructed in leapfrog fashion.

### Contiguous piling

The sides of an excavation can also be supported by piles. These are normally constructed in bored holes lined with steel tubes. Settlement behind the wall can be kept to a minimum, making these walls very suitable for securing excavations directly within the pressure bulb of existing foundations.

Three variations are common (figure 3.3.17):
- In the secant pile method, piles 1, 3, 5 etc. are cast first without reinforcement. The intermediate, reinforced piles are constructed afterwards. In doing so, these interlock with their neighbours already cast to create a continuous concrete wall. These walls can be assumed to be sufficiently watertight for the temporary, excavated condition. Any serious leaks can be sealed by injecting grout or epoxy resin.
- Standard contiguous piling consists of a row of piles spaced about 50-100 mm apart, which is therefore not watertight.
- The piles can also be constructed further part and the spaces between secured with horizontal vaulting.

### Supporting and anchoring diaphragm walls and contiguous piling

Diaphragm walls and contiguous piling must be supported as excavation work proceeds.

In small excavations the walls can be propped against each other across the opening. But in large and deep excavations this propping becomes impracticable and disrupts operations within the excavation. Therefore, we either use ground anchors or support the walls horizontally by means of the floor slabs. Injection anchors (DIN 4125) are normally used. These are tension members made from prestressing steel that are incorporated in drilled holes. The anchorage forms a mechanical interlock with the component to be anchored.

The anchorages are fixed either to walings (figure 3.3.18) or in metal anchor pockets incorporated in the cage of reinforcement (figure 3.3.19). The injection anchor transfers the anchor force to the subsoil.
A cement slurry injected into the drilled hole forms the injection anchor. A simple form of corrosion protection is all that is required for temporary anchors. Permanent anchors must be provided with comprehensive, durable, multiple forms of protection against corrosion. This is achieved by inserting a sheath around the tendons and filling it with cement or a permanently plastic compound. Permanent anchors require building authority approval; current regulations call for an assessment every two years.

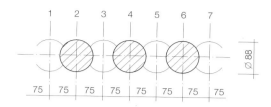

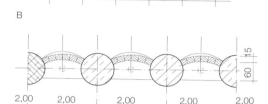

A Secant
B True contiguous
C Spaced with sprayed concrete infills

3.3.17 Contiguous piling

3.3.18 Anchorages fixed to walings

3.3.19 Reinforcement cage around anchorage

# Foundations

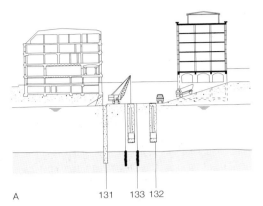

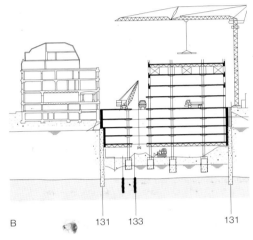

A  Diaphragm wall perimeter, boring and constructing main columns plus permanent tension anchors to prevent flotation
B  Simultaneous construction of floors above ground level and further basement levels

131  Diaphragm wall
132  Main column
133  Ground anchor

3.3.20  Top-down basement construction [63]

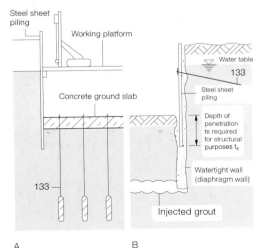

A  with anchored ground slab cast underwater
B  with injected grout at deeper level

3.3.21  Excavation with groundwater control

Diaphragm walls or contiguous piling can also be braced by the basement floor slabs. After constructing the perimeter walls and the piles, which reach to ground level, a flat slab is cast on the (prepared) ground. This slab is supported on the walls and the tops of the piles. After the concrete has reached the necessary strength, the soil beneath the slab is excavated down to the first basement level. The next slab is cast on the prepared ground at this lower level, once again supported on the walls and piles, which have now become columns. This procedure is repeated for all the basement floor levels.

This top-down method of constructing a deep basement has the following advantages:
- It is not necessary to anchor and/or prop the external walls.
- No scaffolding is required in the basement.
- Preparing the ground for casting the slab is less expensive than providing formwork.
- The lower floors can be constructed protected from the weather.
- Work can proceed faster because after constructing the floor slab at ground floor level, construction can proceed both upwards and downwards.

The disadvantages are:
- The limited height for excavation work, especially in basement car parks, where the clear storey height is low.
- Longer spans between columns than is usually the case may be necessary.
- Access openings have to be left in the floors and filled in later.

Figure 3.3.20 illustrates this top-down method of construction with the lowest floor cast on formwork.

## Building below the water table, groundwater control [60]

Lowering the water table is not permitted in many cases for ecological reasons or due to the influence of the foundation on neighbouring structures (risk of settlement, rotting of timber piles). Methods of controlling the groundwater must therefore be used (with the sides and base of the excavation sealed). The walls are almost always constructed as diaphragm walls or contiguous (secant) piling. If the subsoil itself is not more or less watertight at a reasonable depth, it is necessary either to construct a concrete ground slab underwater or to inject grout into the subsoil to form an impermeable layer.

The procedure for constructing an impermeable concrete ground slab is as follows:

After the perimeter walls have been built, the soil is excavated down to the water table. The walls are anchored back into the ground section by section. If the ground slab is to be anchored by piles to prevent flotation, they can be installed from this level (e.g. driven and cast-in-place piles). As the excavation proceeds section by section down to the lowest level of the basement, the walls are braced underwater as required. The plain or reinforced ground slab is then cast underwater.

After the concrete has reached the necessary strength, the water is pumped out of the excavation. If the ground slab is to be anchored with injection anchors, they are installed underwater after concreting the ground slab from a working platform (figure 3.3.21 A).

A watertight base to the excavation can also be constructed by injecting grout into the soil. Injection takes place from the level of the water table after partial excavation. The injected grout is placed as deep as necessary to ensure an adequate factor of safety against flotation (figure 3.3.21 B).

The method chosen to prevent flotation depends on the size of the excavation and the levels of ground slab and water table. It also depends on whether the impermeable concrete ground slab of the structure also has to be anchored. The same anchorages can be used in this case.

Both underwater concrete slabs and injected grout are assumed to remain sufficiently watertight only for the duration of the construction. As a rule, therefore, the ground slab to the structure is also constructed as an impermeable concrete slab. This, too, may need to be anchored against flotation in the final condition.

# References

## Books and articles

### Multistorey structures

[1] Bauarchiv für Architekten und planende Bauingenieure, Bundesverband der Deutschen Transportbetonindustrie, Duisburg 1987

[2] Beck, H., Frenzel, R.: Ausbaudetails im Fertigteilbau, Fachvereinigung Deutscher Betonfertigteilbau, Bonn 1993

[3] Belz, W., Gösele, K., Jenisch, R., Pohl, R., Reichert, H.: Mauerwerk Atlas, Institut für internationale Architektur-Dokumentation, Munich 1984

[4] Betonfertigteile für den Skelett- und Hallenbau, Fachvereinigung Deutscher Betonfertigteilbau, Bonn, Beton-Verlag, Düsseldorf 1993

[5] Betonfertigteile für den Wohnungsbau, Fachvereinigung Deutscher Betonfertigteilbau, Beton-Verlag, Düsseldorf 1994

[6] Bieger, K.-W. (ed.): Stahlbeton und Spannbetontragwerke nach Eurocode, Springer Verlag, Berlin/Heidelberg 1993

[7] Bomhard, H.: BMW-Hochhaus München, Dyckerhoff & Widmann, Munich 1973

[8] Busse, H.-B. v., Waubke, N. V., Grimme, R., Mertins, J.: Atlas Flache Dächer, Institut für internationale Architektur-Dokumentation, Munich 1992

[9] Dierks, K., Schneider, K. J. (eds.): Baukonstruktion, 3rd edition, Werner Verlag, Düsseldorf 1993

[10] Fassaden, Konstruktion und Gestaltung mit Betonfertigteilen (Brandt, J.; Heene, G. V.; Kind-Barkauskas, F. et al.). Beton-Verlag, Düsseldorf 1988

[11] Hart, F., Henn, W., Sontag, H.: Stahlbauatlas Geschossbauten. 2nd edition, Institut für internationale Architektur-Dokumentation, Munich 1982

[12] König, G.: Hochhäuser aus Stahlbeton, Beton-Kalender 1990, part II, pp. 457-539, Ernst & Sohn, Berlin

[13] Kordina, K., Meyer-Ottens, C.: Beton Brandschutz Handbuch, Beton-Verlag, Düsseldorf 1981

[14] Kotulla, B., Urban-Clever, B.-P.: Industrielles Bauen, Fertigteile, 2nd edition, Expert Verlag, Renningen 1994

[15] Mayer, H.: Die Berechnung der Durchbiegung von Stahlbetonbauteilen, Ernst & Sohn, Berlin 1967, DaSt, issue 194

[16] Pfefferkorn, W., Steinhilber, H.: Ausgedehnte fugenlose Stahlbetonbauten: Entwurf und Bemessung der Tragkonstruktion, Beton-Verlag, Düsseldorf 1990

[17] Polónyi, S.: ... mit zaghafter Konsequenz, Bauwelt Fundamente 81, Friedrich Vieweg & Sohn, Braunschweig/Wiesbaden 1987

[18] Polónyi, S.: Rohbaukostenanalyse von Wohngebäuden, memo, Chair of Structural Engineering, Dortmund University 1974

[19] Polónyi, S., Bollinger, K.: Ansätze in der Konzeption des Stahlbetons, Die Bautechnik 4/1983

[20] Polónyi, S., Dicleli, C.: Kosten der Tragkonstruktionen von Skelettbauten, Rudolf Müller Verlag, Cologne 1976

[21] Rösler, W., Stöffler, J.: Beton-Fertigteile im Skelettbau, Fachvereinigung Betonfertigteilbau, Bundesverband Deutsche Beton- und Fertigteilindustrie, Düsseldorf 1982

[22] Scheer, J., Pasternak, H., Hofmeister, M.: Gebrauchstauglichkeit – (k-)ein Problem?, Bauingenieur vol. 69, 1994, pp. 99-106

[23] Schunck, E., Finke, T., Jenisch, R., Oster, H. J.: Der neue Dachatlas – Geneigte Dächer, Institut für internationale Architektur-Dokumentation, Munich 1991

[24] Staffa, M.: Zur Vermeidung von hydratationsbedingten Rissen in Stahlbetonwänden, Beton- und Stahlbetonbau 1994, issue 1, pp. 4-8

[25] Steinle, A., Hahn, V.: Bauen mit Betonfertigteilen im Hochbau, Fachvereinigung Betonfertigteilbau, Bundesverband Deutsche Beton- und Fertigteil-Industrie, Bonn 1991

[26] Suspa: Suspa-Report, company brochure

[27] Weber, H.: Das Poren-Beton-Handbuch, 2nd edition, 1995, Bauverlag, Wiesbaden/Berlin

### Single-storey sheds

[28] Beck, H., Frenzel, R.: Ausbaudetails im Fertigteilbau, Fachvereinigung Deutscher Betonfertigteilbau, Bonn 1993

[29] Betonfertigteile für den Skelett- und Hallenbau, Fachvereinigung Deutscher Betonfertigteilbau, Bonn, Beton-Verlag, Düsseldorf 1993

[30] Born, J.: Praktische Schalenstatik, vol. I.: Die Rotationsschalen, 2nd edition, Ernst & Sohn, Berlin/Munich 1968

[31] Busse, H.-B. v., Waubke, N. V., Grimme, R., Mertins, J.: Atlas Flache Dächer, Institut für internationale Architektur-Dokumentation, Munich 1992

[32] Candela, F.: Structural Applications of Hyperbolic Paraboloid Shells, ACI Journal, 1/1955

[33] Csonka, P.: Theory and Practice of Membrane Shells, Kiado Academy, Budapest, 1987

[34] Die Kunst der leichten Schalen – Zum Werk von Felix Candela, arcus 18, Rudolf Müller Verlag, Cologne 1992

[35] Dyckerhoff & Widmann (eds.): Festschrift Ulrich Finsterwalder – 50 Jahre für Dywidag, Verlag G. Braun, Karlsruhe 1973

[36] Faber, C.: Candela und seine Schalen, Verlag Callwey, Munich 1965

[37] Gestalten in Beton: Zum Werk von Pier Luigi Nervi, arcus 7, Rudolf Müller Verlag, Cologne 1989

[38] Hampe, E.: Statik rotationssymmetrischer Flächentragwerke I-IV, VEB Verlag für Bauwesen, Berlin 1963-73

[39] Isler, H.: Concrete Shells Derived from Experimental Shapes, SEI, Structural Engineering International, 3/94, Journal of the IABSE

[40] Joedecke, J.: Schalenbau, Konstruktion und Gestaltung, Karl Krämer Verlag, Stuttgart 1962

[41] Kaiser, U.: Wiederaufbau der Kongresshalle in Berlin – Die Bauausführung, Die Bautechnik 10/1987, pp. 329-35

[42] Kollár, L.: Schalenkonstruktionen, Beton-Kalender 1974, part II, pp. 327-435, Ernst & Sohn, Berlin

[43] Kordina, K., Meyer-Ottens, C.: Beton Brandschutz Handbuch, Beton-Verlag, Düsseldorf 1981

[44] Mannesmann, Demag materials handling technology, company brochures

[45] Markus, G.: Theorie und Berechnung rotationssymmetrischer Bauwerke, Werner Verlag, Düsseldorf 1976

[46] Otto, F.: Das hängende Dach (The Suspended Roof), Bauwelt Verlag, Berlin 1954 (reprint: dva 1990)

[47] Patzkowsky, K.: Bewehren von Stahlbetonbalken mit ausgeklinkten Auflagern, dissertation, Dortmund University, 1990

[48] Polónyi, S.: Berechnung der hyperbolischen Paraboloidschalen über beliebigen Viereckgrundrissen, Beton- und Stahlbeton, 6/1962

[49] Polónyi, S., Stein, H.: Hallen, Rudolf Müller Verlag, Cologne 1986

[50] Polónyi, S.: Neue Aspekte im Stahlbeton-Schalenbau, Bauwelt 32/1965

[51] Polónyi, S., Koch, P.: Hautartige Schalen für Überdachungen, Die Bautechnik 6/1973

[52] Ramm, E., Schunck, E.: Heinz Isler Schalen, Karl Krämer Verlag, Stuttgart 1986

[53] Rösler, W., Stöffler, J.: Beton-Fertigteile im Skelettbau, Fachvereinigung Betonfertigteilbau, Bundesverband Deutsche Beton- und Fertigteilindustrie, Düsseldorf 1982

[54] Rühle, H.: Räumliche Dachtragwerke, Konstruktion und Ausführung, vol. 1: Beton, Holz, Keramik, Rudolf Müller Verlag, Cologne 1969

[55] Schlaich, J., Kordina, K., Engell, H.-J.: Teileinsturz der Kongresshalle in Berlin, Schadensursachen, Zusammenfassendes Gutachten, Beton- und Stahlbetonbau 12/1980, pp. 281-94

[56] Schunck, E., Finke, T., Jenisch, R., Oster, H. J.: Der neue Dachatlas – Geneigte Dächer, Institut für internationale Architektur-Dokumentation, Munich 1991

[57] Siegel, C.: Strukturformen der Modernen Architektur, Callwey Verlag, Munich 1960

[58] Steinle, A., Hahn, V.: Bauen mit Betonfertigteilen im Hochbau, Fachvereinigung Betonfertigteilbau, Bundesverband Deutsche Beton- und Fertigteilindustrie, Bonn 1991

[59] Weber, H.: Das Poren-Beton-Handbuch, Bauverlag, Wiesbaden, Berlin

### Foundations

[60] Arz, P., Schmidt, H. G., Seitz, J., Semprich, S.: Grundbau, Beton-Kalender 1994, part II, Ernst & Sohn, Berlin

[61] Bollinger, K.: Bewehren nach neuer Stahlbeton-Konzeption II, Tragverhalten von rotationssymmetrisch beanspruchten Stahlbetonplatten, Bautechnik 11/1985

[62] Grundbau-Taschenbuch, 4th edition, Ernst & Sohn, Berlin 1991

[63] Pause, H.: Deutsche Bank AG, Düsseldorf, technical report, Philipp Holzmann company, June 1985

[64] Polónyi, S., Block, K., Bollinger, K.: Bewehren nach neuer Stahlbetonkonzeption I; Der Balken, Die Kreisplatte, Die innere Steifigkeit; Die Bautechnik 12/1984, pp. 422-31

[65] Staffa, M.: Zur Vermeidung von hydratationsbedingten Rissen in Stahlbetonwänden, Beton- und Stahlbetonbau 1994, issue 1, pp. 4-8

[66] Styn, E.: Unterschiedlich gelagerte Kreisplatten mit Ringbewehrung, dissertation, Dortmund University, 1991

[67] Walochnik, W.: Fertigteilstützen aus Schleuderbeton, Bautechnik 8/1995

## Standards and directives

Eurocode 1 (EC 1) Basis of design and actions on structures

Eurocode 2 (EC 2) Design of concrete structures

DIN 1055 parts 1-5: Design loads for buildings, edition 1994

DIN 1045 Structural use of concrete; design and construction, edition 1994

DIN 1054 Subsoil; permissible loading of subsoil, edition 1994

DIN 4227 parts 1-6: Prestressed concrete, edition 1994

DIN 4219 Lightweight concrete and reinforced lightweight concrete of dense structure
part 1 Properties, manufacture and inspection, edition 1984
part 2 Design and construction, edition 1984

DIN 4232 No-fines lightweight concrete walls; design and construction, edition 1989

DIN 4028 Reinforced concrete slabs made of lightweight concrete with internally porous texture, edition 1984

DIN 18806 Composite steel and concrete structures; composite columns, edition 1984

Directive for composite steel and concrete beams, edition 1993

DIN 4014 Bored cast-in-place piles; formation, design and bearing capacity, edition 1994

DIN 4026 Driven piles; manufacture, dimensioning and permissible loading, edition 1988

DIN 4126 Cast-in-in situ concrete diaphragm walls; design and construction, draft edition with explanatory text, 1994

DBC-Merkblatt collection: data sheets, status reports, directives, edition 1991, Deutscher Beton-Verein

## 4.1 Loadbearing external wall, single-leaf
## B Precast concrete

In this example our single-leaf loadbearing wall consists of precast concrete elements. As with the in situ concrete construction (4.1 A), an additional layer of thermal insulation with suitable protection against external influences is required.

The (heated) basement consists of an in situ concrete ground slab and precast concrete wall elements with the waterproofing layer on the outside. The lightwell is hung on the external wall at the top and is thus secured against overturning. The thermal insulation to the wall is placed on the outside, on top of the waterproofing, and should therefore comprise closed-cell, moisture-resistant material. Isolating the components at the balcony by way of suitable details avoids the formation of a thermal bridge. The balconies are supported on projecting wall panels, which in turn are connected to the loadbearing construction by anchors to provide security against overturning. This solution is preferable to the arrangement shown in figure 4.1 A. As the rendering can only be applied after erecting the precast concrete panels, this solution requires scaffolding to the facade.

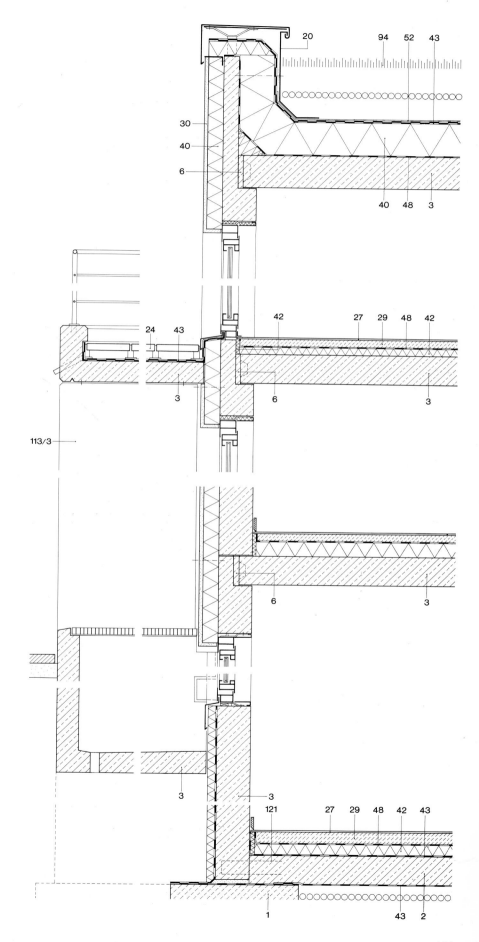

4.1.B

| | |
|---|---|
| 29 Screed | 88 Gravel |
| 30 Rendering | 92 Water bar |
| 40 Thermal insulation | 94 Roof planting |
| 42 Impact sound insulation | 113 Wall |
| 43 Bitumenised felt (waterproofing) | 121 Reinforcement |
| 48 Interlayer | |
| 52 Root penetration protection mat | |

# Construction details

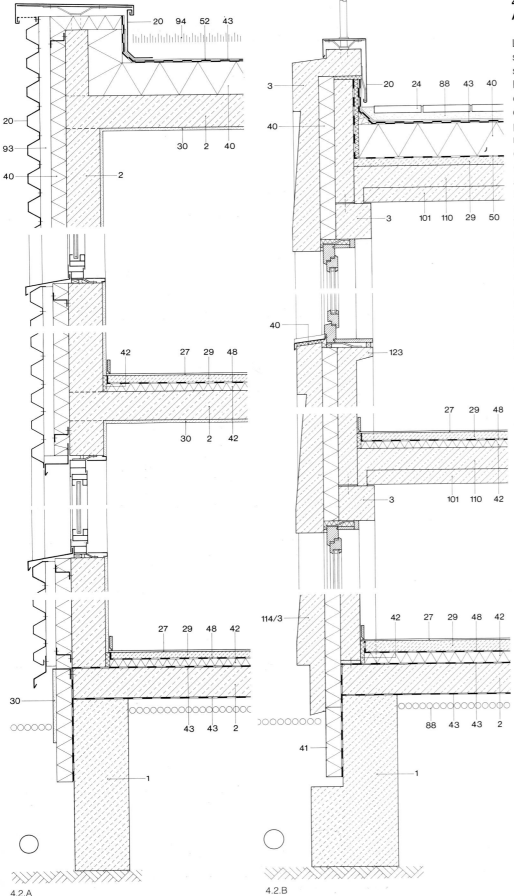

## 4.2 Loadbearing external wall, double-leaf
### A In situ concrete

Loadbearing external walls are frequently constructed with more than one leaf for building science, structural and architectural reasons. Here, the loadbearing construction is of in situ concrete. The facing leaf can make use of various materials; this example shows trapezoidal profile metal cladding; other materials such as reconstituted stone panels, natural stone panels or masonry are also feasible. The concrete ground slab is placed directly onto the subsoil. A compacted layer of gravel is therefore required beneath the slab in order to interrupt the capillary action, plus a sufficiently thick layer of thermal insulation below the floating screed and a secure connection between horizontal and vertical waterproofing materials. The external insulation in front of the strip footing continues below ground level for a distance sufficient to prevent a thermal bridge being formed between ground slab and wall. A layer of rendering at the base of the wall protects the closed-cell insulating material from splashing water. An airspace is included in the lightweight facade to improve sound insulation and diffusion processes. The cavities are filled with a foam insulating material around the windows (thermal bridges). Care should be taken to ensure there are no gaps in the thermal insulation at the junction between wall and roof.

## 4.2 Loadbearing external wall, double-leaf
### B Precast concrete

This loadbearing external wall consists of precast concrete elements with a central core of thermal insulation without air space (sandwich construction). The inner leaf provides the loadbearing function, the outer one protection against the weather. Allocating the wall functions in this way means that the concrete can be used as an architectural element with an effective surface texture. The thermal insulation must continue the full height of the wall without interruptions. Special care must be taken at the junctions between the individual precast units. The impact sound insulation on the floors is guaranteed by a floating screed, but it is necessary to isolate the screed on all sides from the walls. The waterproofing to the (accessible) roof must continue min. 300 mm up the parapet and be properly flashed.

| | | | |
|---|---|---|---|
| 1 | Concrete | 27 | Synthetic floor covering / carpet |
| 2 | In situ reinforced concrete | 29 | Screed |
| 3 | Precast concrete element | 30 | Plaster |
| | | 30a | Rendering |
| 20 | Sheet metal capping / flashing | 40 | Thermal insulation |
| | | 41 | Closed-cell insulation |
| 20a | Trapezoidal profile metal cladding | 42 | Impact sound insulation |
| 24 | Paving slabs | 43 | Waterproofing |

## 4.11 Stairs

### 4.11.1 In situ concrete
For reasons of impact sound insulation, the in situ concrete stair is supported on discrete, soft bearings. The open joint between wall and flight also serves to improve the sound insulation.

### 4.11.2 In situ concrete with precast concrete flights
The landings to this stair are supported on discrete, soft bearings at the sides for reasons of impact sound insulation. Again, the open joint between internal wall and flight improves the sound insulation. The landings are isolated from the external wall by means of strips of non-rigid insulation.

### 4.11.3 Precast concrete (dog-leg)
The precast concrete element consists of landings and connecting flight. It is supported on discrete, soft bearings at each end. The half-landing is supported on nibs seated in bearing pockets. The support at floor level is formed by means of nibs bearing on the floor beam.

### 4.11.4 Precast concrete (half-turn with landings)
One precast concrete element consists of one flight plus the intermediate landings. It is supported on discrete, resilient bearings at the sides. The other precast concrete elements consist of one flight and the landing at floor level. Support is by means of nibs, or the precast concrete elements supported at floor level on nibs on resilient bearings within pockets in the floor beams. The open joint between the stair and the adjoining walls improves the impact sound insulation.

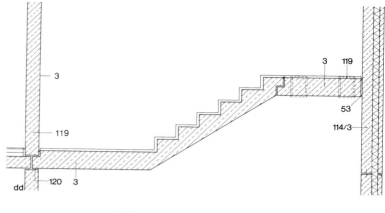

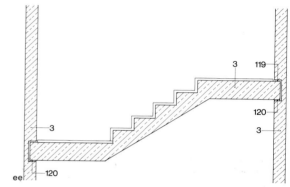

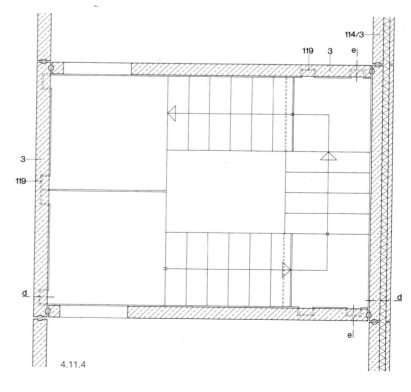

| 2 | In situ reinforced concrete |
| 3 | Precast concrete element |
| 30 | Rendering |
| 40 | Thermal insulation |
| 42 | Impact sound insulation |
| 53 | Sealing compound |
| 114 | Sandwich panel |
| 119 | Pocket for bearing |
| 120 | Elastomeric bearing |

# Part 5 · Built examples in detail

Bruno Kauhsen

The buildings in this chapter were selected on the basis of their architecture, form and surface finishes. If one or other of the details does not comply with the latest German energy economy requirements, this is primarily for the following reasons:

The requirements concerning thermal insulation and energy efficiency of buildings have been regularly tightened since the first Thermal Insulation Act was passed in 1977. Since the Energy Economy Act came into force, passed by the German parliament in July 2001, the previous acts covering thermal insulation and heating systems have been combined. Although a detailed account of energy requirements is now necessary, this does not enable us to make any direct statements about the thickness of insulation currently being used. However, we can assume that increased thicknesses are now being called for. The designer of a building can now decide for himself which energy-saving measures should be employed to achieve the necessary targets. He may decide to compensate for possible weak points in other ways. Energy-saving measures include:

- better thermal insulation
- efficient heating system technology
- the use of renewable energy sources
- heat recovery concepts

Depending on the type of building, its utilisation, type of construction and the local market situation, this usually leads to a combination of these measures in order to obtain the required energy economy targets cost-effectively (see p. 90).

Some of the structures described here are located in different parts of the world and have been planned and erected according to the particular regulations of the respective country. The degree of thermal insulation required depends on the utilisation (internal climate) and the location of a building (external climate). In this context, heating and ventilation in winter are equally important in terms of avoiding condensation on the internal surfaces of components, as is the prevention of thermal bridges.

## Built examples in detail – overview

| Number | Page | Architect(s) | Project | Features |
|---|---|---|---|---|
| 1 | 196 | Ando, Osaka | Church of the Light, Ibaraki, J | Quality of fair-face concrete |
| 2 | 198 | Drexler + Tilch, Riederau | Detached house, Riederau, D | Lightweight concrete |
| 3 | 200 | Alder, Basel | Apartment block, Stuttgart, D | Fair-face concrete |
| 4 | 202 | Hopkins, London | Mixed commercial and residential block, London, GB | Quality of fair-face concrete |
| 5 | 206 | Schultes, Berlin | Art gallery, Bonn, D | Spun concrete columns, column-roof connection |
| 6 | 210 | Galfetti, Bellinzona | Castle restoration, Bellinzona, CH | Fair-face concrete in refurbishment work |
| 7 | 212 | Ciriani, Paris | Archaelogy museum, Arles, F | Fair-face concrete |
| 8 | 216 | Olgiati, Zurich | School, Paspels, CH | Fair-face concrete |
| 9 | 218 | Atelier 5, Bern | Training centre, Thun, CH | Fair-face concrete |
| 10 | 220 | Ando, Osaka | Conference pavilion, Weil am Rhein, D | Quality of fair-face concrete |
| 11 | 224 | Reichel, Kassel/Munich | Mixed office and residential block, Kassel, D | Modular reinforced concrete frame |
| 12 | 226 | Auer + Weber, Stuttgart | Health spa treatment centre, Bad Salzuflen, D | Concrete canopies on single column |
| 13 | 228 | Fuses + Viader, Girona | Central university building, Girona, E | Fair-face concrete in refurbishment work |
| 14 | 230 | Baller, Berlin | Two-tier sports hall, Berlin-Charlottenburg, D | Roof construction |
| 15 | 232 | OIKOS, Berlin | Detached house, Berlin, D | Fair-face concrete |
| 16 | 234 | Galfetti, Bellinzona | Tennis courts, Bellinzona, CH | Profiled fair-face concrete wall |
| 17 | 236 | Kochta + Lechner, Munich | Rhinoceros and tapir house, Munich, D | In situ concrete shell, sprayed concrete cladding |
| 18 | 238 | Böbel + Frey, Göppingen | Works forum, cement works, Dotternhausen, D | Lightweight concrete roof shell |
| 19 | 240 | von Gerkan + Marg, Hamburg | Multistorey car park, Hamburg, D | Fair-face concrete |
| 20 | 242 | Dorn + Marg, Hamburg | Car parking facility, Paderborn, D | Prestressed concrete deck without additional waterproofing |
| 21 | 244 | Schultes, Berlin | Crematorium, Berlin, D | Fair-face concrete |
| 22 | 248 | Hertzberger, Amsterdam | Primary school and kindergarten, Amsterdam, NL | Concrete masonry |
| 23 | 250 | Schürmann, Cologne | Letter sorting centre, Cologne, D | Concrete masonry |
| 24 | 254 | Calatrava, Zurich, Paris | Railway station, Lyon, F | Expressive in situ concrete loadbearing structure |
| 25 | 256 | Böhm, Cologne | University library, Mannheim, D | Coloured profiled precast concrete elements |
| 26 | 258 | Böhm, Cologne | Office building, Stuttgart, D | Self-supporting precast concrete facade |
| 27 | 262 | Seidler, Sydney | Office building, Canberra, AUS | Prestressed concrete |
| 28 | 264 | Gerber, Dortmund | Office building, Dortmund, D | Precast concrete facade |
| 29 | 266 | Piano, Paris | Museum, Houston, USA | Roof louvres of ferrocement |
| 30 | 268 | Hopkins, London | Administration centre, Nottingham, GB | Precast concrete vaulting |
| 31 | 270 | Mangiarotti, Milan | Industrial building, Bussolengo Barese, I | Precast concrete system building |
| 32 | 272 | Piano, Paris | Sports stadium, Bari, I | Precast concrete structure |
| 33 | 276 | Foster, London | Underground station, Canary Wharf, London, GB | Reinforced concrete loadbearing structure |

Example 1

**Church of the Light, Ibaraki, Japan**

1989

Architects:
Tadao Ando & Associates, Osaka

Structural engineers:
Ascoral Engineering Associates, Osaka

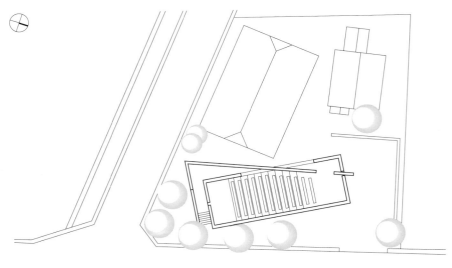

Site plan    scale 1:500

This Protestant church has been sensitively integrated to augment an existing timber church and vicarage in quiet residential surroundings between Osaka and Kyoto. The extremely strict structure reflects the link between traditional Japanese notions of space and the modern architecture so typical of Ando. In this modest room measuring 6.28 x 18.0 m, the most striking architectural device is the diagonal wall, which penetrates the room at an angle of 15°. In doing so, it creates the entrance lobby and also separates the church from the administrative functions. The visitor enters the church on the side with the slits cut in the concrete wall to form the cross of light, from which the church gets its name, and follows the diagonal wall, which is 180 mm lower than the building itself. The ribbon of light entering over the wall provides an additional source of light for the interior. The simple pews are progressively shortened to follow the line of the diagonal wall and, as in a theatre, the floor is raked down to the lowest point of the church, the altar. High-quality fair-face concrete was used for the walls of the church. No particular thermal insulation measures are necessary owing to the climate of the region. The outcome of the special concrete mix is an extremely smooth surface that reflects the light. The individual concrete panels were cast in situ with the help of painted formwork panels matching the size of the traditional Japanese tatami mats. The joints between the panels form a grid and the holes for the accurately positioned formwork ties have not been made good and thus lend the walls additional texture. The ascetic ambience of the interior is emphasised by the simple furniture and floorboards of cedar. However, the true impression of the interior is determined by the lighting effects. The slits forming the cross in the wall behind the altar, as well as the 180 mm high ribbon of glass above the diagonal wall, have intentionally been kept narrow to intensify the experience of the contrast between daylight and darkness. The light penetrating from outside illuminates the cross, and the size and position of its reflections within the room change with the hour and the seasons. This satisfies the architect's wish to create a relationship between people, nature and architecture.

Church of the Light, Ibaraki

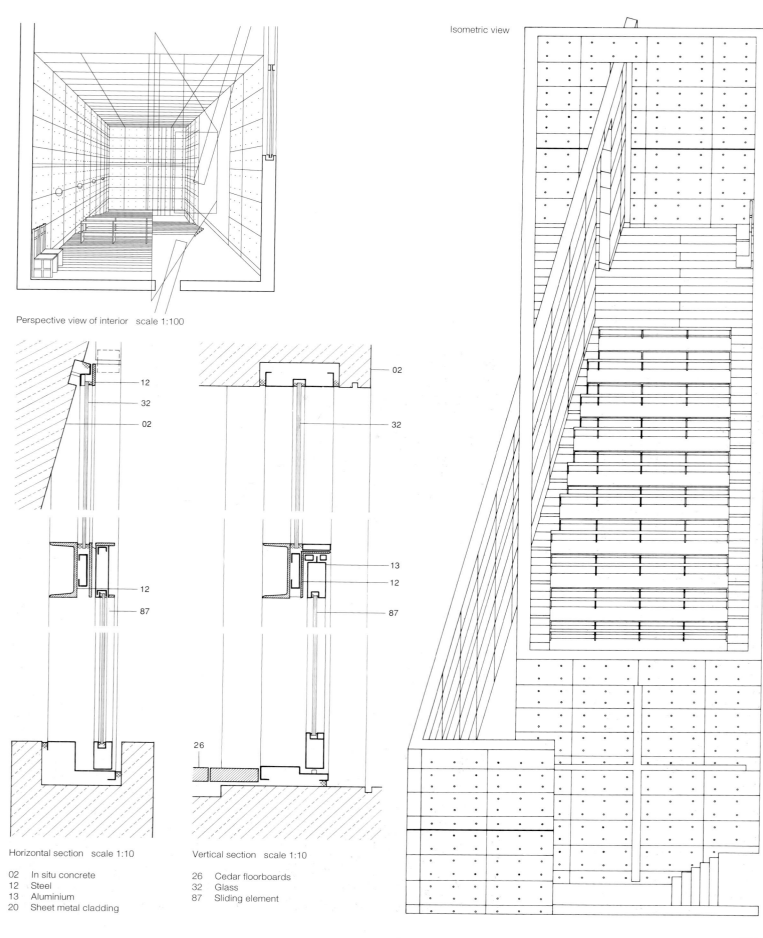

Perspective view of interior   scale 1:100

Horizontal section   scale 1:10

Vertical section   scale 1:10

Isometric view

| 02 | In situ concrete |
| 12 | Steel |
| 13 | Aluminium |
| 20 | Sheet metal cladding |
| 26 | Cedar floorboards |
| 32 | Glass |
| 87 | Sliding element |

# Example 2

**Detached house, Riederau, Germany**

1987

Architects:
Axel Tilch, Gisela Drexler, Riderau

Structural engineer:
Rudolph Meyer, Dettenschwang

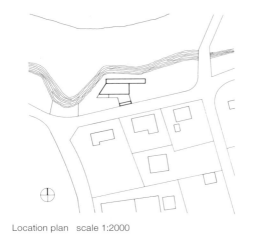

Location plan   scale 1:2000

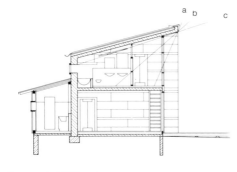

Section   scale 1:200

Sun's elevation
a   66° on 22 June
b   42° on 22 March and 2 November
c   18° on 22 December

The house lies on the edge of a residential district. The plot itself is bounded to the north by a river. The exploitation of passive solar energy played a key role in the planning. The plan shape, a rectangle facing south, is varied by a projecting conservatory at the south-west corner. The fully glazed south elevation contrasts with the northern side, which, with its ancillary rooms and few openings, acts as a buffer zone. The angle of the monopitch roof and the overhang of the eaves above the living quarters are based on the sun's elevation in summer and winter. Internal walls, concrete slabs and the dark natural stone floor covering can store the incoming heat and so make a major contribution to the passive energy concept. Small openings in the solid east and west elevations allow cross-ventilation. A lightweight concrete with a dense microstructure was employed for internal and external walls, floors and roof.

The single-leaf external walls are 400 mm thick and exhibit a low gross density (grade LB 8) in order to increase the thermal insulation effect. However, the latest Thermal Insulation Act would require thicker walls or additional insulation. The internal walls, 200-300 mm thick, employ grade LB 15 concrete, the floors and roof grade LB 25. The roof is provided with a layer of thermal insulation 120 mm thick.

To ensure a clear demarcation between the concrete walls, there are no chamfers on the edges. The timber formwork panels (500 x 2000 mm) were arranged in a "bond" pattern, and after every two rows of panels the inclusion of 120 mm boards can be clearly seen. The wires of the formwork ties passed through these boards and the positions have been left visible. Battens attached to the inside of the formwork create deep rebates, indicating the position and depth of the upper floor slab.

Detached house, Riederau

A  Plan of ground floor   scale 1:200
1  Conservatory
2  Living / dining, kitchen
3  Bedroom
4  Ancillary rooms

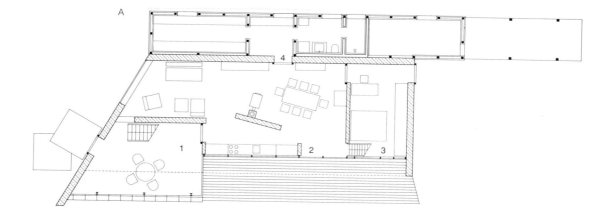

B  Section through conservatory   scale 1:50

04  Lightweight concrete
12  Steel frame
24  Concrete flags
33  Double glazing

90  Roof construction:
    titanium-zinc metal roof covering
    bitumenised roofing felt
    timber boarding
    air space
    thermal insulation
    vapour barrier

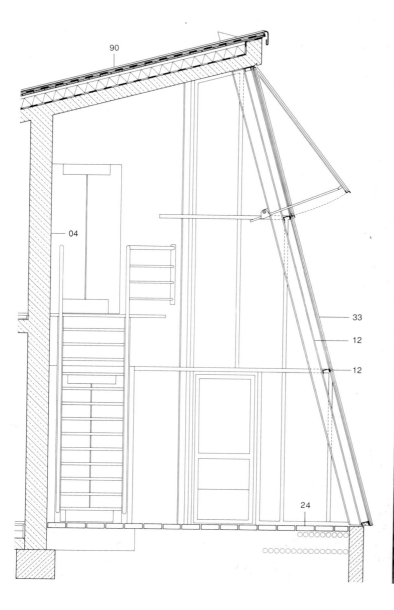

Example 3

**Apartment block, Stuttgart, Germany**

1993

Architect:
Michael Alder, Basel
Partner:
Hanspeter Müller
Assistant:
Roland Fischer

Structural engineers:
Greschick & Falk, Lörrach

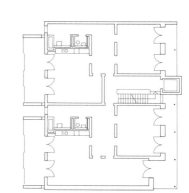

Location plan   scale 1:3000

Plan of ground floor   scale 1:400
Plan of 1st to 5th floors   scale 1:400

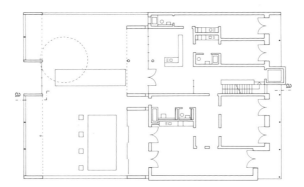

This five-storey residential block stands out from its heterogeneous surroundings thanks to its restrained, meticulous facade design. Even the entrance is shielded from the public gaze by a separate structure. The intervening forecourt creates the right distance to the main building; the ground-floor apartments also benefit from this arrangement. The generously sized entrance zone within the building can also be used as a central common room, as a kitchenette has been included. Another advantage is that all apartments are fitted out suitably for disabled occupants. The open-plan individual apartment layouts can be modified by adding partitions and so provide ample scope for meeting the needs of individual tenants. The arrangement of doors and windows creates an all-round transparency throughout each apartment.
When the weather permits, the fully glazed veranda on the south side extends the usable area of the living/dining room. At the same time, it serves as a noise buffer and also contributes to passive solar heat gains. The external clay brickwork, 500 mm thick, provides adequate storage capacity and thermal insulation. The overhanging concrete roof slab and the use of fair-face concrete in the entrance area emphasise the restrained and unambiguous appearance of this building.

# Apartment block, Stuttgart

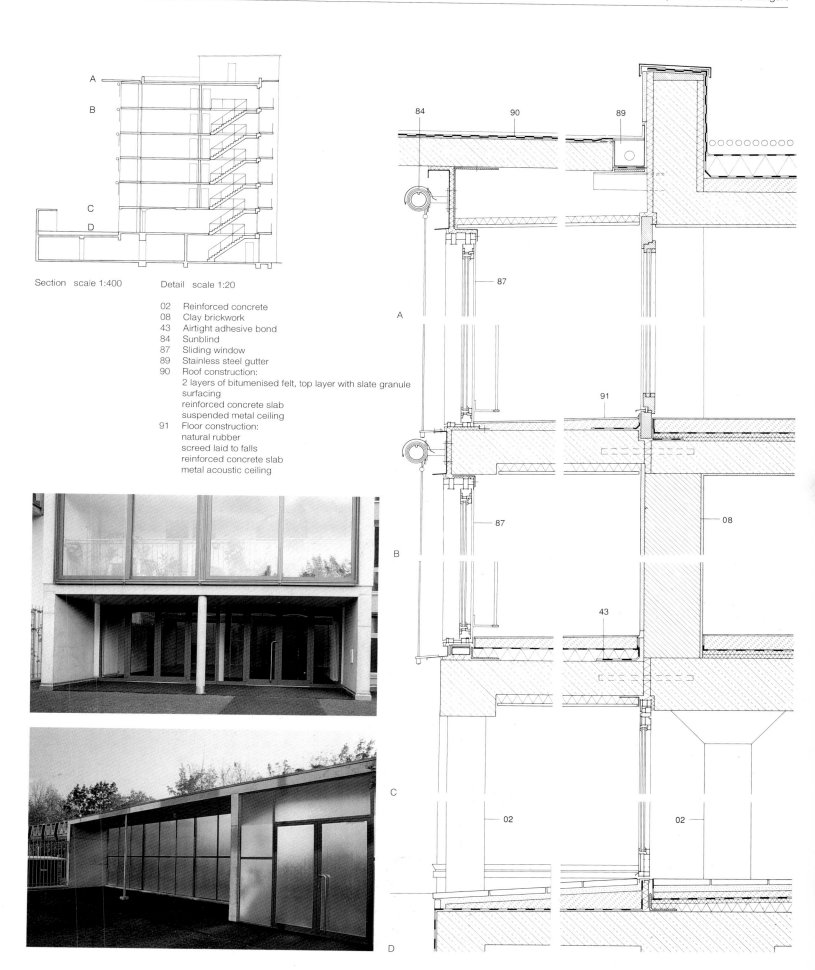

Section scale 1:400    Detail scale 1:20

- 02 Reinforced concrete
- 08 Clay brickwork
- 43 Airtight adhesive bond
- 84 Sunblind
- 87 Sliding window
- 89 Stainless steel gutter
- 90 Roof construction:
  2 layers of bitumenised felt, top layer with slate granule surfacing
  reinforced concrete slab
  suspended metal ceiling
- 91 Floor construction:
  natural rubber
  screed laid to falls
  reinforced concrete slab
  metal acoustic ceiling

Example 4

**Mixed commercial and residential block, London**

1991

Architects:
Michael Hopkins & Partners, London
with John Pringle, Bill Dunster,
Ernest Fasanya, Lucy Lavers, Neno Kezic

Structural engineers:
Büro E. Happold, London

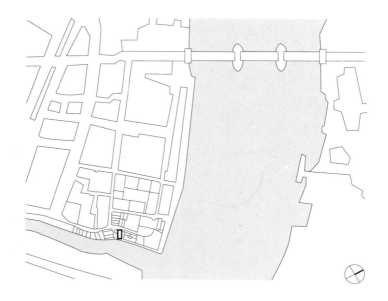

Location plan
scale 1:10 000

This six-storey mixed-use block is situated at St Saviour's Dock on the south side of the River Thames not far from Tower Bridge. Buildings with new uses are springing up on the old, narrow plots of this district, dominated by warehouses and abandoned dockyard buildings. This building, erected for a designer, manufacturer and distributor of steel products, contains sales premises and a showroom on the ground floor, three storeys of offices, and a maisonette apartment above them. The top storey of the apartment is set back to create spacious rooftop patios. Two service towers, containing building services, sanitary facilities and a second staircase, flank the clean lines of this rectangular structure.

The structure consists of a reinforced concrete frame with fair-face circular columns carrying flat slabs. The facades to the showroom are fully glazed and those to the office storeys have full-height sliding windows with anodised aluminium frames, while the lower floor of the maisonette has cantilevering steel balconies. The infill panels of the side walls are clad externally with lead sheet; the ensuing thermal bridges around the ends of the floor slabs would not be permitted according to German standards.

The architects placed great emphasis on obtaining a fair-face concrete surface of the best possible quality, as though it had been "moulded" in one piece. Special attention was paid to the formwork. Plywood panels with chamfered edges were employed for flat surfaces; the result of this was that the joints are given extra prominence. The structural nodes were cast in a special aluminium form and the circular columns in steel forms. All traces of rust were removed from the reinforcement and the entire formwork carefully cleaned out with compressed air prior to concreting. The architects also gave thought to the correct concrete mix and most effective means of compaction. Manual sanding gave concrete surfaces their final "polish", as it were, resulting in a reflective surface finish.

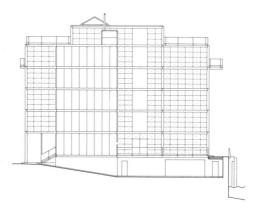

Elevation
scale 1:500

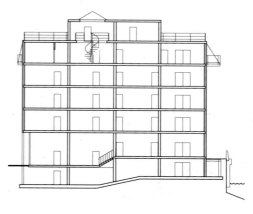

Section
scale 1:500

# Mixed commercial and residential block, London

Example 4

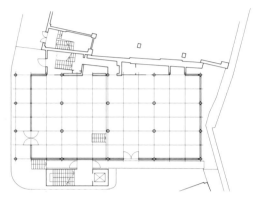

Plan of ground floor   scale 1:500

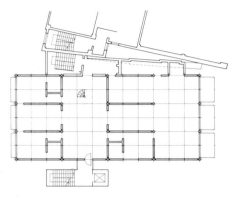

Plan of 4th floor   scale 1:500

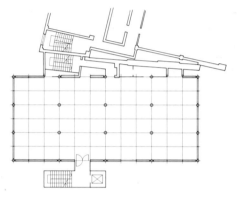

Plan of 2nd floor   scale 1:500

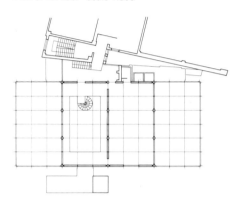

Plan of 5th floor   scale 1:500

204

Mixed commercial and residential block, London

Detail of roof construction    scale 1:20

| | | | | | |
|---|---|---|---|---|---|
| 02 | Reinforced concrete | 20 | Lead sheet cladding with fibre-reinforced concrete filling | 24 | Concrete flags |
| 04 | Lightweight concrete laid to falls | | | 40 | Thermal insulation |
| | | 21 | Plasterboard panels | 41 | Rigid thermal insulation |
| | | | | 43 | Bitumenised roofing felt |
| 50 | Vapour barrier | | | | |
| 87 | Sliding window with double glazing | | | | |

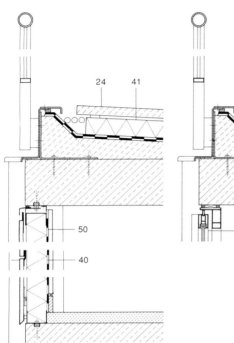
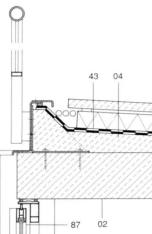
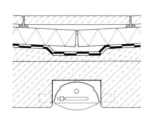
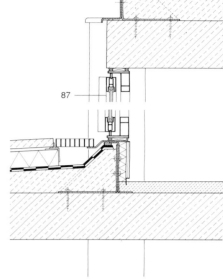
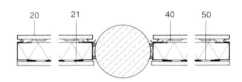

Example 5

**Art gallery, Bonn, Germany**

1992

Architect:
Axel Schultes, Berlin
of Bangert, Jansen, Scholz, Schultes
with Jürgen Pleuser
Assistants:
Georg Bumiller, Michael Bürger,
Margret Kister, Enno Maass, Heike Nordmann,
Volker Staab

Structural engineers:
Polónyi & Fink, Cologne
Spitzlei & Jossen, Siegburg
Stettner & Wald, Bonn

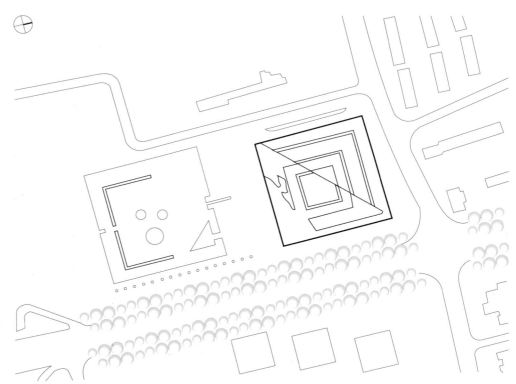

Location plan   scale 1:3000

Bonn Art Gallery is one of the elements of the new "cultural mile" along the Friedrich-Ebert-Allee. It contains attractive collections, with the focus on post-war German art and "August Macke and the Rhenish expressionists".
The plan layout of the building is a square measuring 100 x 100 m, half of which develops with projections and setbacks along the diagonal. The gallery itself is screened from the road by a narrow administrative block, which is given a visual link to the main building by the extended roof slab. The intervening space acts as a large open, covered extension to the entrance foyer. The entrance itself leads diagonally into the actual foyer.
Once inside, the visitor is presented with an interesting sequence of exhibition rooms of different sizes. The simple succession of rectangular spaces and their open relationship with each other enables the visitor to move around at will without any loss of orientation, repeatedly presented with interesting views. The lighting has been very carefully planned. On the upper floor, intricately designed rooflights provide illumination. Further daylight is able to enter via internal courtyards and storey-height windows in the external walls.
The entire building employs double-leaf normal-weight concrete. The many fair-face concrete surfaces remain the dominant impression despite the light grey internal paint finish in the exhibition areas and the sandstone facing leaf opposite the road. The unclad concrete demonstrates that this smooth surface contributes to the interior ambience and underlines the exacting, elegant architecture.
The detail at the top of the external, 13 m tall, spun concrete columns is interesting. An opening has been left in the fair-face in situ concrete roof slab and the columns connected via four steel webs.

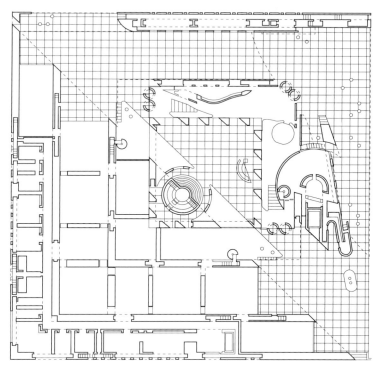

Plan of ground floor   scale 1:1000

Plan of 1st floor   scale 1:1000

Example 5

03 Precast concrete element
20 Aluminium sheet, painted white
28 Textile sunshade
32 Double glazing
40 Thermal insulation
82 Smoke vent
84 Sunshade
86 Ventilation

Section   scale 1:1000

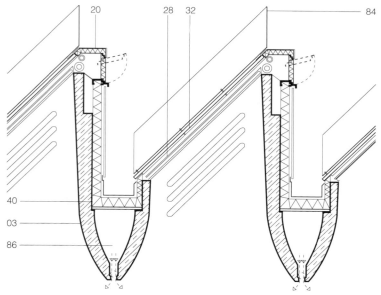

Detail   scale 1:50

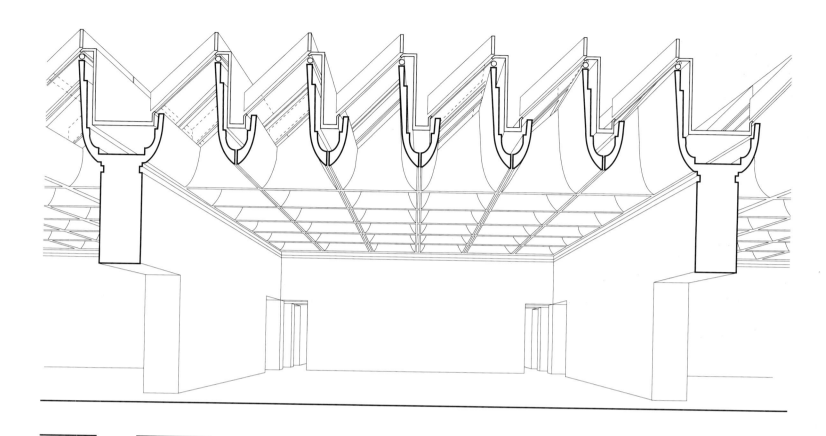

Perspective view of exhibition hall

**Castle restoration, Bellinzona, Switzerland**

1989

Architect:
Aurelio Galfetti, Bellinzona
Assistants:
Valentino Mazza, Luigi Pellegrini,
Rolf Lauppi, Renato Regazzoni,
Ernesto Bomger, José Ormazabal

Structural engineer:
Enzo Vanetta, Lugano

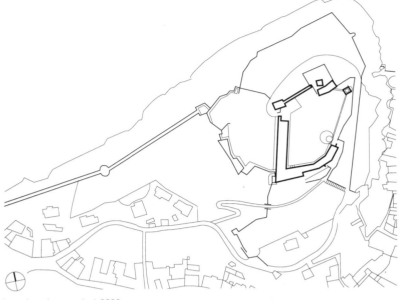

Location plan   scale 1:3500

Restoration work on the Castelgrande, the largest of the three castles in Bellinzona, began in 1980. This castle stands on basalt rocks and is supported by flanking walls, which extend right down into the town. During restoration work, the overgrowth on the rocks was cleared to again reinforce the bleak, fortified impression of the castle. The new entrance to this hilltop fortification starts at the level of the town, directly on the Piazza del Sole. A natural wedge-shaped gap in the rock – lined with concrete, however – leads to a concrete-lined, dome-shaped rotunda deep within the rock. This half-sphere with its "as struck" finish concrete walls forms the entrance lobby for the lifts taking visitors up to the castle. The emergency stairs, also of concrete, are clearly separated from the rock and thus portray a space claimed by people in their fight against nature. Besides the lift shaft there is also a lightwell which permits a glimpse of the sky through a small window at the top of the 40 m shaft. From the top of the lift shaft the visitor enters the castle's inner courtyard. The sloping, paved area, enclosed by a new straight wall and an old curved one, creates a new space. From the outside, the strict horizontal lines of the walls emphasise the vertical line of the towers.

Castle restoration, Bellinzona

Stairs and lifts to Castelgrande   scale 1:1000

Plan of and section through new entrance   scale 1:200

1 Subterranean passage
2 Chamber
3 Lifts
4 Lightwell

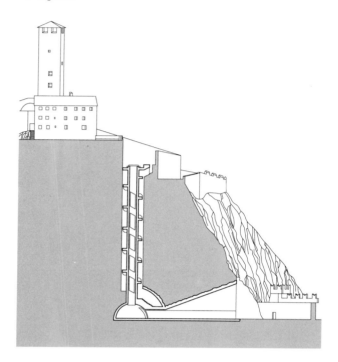
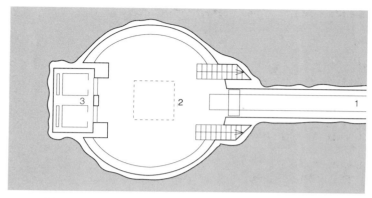
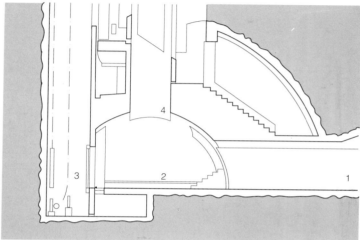

## Archaeology museum, Arles, France

1992

Architect:
Henri Ciriani, Paris
Assistant:
Jackie Nicolas

Structural engineers:
BET Scobat, Paris
Cesba, Aix-en-Provence

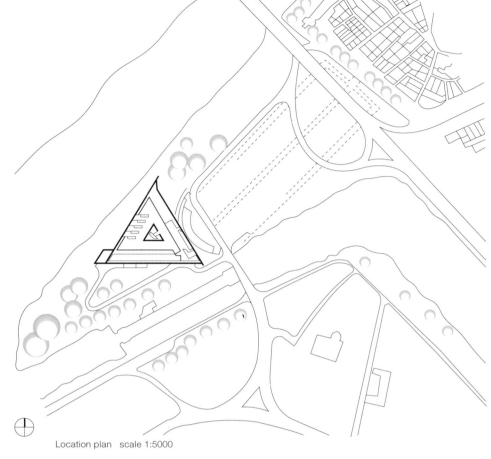

Location plan scale 1:5000

Between the River Rhône and the Canal du Midi, sited on a spit of land in the south-west corner of the old quarter of Arles, stands the triangular edifice of the archaeology museum. Two sides of the building face the water, the third overlooks the oval of a Roman forum that was discovered here. The unusual shape of the museum was dictated by both urban planning and functional requirements.
The museum is divided into research, exhibition and education zones. There are two long blocks with offices, laboratories and classrooms, and these frame the actual museum. The two exhibition areas for permanent and temporary presentations are located on the same level and extend the full height of this two-storey building. The architect conceived a long and a short route though the exhibition; both tours circulate around a central staircase and terminate at the viewing platform.
In situ concrete columns extending over both storeys, some of which also support galleries projecting into the space, plus a northlight roof and walls of fair-face concrete with a wide formwork panel grid characterise the interior atmosphere of the museum. The exhibition hall, making use of natural stone, is flooded with light from the overhead windows and large expanses of glazing. Externally, the blue enamel facade panels dominate the architecture.

Archaeology museum, Arles

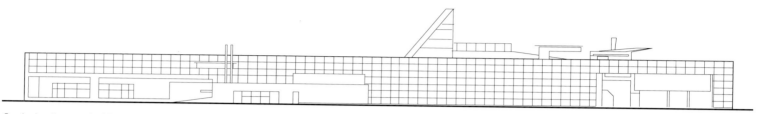

South elevation   scale 1:750

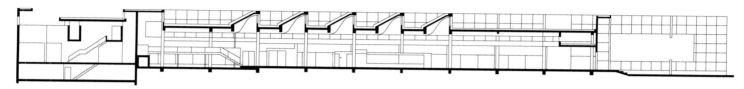

Section   scale 1:750

Example 7

Plan scale 1:1200

1 Entrance
2 Foyer
3 Permanent exhibition
4 Gallery
5 Temporary exhibitions
6 Workshops
7 Deliveries
8 Store

Detail scale 1:100

02 In situ concrete
20 Aluminium sheet, enamelled
32 Glass
40 Thermal insulation
90 Roof construction:
 concrete flags
 synthetic roofing felt
 rigid thermal insulation
 vapour barrier
 reinforced concrete slab

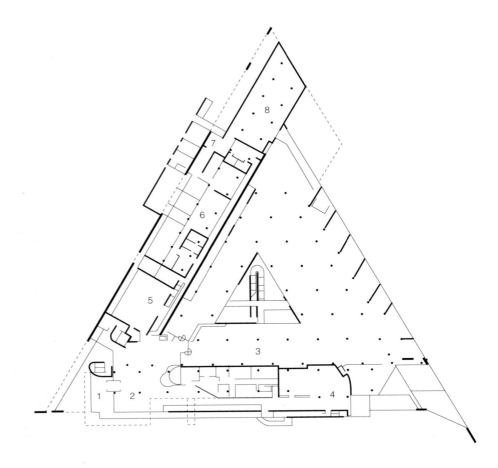

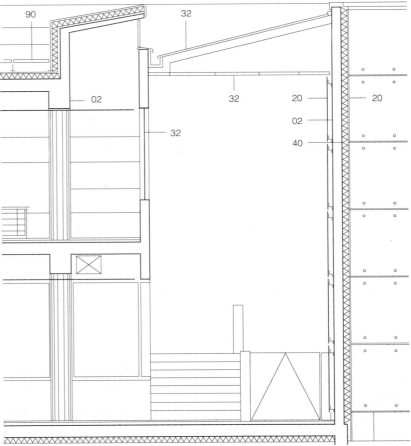

Archaeology museum, Arles

Example 8

## School, Paspels, Switzerland

1999

Architect:
Valerio Olgiati, Zurich
Assistants:
Iris Dätwyler, Gaudenz Zindel,
Raphael Zuber

Structural engineer:
Gebhard Decasper, Chur

Location plan
scale 1:5000
Section aa
Plans
scale 1:500

1 Cloakroom
2 Multipurpose room
3 Staff room
4 Classroom

The little village of Paspels, about 20 km south of Chur, has enriched the Swiss architectural landscape with its new school building. Here in an Alpine setting, this is a remarkable example of contemporary building design. The viewer does not suspect that the client of this school building, with its amazingly simple, unequivocal design by Valerio Olgiati, is a tiny mountain village with a population of just 400. The architect has created a building that clearly stands out from its counterparts in neighbouring communities. Apart from its very economical footprint, it employs only materials familiar to the local inhabitants from the local agricultural structures. The three-storey fair-face concrete building rises like an outcrop of rock from the surrounding meadow. It is only the fenestration that discloses the fact that this is a building for use by humans. The varying treatment of circulation and working zones within the building is also reflected in the facade. The wide windows of the wood-lined classrooms are set in deep reveals, flush with the inner face. Each window projects a different but clearly defined extract of the surroundings into the classroom. Windows positioned flush with the outer face are located at all places where light is admitted to the corridors and stairwells constructed exclusively in concrete. The cruciform access layout enables daylight to enter from all directions, which leads to ever-changing impressions of the interior over the course of a day. In structural terms, the concrete internal walls, cast monolithically with the floor slabs, form an independent loadbearing frame, which is connected to the enclosing fair-face concrete facade only by means of individual shear pins. All the junctions between walls and floors/ceilings are clearly delineated by way of deep rebates.

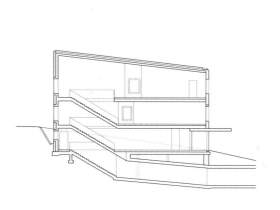

Section aa

Ground floor

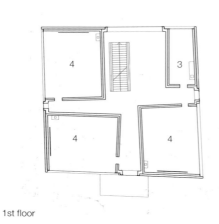

1st floor

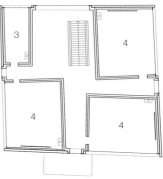

2nd floor

School, Paspels

Detail
scale 1:20

02  In situ reinforced concrete
12  Steel
13  Aluminium
40  Thermal insulation, 120 mm
53  Sealing compound, resistant to UV radiation
87  Window element
90  Roof construction:
    copper roof covering
    bitumenised felt
    30 mm timber boarding
    100 mm battens / air space
    roofing felt, bonded over entire surface
    2 layers of 100 mm thermal insualtion
    vapour barrier
    260 mm reinforced concrete slab
91  Floor construction:
    20 mm granolithic concrete
    80 mm cement screed incorporating pipes for
    underfloor heating
    separating layer
    40 mm impact sound insulation
    280 mm reinforced concrete fair-face slab

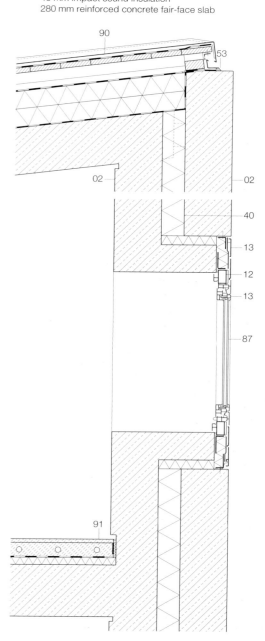

Example 9

**Training centre, Thun, Switzerland**

1991

Architects:
Atelier 5, Bern

Structural engineers:
Finger & Fuchs, Thun

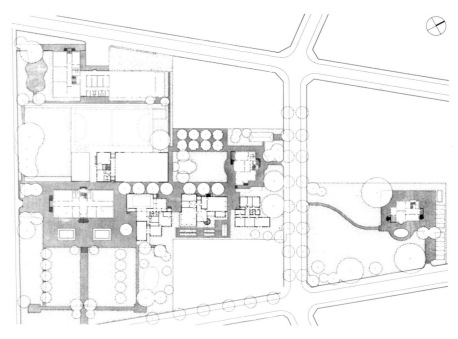

Location plan   scale 1:2000

The Thun Training Centre is a school, whose operations forced it to expand to accommodate two large old houses nearby. The former residential premises had been appropriated because the necessary funds to adapt the old houses to suit their new function had not been available for a long time. The plan for expanding the training centre provided for retaining and extending precisely this principle of dividing up the school facilities among various larger and smaller buildings.
Basing the buildings on the architecture of the period around 1900 or 1920-30 was never an issue. Instead, precisely because of this integration within an existing setting, it was decided that these new buildings should clearly reveal their 1980s heritage. But they should embody an exemplary simplicity in terms of form, material and colour, both inside and outside.
The reinforced concrete is left untreated as fair-face concrete. Internally, the 60 mm layer of vapour-proof insulation to the external walls is covered with plasterboard. Steel balustrades and generously dimensioned areas of glazing relieve the in situ concrete facade, and unbroken expanses of concrete are broken up with planting.
The architects have succeeded in using reinforced concrete to highlight the simplicity. However, the extension to Thun Training Centre is a practical solution. An internally insulated design for buildings used only temporarily appears sensible.

Training centre, Thun

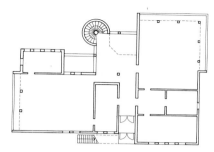

Plan scale 1:500

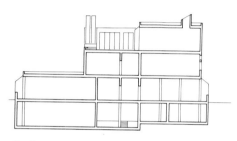

Section scale 1:500

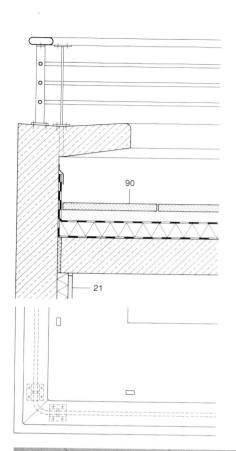

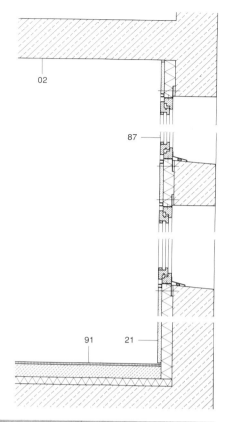

Details scale 1:20

02  Reinforced concrete
21  Plasterboard, 20 mm
87  Window element
90  Terrace construction:
    paving slabs in mortar bed
    waterproofing
    thermal insulation
    vapour barrier
91  Floor construction:
    floor covering
    cement screed
    thermal insulation

### Conference pavilion, Weil am Rhein, Germany

1993

Architects:
Tadao Ando & Associates, Osaka

Detailed design and site management:
Günter Pfeifer, Roland Mayer, Lörrach
Project Manager: Peter M. Bährle
Assistant: Caroline Reich

Structural engineer:
Johannes C. Schuhmacher, Bad Krozingen

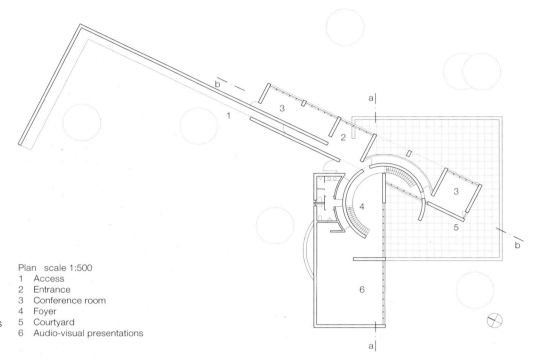

Plan scale 1:500
1 Access
2 Entrance
3 Conference room
4 Foyer
5 Courtyard
6 Audio-visual presentations

This two-storey conference and training centre for an office furniture manufacturer was built in an open field among cherry trees on the edge of the company's site. Contrasting with the neighbouring deconstructionist museum, Ando's architecture is characterised by its austerity and clarity. Visitors approach the pavilion along a wall laid to an L-shaped plan and enter a narrow rectangular tract that represents a continuation of the wall. This long access route is reminiscent of a Japanese path of meditation. The building's layout is ingenious: two rectilinear volumes intersecting at an angle, penetrated by a hollow cylindrical element at their point of intersection. This latter forms the foyer with the stairs.

The concrete-paved courtyard lies below the surrounding ground level. This means that, from the encircling field, only one storey of the pavilion is visible and the cherry trees tower above the building.

The rooms, of different sizes and spread over two levels, comprise large and small conference rooms, accommodation for guests, a lobby and library.

Tadao Ando again makes full use of fair-face concrete in this building in order to emphasise the restrained rigour of his architecture. The walls were constructed using formwork panels matching the size of Japanese tatami mats (910 x 1820 mm). The double-leaf walls were necessary to comply with German thermal insulation standards, and this led to walls up to 400 mm thick. Only in the case of the external walls against the ground was a single-leaf wall with perimeter insulation permitted, or double-leaf reinforced concrete walls with cavity insulation. Here again, the internal faces are fair-face concrete.

The external walls finished in fair-face quality both sides consist of two leaves of reinforced concrete with cavity insulation; the inner leaf is loadbearing. The accuracy with which the junctions of three different concreting levels have been constructed is a magnificent example of in situ fair-face concrete.

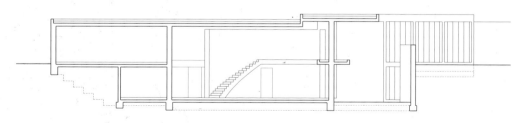

Section aa   scale 1:300

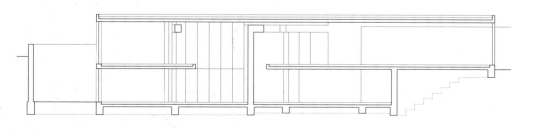

Section bb   scale 1:300

Conference pavilion, Weil am Rhein

Example 10

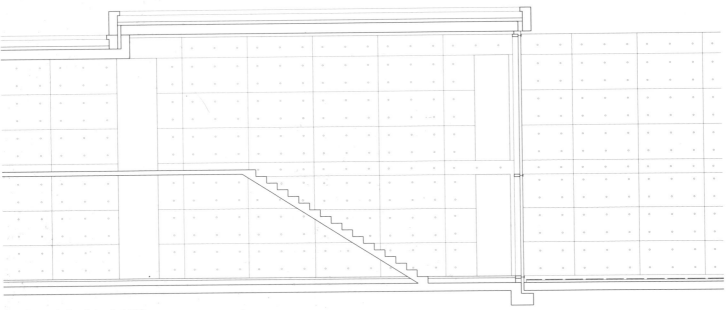

Development of wall  scale 1:100

| 02 | In situ reinforced concrete | 44 | Bitumen coating |
| 03 | Precast concrete element | 50 | Vapour barrier |
| 14 | Timber | 66 | Retaining anchor |
| 29 | Screed | 87 | Window element |
| 40 | Perimeter insulation | 88 | Gravel |
| 41 | Closed-cell thermal insulation | 92 | Water bar |
| 43 | Bitumenised roofing felt | | |

Detail  scale 1:5

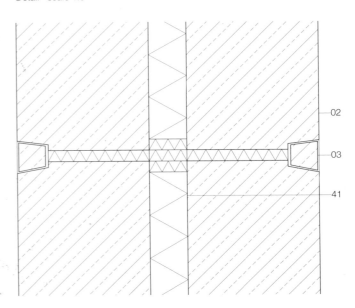

# Conference pavilion, Weil am Rhein

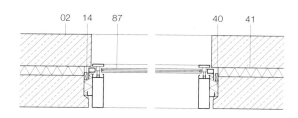

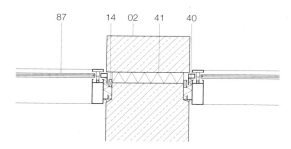

Horizontal sections
Vertical section  scale 1:20

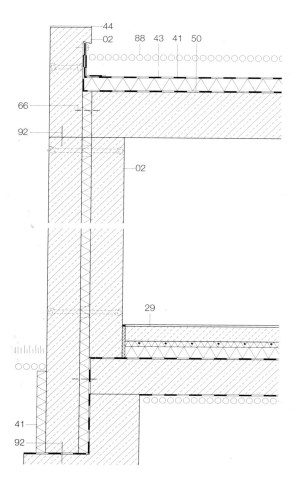

Example 11

**Mixed office and residential block, Kassel, Germany**

1999

Architect:
Alexander Reichel, Kassel/Munich
Assistants:
Johanna Reichel-Vossen, Stefan Seibert,
Caroline Ossenberg-Engels, Elke Radloff

Structural engineers:
Hobein, Kleinhans, Marx,
Hochtief AG, Kassel

Plans
Ground floor · Upper floor
Section   scale 1:500

1 Entrance
2 Bicycle shed
3 Garage
4 Office
5 Storage
6 Apartment

Location plan   scale 1:3000

The modular principle of this town house is based on the brief for an architectural competition. The task was to design a building type for the eight different plots of this residential development on the outskirts of Kassel's Unterneustadt. Starting with a column grid of about 3.0 x 3.3 m, this town house can be extended or modified to suit different uses and topographical conditions. One prototype was built as a straightforward cube measuring 13.52 x 12.30 x 15.40 m; the other seven town houses were the responsibility of other prize winners. The building is set amid idyllic park-like surroundings not far from the River Fulda with its boat moorings and historic suspension bridge.
A southern flair is obtained by the use of full-height glazing to the living rooms, from where occupants enjoy a view over the pleasant surroundings. Also through the untreated larch wood infill panels between the reinforced concrete frame members. The structure and the solid sections of the external walls are clad with precast glass fibre-reinforced concrete panels; this artifice helps to indicate the different internal uses. To reinforce the character of a detached villa, ancillary buildings were omitted and eight parking spaces accommodated

Mixed office and residential block, Kassel

within the building itself by means of a mechanical car stacking system. A maisonette with a floor area of 120 m² plus a low-level yard occupies the semi-basement and ground floor. This can be used as office or apartment. The accommodation above can be divided to create two- or three-room apartments (plus kitchens and bathrooms). The top two floors are again maisonettes, and have a generous rooftop patio overlooking the river.

In order to achieve the desired variety in the facade and the necessary structural clarity, the building was divided into various systems: the loadbearing construction of reinforced concrete frame with precast concrete plank floors and walls, the timber framing elements and the cladding to the structural members. These individual systems are designed to remain visible in the facade and hence structure the building's appearance. However, leaving a concrete structure exposed in Germany creates a building science problem. Owing to its good thermal conductivity, concrete must be insulated to prevent energy losses and damage caused by moisture. The concrete loadbearing structure was therefore clad with insulated precast elements. Glass fibre-reinforced units just 30 mm thick were chosen. Besides their slim design and low weight, they are also easy to erect and work. The material and pattern of the joints of these accurate panels convey the structural rhythm of the concrete frame to the observer. Glass fibre-reinforced concrete is normally used as permanent formwork, as textured formwork or for rebuilding reliefs and cornices on older buildings. It consists of fine-aggregate concrete, aggregate size <4 mm, to which the alkali-resistant glass fibres, approx. 2-4 mm long, are added. These act as tension and anti-crack reinforcement. Each precast component is coated with a hydrophobic fluid at the works to produce a consistent, water-repellent outer surface. This gives the surface a "milky" shade, which lends the material a vibrant quality.

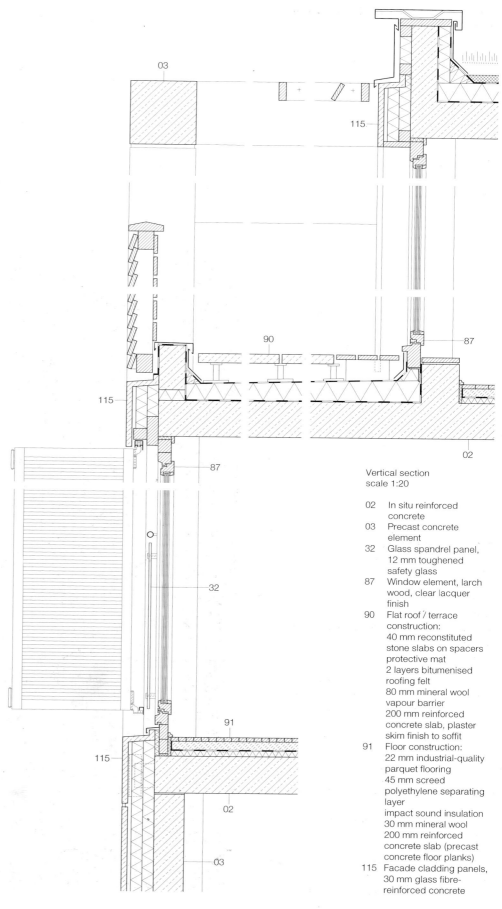

Vertical section
scale 1:20

02  In situ reinforced concrete
03  Precast concrete element
32  Glass spandrel panel, 12 mm toughened safety glass
87  Window element, larch wood, clear lacquer finish
90  Flat roof / terrace construction:
    40 mm reconstituted stone slabs on spacers
    protective mat
    2 layers bitumenised roofing felt
    80 mm mineral wool
    vapour barrier
    200 mm reinforced concrete slab, plaster skim finish to soffit
91  Floor construction:
    22 mm industrial-quality parquet flooring
    45 mm screed
    polyethylene separating layer
    impact sound insulation
    30 mm mineral wool
    200 mm reinforced concrete slab (precast concrete floor planks)
115 Facade cladding panels, 30 mm glass fibre-reinforced concrete

Example 12

**Health spa treatment centre, Bad Salzuflen, Germany**

1988

Architects:
Auer & Weber, Stuttgart

Structural engineers:
Schlaich Bergermann & Partner, Stuttgart

Location plan   scale 1:5000

The centre for guests of this health spa is located on the edge of the town centre. Besides the spa water spring, first tapped in the 1950s, it accommodates the spacious reception hall, offices, treatment rooms, shops, cafés, workshops and activity rooms. Situated at the transition point between the town and the park on the River Salze, the complex – together with the concert hall – forms the focal point for guests "taking the waters".
The centre integrates sensitively into its urban surroundings, but without adopting the existing forms. Although the park has clear boundaries and reference points, emphasised by the paving layout, there is no sudden transition between old and new. Especially striking in the new development are the smooth-finish concrete "mushroom" canopies. These continue outside as a stylised tree group.
The rooflights, created by the fact that the one- and two-storey-high concrete canopies only make contact tangentially, plus the planting to the flat roofs of certain canopies, give the hall an airy but nonetheless protective ambience. The topography of the land has been skilfully incorporated into the architecture of the construction. The planting, to the roofs in particular, plays an especially important role in this interweaving of landscape and structure. Concrete members were kept as slim as possible and their appearance improved through the use of light-coloured aggregates.

# Health spa treatment centre, Bad Salzuflen

1 Hall
2 Cashier
3 Reception
4 Exhibitions, presentations
5 Café
6 Terrace
7 Shop
8 Health spa treatment and tourist information
9 Computer room
10 Offices
11 Spa water spring

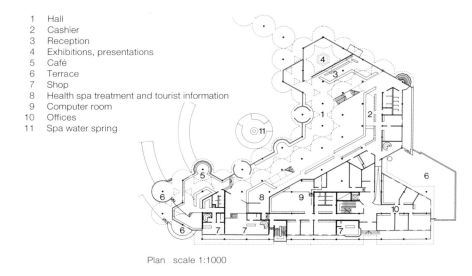

Plan   scale 1:1000

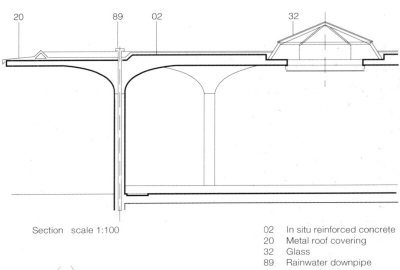

Section   scale 1:100

02  In situ reinforced concrete
20  Metal roof covering
32  Glass
89  Rainwater downpipe

Section   scale 1:100

Example 13

**Central university building, Girona, Spain**

1993

Architects:
Josep Fuses, Joan Maria Viader, Girona

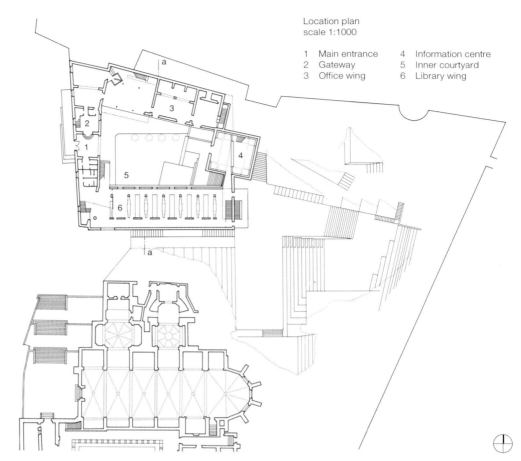

Location plan
scale 1:1000

1 Main entrance
2 Gateway
3 Office wing
4 Information centre
5 Inner courtyard
6 Library wing

"Les Aligues" was built in the 16th century and, in its heyday, was the most noteworthy public building in the town. It served as a university building into the 17th century and was later used for various purposes until it was finally vacated and left standing empty. Only the principal facade overlooking the Plaza de San Domènec and the facade facing the old church still bore their original appearance when it was decided to restore the ruin. The L-shaped layout of the old building has been restored in order to accommodate the university's most important administrative functions. The chapel now serves as an information centre for students, while the inner courtyard to the south is bounded by a new building housing further university facilities. The inner courtyard linking the different parts of the complex has the character of a public square. The inclusion of a number of concrete stairs illustrates the different levels of the various buildings. The stairs also enable various old walls and arches to be preserved, but do force the new construction to be separated from the main building.
The gable wall of the cube-like chapel was demolished and replaced by a glass storey, to admit daylight, and a concrete roof. The four sides of the "as struck" finish concrete roof curve inwards in order to direct light into the interior. The interior of the old wing is designed with consideration for the preserved elements. The use of fair-face concrete plays a key part in the clear separation between old and new. Stainless steel, welded, polished and painted metal plates and coloured inserts, attached directly to the stone or embedded in the rendering, have also been employed.

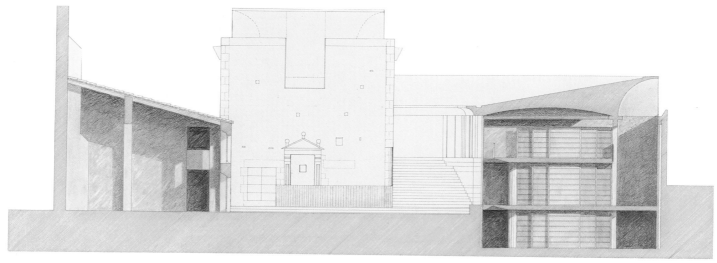

Section aa

Example 14

**Two-tier sports hall, Berlin-Charlottenburg, Germany**

1988

Architects:
Hinrich & Inken Baller, Berlin

Structural engineer:
Gerhard Pichler, Berlin

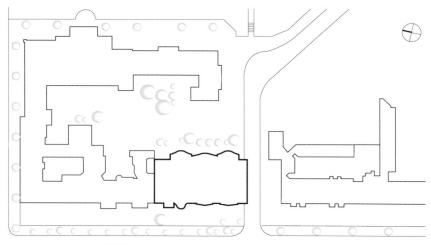

Location plan   scale 1:2000

The piggyback sports facility forced by the lack of space in densely developed conurbations is not new. Two sports halls, one above the other, were erected on Berlin's Schlossstrasse and integrated in the block developments of a district dominated by the architecture of the late 19th century.
A staircase and adjoining "open walkway" link the parking area at ground level with the two sports hall levels. Ancillary rooms are positioned at the side of each floor on the north gable. The equipment rooms project from the rear of the building, providing relief. With its large and small gables and balconies, the facade facing the street is akin to a traditional residential development. But the metal-covered barrel vault roof, large areas of glazing to the walls and the projecting helical staircase indicate the special utilisation of the building. Besides the posts and rails of the timber-and-glass facade, profiled columns and beams of reinforced concrete in fair-face quality enrich the street-side facade. Bracing to resist horizontal loads in the transverse direction is provided by the stiffened north gable, which also acts as a fire wall. Further transverse bracing to this structure without any joints is provided by the steel wind girder in the south gable. The crosswalls of the staircases also contribute to stability. In the longitudinal direction the walls to the stairs and the columns they restrain provide the necessary stability. Foundations, columns, walls, floor slabs and composite columns are cast in concrete grades B 25 to B 45, the composite beams in the facade in lightweight concrete grade LB 25.
Slab soffits and walls have an "as struck" board finish. The tapering columns and wind girders are smooth. Prefabricated formwork elements were used for the floor slabs to the large sports halls.

Two-tier sports hall, Berlin-Charlottenburg

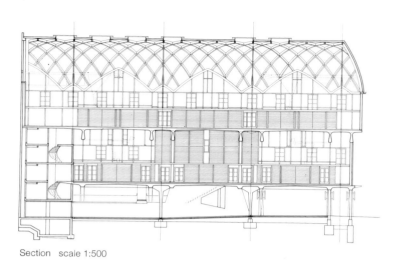

Section scale 1:500

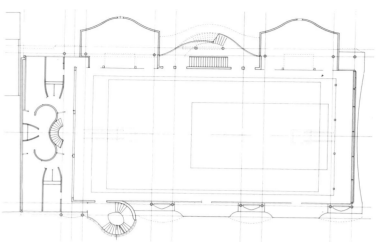

Plan scale 1:500

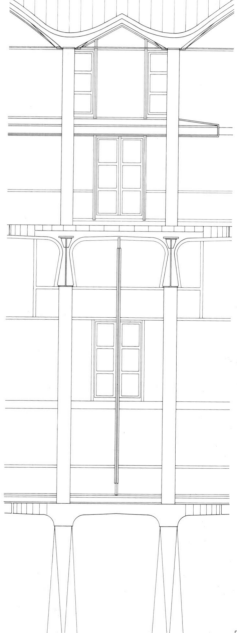

Part-elevation
scale 1:100

Example 15

**Detached house, Berlin, Germany**

1999

Architects:
OIKOS, Peter Herrle and
Werner Stoll, Berlin
Assistants:
Amun Bieser, Tobias Schmachtel

Structural engineers:
Wilhelm & Wulle, Stuttgart

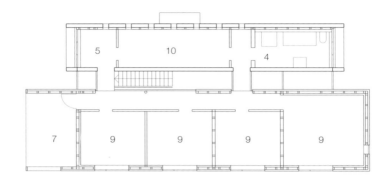

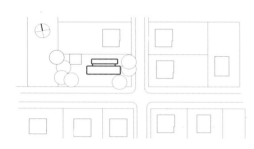

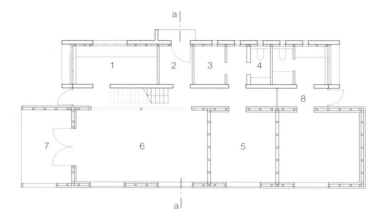

Location plan
scale 1:2000
Plans
scale 1:200
1 Kitchen
2 Lobby
3 Incoming services
4 Bathroom
5 Study
6 Living room
7 Terrace
8 Separate apartment
9 Room
10 Dressing room

This private house on the outskirts of Berlin consists of two halves joined by a glass link. There is a separate apartment on the ground floor with its own entrance at one end of the glass link. The narrow concrete block on the north side contains the utility rooms and the entrance, in the long north facade. The living quarters are situated on the south side, in the timber block with pitched roof.
The ends of the concrete block are closed off with inset timber facades. Their horizontal battens of untreated larch wood match the timber facade of the southern block. This facade consists of prefabricated timber-frame elements 8 m high, clad both inside and outside with the horizontal larch wood battens. The timber battens also enclose the two terraces at the western end of the southern block, and are omitted on the narrow side facing the garden only.
Concrete and timber dominate the interior. The steps of the fair-face concrete stair cantilever from the face of the – likewise – fair-face concrete wall to the northern block. The stair leads to a gallery on the upper floor, which permits access to the accommodation in the southern block.
The unequivocal external architecture of the house is reflected in the interior.

Detached house, Berlin

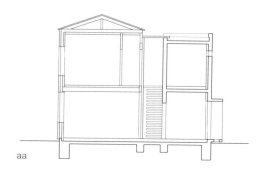

aa

Section aa
scale 1:200

Detail
scale 1:20
- 02 Reinforced concrete, 200mm
- 21 Plasterboard, 12.5 mm
- 29 Screed
- 32 Glass
- 40 Thermal insulation, 100 mm
- 43 Bitumenised felt (waterproofing)
- 48 Separating layer
- 50 Vapour barrier
- 87 Window element
- 90 Roof construction: synthetic waterproofing layer
  separating layer
  100 mm polyurethane thermal insulation
  vapour barrier
  160 mm fair-face concrete
- 91 Floor construction:
  15 mm bamboo parquet flooring
  2 layers of 19 mm composition board
  separating layer
  100 mm polyurethane thermal insulation
  vapour barrier
  160 mm fair-face concrete

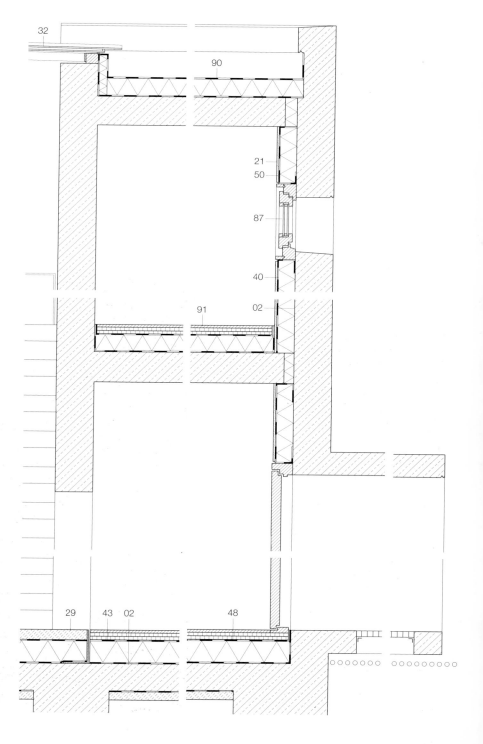

Example 16

**Tennis courts, Bellinzona, Switzerland**

1986

Architect:
Aurelio Galfetti, Bellinzona
Assistants:
Walter Büchler, Piero Ceresa

Location plan   scale 1:4000

The local tennis courts form part of a sports centre which includes an indoor pool and an ice hockey rink. They are situated near the river on the edge of a residential district.
This amenity is divided into two symmetrical halves and is bounded on the north-east side by a long wall extending over two storeys. The entrance is located in the middle of this wall. Entering the facility is reminiscent of entering a town through its town gate. The ancillary rooms are placed directly adjacent to this wall on two levels. Those on the ground floor are fully glazed on the side facing the tennis courts. The left-hand wing contains offices and services, the right-hand wing a restaurant with kitchen and bar.
The walls are of fair-face concrete. Their ornamental texture was achieved by placing horizontal battens in the formwork. This leads to impressive effects of light and shadow on the surface of the wall, even though the recess is only 25 mm deep. At the top of the wall there is a barrel vault roof of transparent plastic supported by delicate steel framing.

Tennis courts, Bellinzona

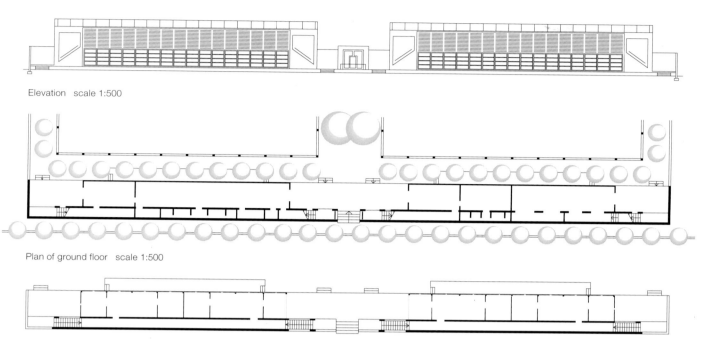

Elevation scale 1:500

Plan of ground floor scale 1:500

Plan of upper floor scale 1:500

Detail scale 1:20

| 02 | Reinforced concrete | 33 | Transparent plastic | | insulation |
| 12 | Steel tie | 40 | Thermal insulation | 43 | Bitumenised roofing felt |
| 22 | Board of timber derivative | 41 | Rigid thermal | 88 | Gravel |

Example 17

**Rhinoceros and tapir house, Munich**

1992

Architects:
Kochta & Lechner, Munich
Assistants:
Stefan Endl-Storek (Project Manager),
Reinhard Nägele

Structural engineer:
Dieter Herrschmann, Munich

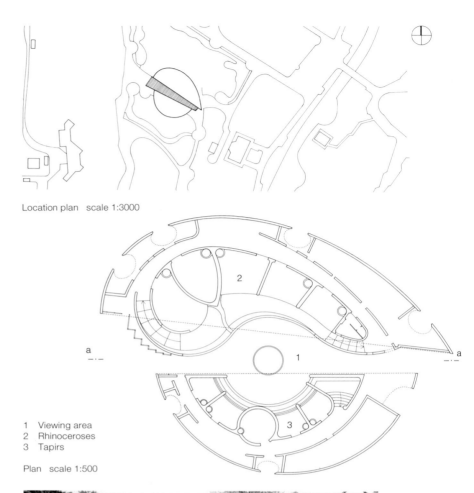

Location plan   scale 1:3000

1  Viewing area
2  Rhinoceroses
3  Tapirs

Plan   scale 1:500

The rhinoceros and tapir house in Munich's Hellabrunn Zoo is integrated into the forest and flood plain landscape. The shallow, rounded lines of the building accentuate its synthesis within the surrounding animal pen.
The roof consists of two half-shells of different lengths (34 and 47 m) facing each other at an angle of 7°. Each shell covers five pens for tapirs and rhinoceroses. A tapering glass lantern light (span: 3-10 m) creates a weatherproof, transparent link between the two shells. As a freely modelled three-dimensional form, the geometric plan shape, based on successive spiral and circular segments, is intended to resemble the natural habitat of the animals. The contour lines of the rhinoceros half are concentric ellipses, those of the tapir half concentric circles.
The entire formwork for the tapir shell, circular on plan, was made from prefabricated segments approx. 8.50 m long along the circumference and assembled on site. Formwork to the top surface was required in the steep areas exceeding about 30°. The in situ shell 180-400 mm thick is constructed as a warm deck, with vapour barrier, mineral wool and mesh-reinforced plastic sheeting on the outside. The pre-assembled roof covering is additionally fixed with individual screw fixings at the junctions at ridge and base.
The internal walls were concreted and compacted from outside through access openings in the half-shells.
The two half-shells are founded on strip footings whose ends are connected by means of a reinforced concrete tie.
In order to avoid shrinkage cracks, concrete grade B 35 with aggregate up to 32 mm dia. was employed.
The rough, exposed concrete surfaces are intended to resemble the inside of a cave. They were created by treating the surface with a retarder and brushing the new concrete immediately after striking the formwork. In addition, sprayed concrete was used on the inner face.

# Rhinoceros and tapir house, Munich

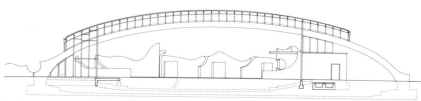

Section aa  scale 1:500

South-east elevation  scale 1:500

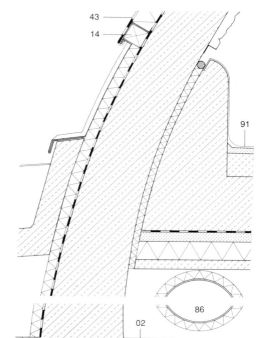

Details  scale 1:20
01  Sprayed concrete
02  In situ reinforced concrete
14  Timber
20  Sheet metal
29  Screed
30  Plaster
41  Closed-cell thermal insulation
43  Bitumenised roofing felt
50  Vapour barrier
86  Main ventilation duct
87  Sliding steel door
88  Gravel
91  Floor construction:
    mastic asphalt
    screed
    concrete laid to falls
    reinforced concrete slab with
    underfloor heating pipes
    separating layer
    protective screed
    thermal insulation
    blinding layer

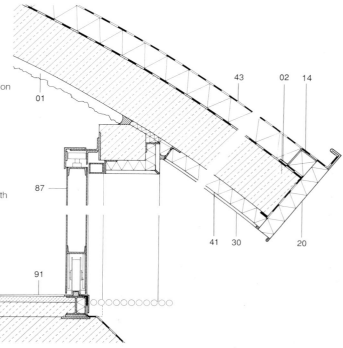

237

Example 18

**Works forum, cement works, Dotternhausen, Germany**

1990

Architects:
Böbel & Frey, Göppingen

Structural engineer:
Ulrich Otto, Stuttgart

For more than 15 years, fossils have been uncovered in Dotternhausen while mining oil shale for use in cement production. In order to exhibit this now quite extensive collection of fossils near their place of discovery, the cement works decided to add a fossils museum to the new works entrance it was planning. An information pavilion, training facilities, a computer centre and offices would be included.
The idea behind the project was to create an imposing entrance for the plant, position the offices closer to the existing administration block and locate the museum closer to the perimeter of the site. In line with its functions, the complex is divided into three levels:
Level 1 on the base of the test pit is constructed in in situ concrete and comprises the museum, the open-air preparation yard and the presentations room.
Level 2 contains the highly effective new works entrance – with foyer, preparation room, training facilities – and computer centre.
Level 3 is a bridge arrangement employing a steel-reinforced concrete composite construction; it is supported on four columns and contains the offices. The roof shell of precast lightweight concrete elements forms the compression zone, the steel girders the tension zone. The rotunda that penetrates the roof of the hall houses the presentations room with gallery. The air-conditioning plant is accommodated above the ceiling. The "shale quarry" is also part of the museum, with the formation of a fossil presented in its natural environment to the visitor in an attractive, understandable manner. The close link between museum and test pit fulfils the museum's goal of an educational and tangible presentation. The route taken by employees every day, as well as visitors, therefore becomes a real experience of an unusual nature. To compensate for the areas of landscaping now claimed by the new complex, planted areas have been included on parts of the flat roofs. The sparingly planted shale tip, which resembles the crannies in a "real" shale quarry, has been returned to nature and is now a valuable biotope. The use of in situ concrete, precast lightweight concrete elements and coloured concrete bricks has been realised sensibly and does justice to the materials. Furthermore, this also displays the range of products manufactured at the plant.

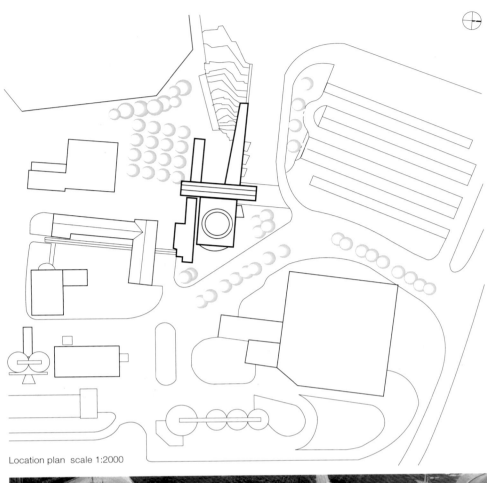

Location plan  scale 1:2000

238

Works forum, cement works, Dotternhausen

1 Exhibition
2 Foyer
3 Hall
4 Presentations
5 Walkway
6 Training
7 Computer room
8 Preparation
9 Meeting room
10 Courtyard

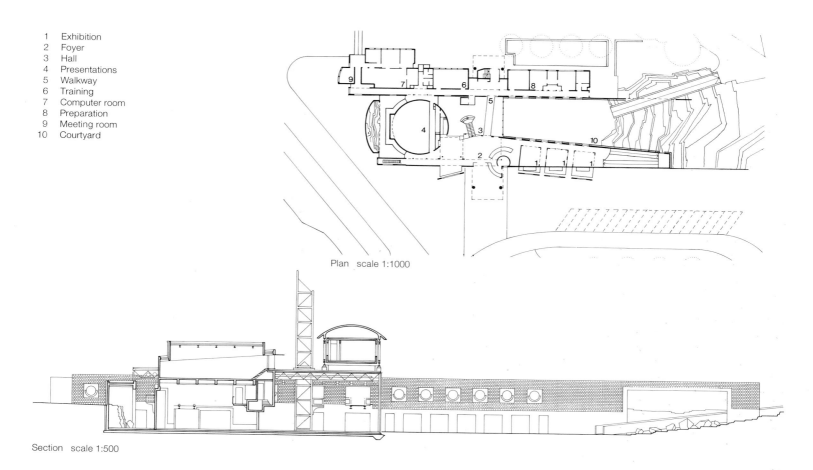

Plan  scale 1:1000

Section  scale 1:500

Detail  scale 1:50

04 Lightweight concrete
12 Steel stanchion
87 Window element
89 Rainwater downpipe
90 Roof construction:
   roof covering
   thermal insulation
   vapour barrier
   reinforced concrete

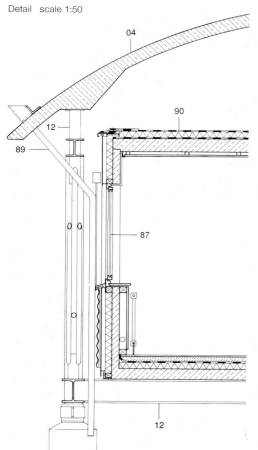

239

Example 19

**Multistorey car park, Hamburg, Germany**

1990

Architects:
von Gerkan, Marg & Partner, Hamburg
Project manager: Karsten Brauer
Assistant: Klaus Hoyer

Structural engineers:
Schwarz & Weber, Hamburg

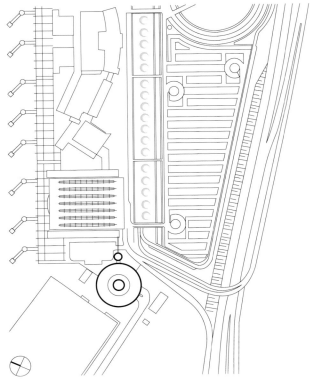

Location plan   scale 1:5000

This circular multistorey car park structure has about 800 spaces and a circular stair tower linked by means of delicate steel bridges. Its position and architecture allow it to assume the role of an urban fulcrum between the various airport buildings and facilities.
The essential features of the car park are the division of the storeys by the contrast between open areas with thin concrete edge and tubular balustrading, as well as a lattice facade suspended in front of the parking decks. The primary structure is a reinforced concrete construction with circular parking decks measuring 61 m in diameter. Flat slabs are used for the floors, with annular haunches. They are between 250 and 450 mm deep depending on structural requirements. Loads are transferred inwards via a ring of 20 columns measuring 500 x 1200 mm (the narrow side is formed as a semicircle), and outwards via a cylindrical wall with a thickness of 450 mm. The access ramps are constructed as slabs cantilevering from the inside face of this reinforced concrete cylinder. They are likewise between 250 and 450 mm deep. The services level is a concrete slab on the cylindrical wall and is stiffened by ribs 250-750 mm deep and annular upstand beams. The parking decks are inclined 2% to the inside, the ramps 3%. These slopes are always transverse to the direction of travel in order to ease driving in a circle. The stair tower is divided into a semicircular reinforced concrete shaft for the lifts and electrics and a reinforced concrete frame for the stairs. The latter is stiffened by the floor slabs. The infill panels are made of glass blocks set in storey-height steel frames. All concrete surfaces have a smooth finish and are defined by cast-in blocks. The circular geometry favoured the reuse of formwork sections. The positions of formwork ties were planned in order that they could remain visible in the final surface. Steel balustrading and bridges as well as the infill panels of steel frame with glass blocks set the architectural accent.

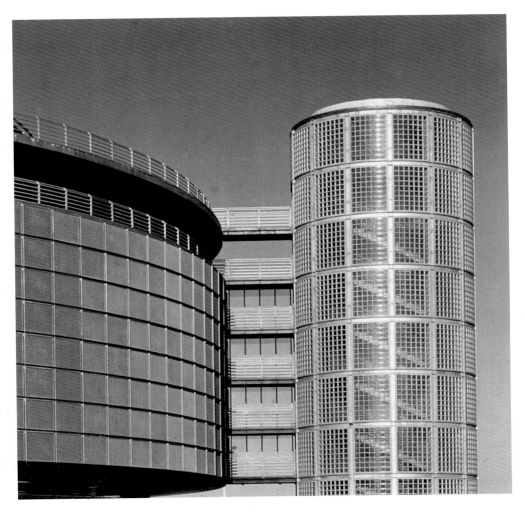

# Multistorey car park, Hamburg

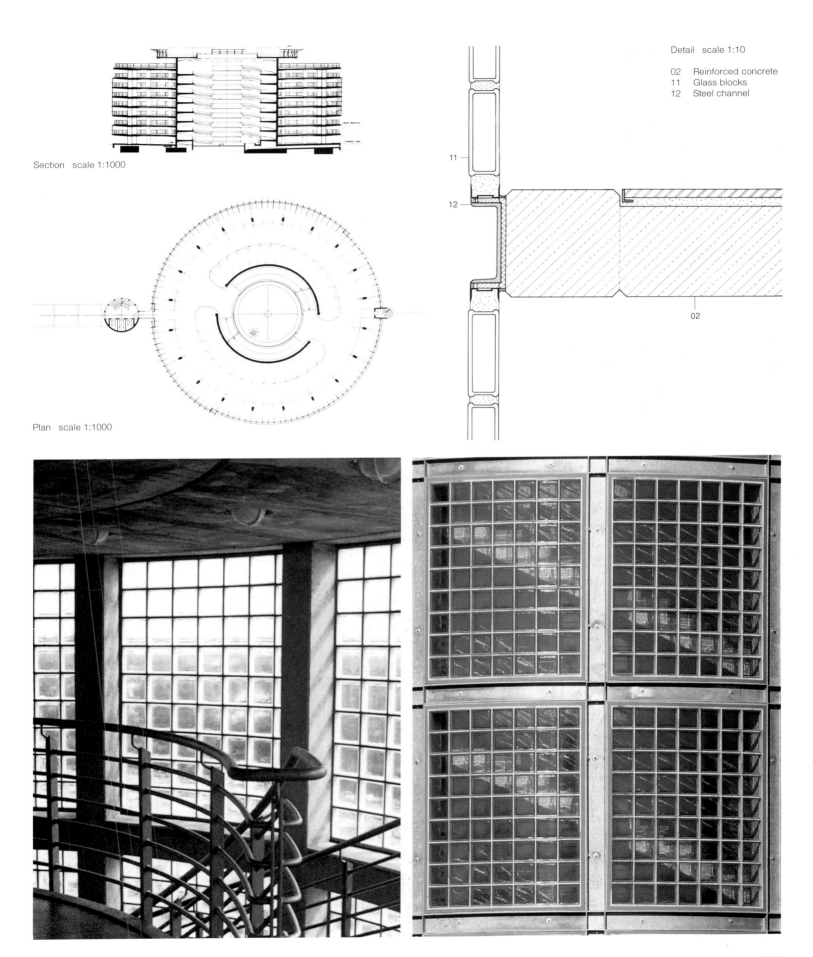

Section scale 1:1000

Plan scale 1:1000

Detail scale 1:10

02 Reinforced concrete
11 Glass blocks
12 Steel channel

Example 20

**Car parking facility, Paderborn, Germany**

1993

Architects:
Roland Dorn, Eva Matern, Paderborn/Cologne
Assistants: Bernhard Gieselmann,
Klaus Hovestadt

Structural engineers:
Thormälen & Peuckert,
Paderborn/Aachen
with Domostatik, Zurich

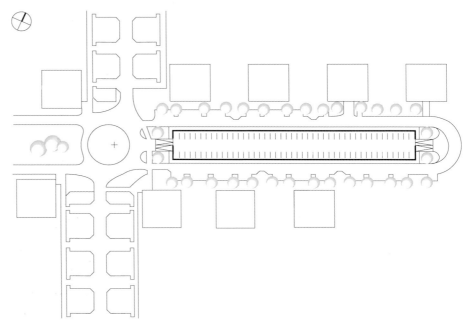
Location plan   scale 1:2000

This prototype for a two-storey car parking facility for 207 cars was developed according to the principles of simple and economical construction. However, it also takes account of landscaping aspects, particularly in terms of ecological processes.
A curving monolithic concrete slab extends – following the topography – with a fall of approx. 4% over a length of 127.40 m down the slope. Due to the simultaneous longitudinal and transverse falls, rainwater rapidly runs off into the lateral drainage system and is distributed over 40 precast concrete planting tubs arranged around the perimeter embankments. Tall trellises ensure vegetation related to the architecture. They highlight the column grid three parking spaces wide and define the extent of the upper deck. The arrangement of the formwork, made from graded individual boards placed lengthwise, reinforces the elongated overall impression of the construction. Cement paste was spread over the formwork prior to concreting in order to achieve a consistent surface finish. The light colour was obtained not by using pigments but exclusively by choosing the right concrete mix. The prestressed concrete flat slab forming the upper deck is supported on slender columns and spans without joints or beams.
The extremely small cross-section of the upper deck matches the bending moment diagram for a single-span beam (l = 10.00 m) with two cantilevers (l = 3.25 m): Its depth decreases continuously from 320 to 260 mm. Bracing is provided by the fixed-base columns – all columns assisting in the transverse direction but only the four central columns in the longitudinal direction (core effect). The longitudinal expansion from the middle to the ends is accommodated with the help of carriageway movement joints at the ramps. Single strands protected against corrosion provide the prestress in both the transverse and longitudinal directions. The compressive force generated by the prestress limits the width of cracks and thus assists the impermeable construction without the need for additional waterproofing measures.

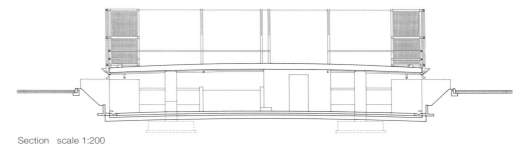
Section   scale 1:200

Car parking facility, Paderborn

A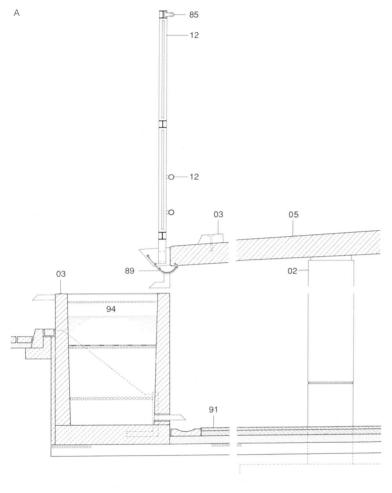

B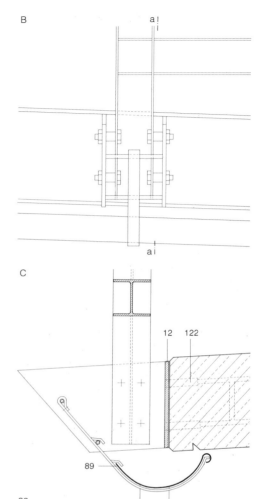

A   Detail of edge of slab showing drainage and planting tub
Section   scale 1:50

Detail of trellis fixing and gutter
scale 1:10

B   Elevation
C   Section

02   In situ concrete
03   Precast concrete element
05   Prestressed concrete
12   Steel
85   Lighting
89   Rainwater gutter
91   Floor construction:
     30 mm bituminous surfacing
     100 mm loadbearing layer
     100 mm sub-base of chippings
     ballast subgrade
94   Planting
122  Prestressing tendon

Reinforcement over column head

## Crematorium, Berlin, Germany

1998

Architects:
Axel Schultes Architekten, Berlin
Frank, Schultes, Witt
Design:
Axel Schultes, Charlotte Witt
Project managers:
Margret Kister, Christoff Witt

Structural engineers:
GSE Saar Enseleit & Partner, Berlin
IDL, Berlin

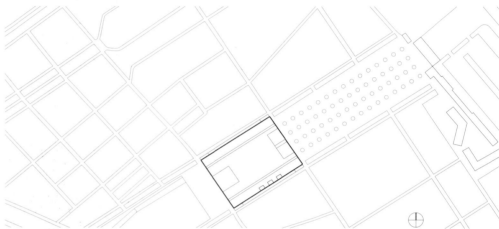

The new crematorium for Germany's capital city has been erected in the Treptow district. This monolithic fair-face concrete structure follows in the tradition of Gunnar Asplund in Stockholm and Fritz Schumacher in Hamburg-Ohlsdorf. The objective of such a building must be to provide the grieving friends and relatives with a worthy framework for the funeral service. Guests enter the crematorium grounds via an existing gatehouse and approach the symmetrical front elevation of the building across a long open area interspersed with rows of trees. Three covered forecourts cut deep into the building and lead to the entrance to the spacious vestibule. This square hall with its 29 towering circular columns with their "capitals of light" forms the key spatial event. It can accommodate up to 1000 persons. From here, mourners can move on to one of the ceremonial halls – one large, two smaller – where the actual funeral services take place. The structural properties of concrete were taken to the limit for the capitals of light on the columns. Supported on the columns by means of brackets, the "as struck" concrete roof slab seems to float – separated by a transparent ceiling rose. The pattern of the joints and the large cones for the formwork ties lend the walls of the building a harmonising structure. The chamfered edges to the steel climbing formwork have created a regular pattern of raised joints on the surface of the concrete.

# Crematorium, Berlin

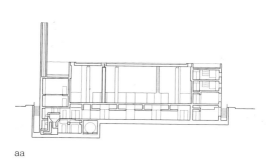

aa

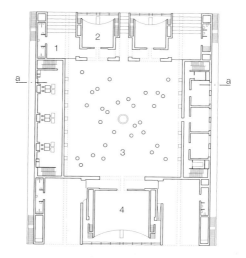

Location plan
scale 1:3000
Section aa
Plans
Ground floor
Basement
scale 1:1000

1 Covered forecourt
2 Small ceremonial hall
3 Vestibule
4 Large ceremonial hall (approx. 250 persons)
5 Cremation furnaces
6 Laying-out cells

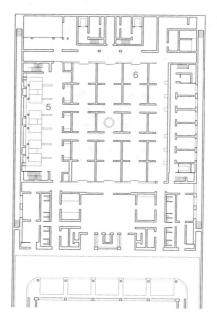

Crematorium, Berlin

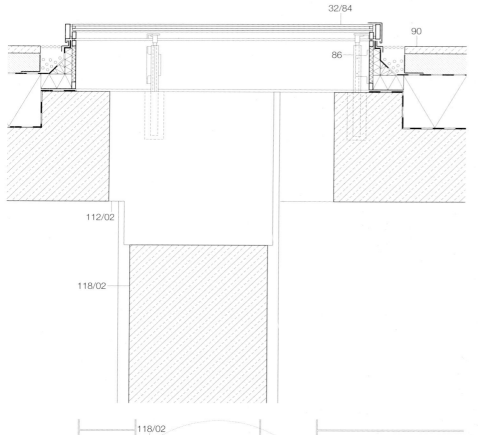
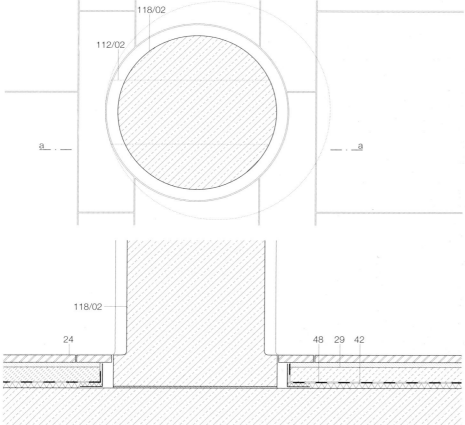

Vertical section aa
Top of column, rooflight ("capital of light")
Horizontal section through column, base of column
scale 1:20

02  In situ reinforced concrete
24  Paving slabs, serpentine stone, sandblasted
29  Screed
32  Glass
42  Impact sound insulation
48  Separating layer
84  Sunshading
86  Ventilation
90  Roof construction:
    40 mm reconstituted stone slabs
    110 mm loose expanded clay
    separating layer
    bitumenised roofing felt
    295 mm rigid expanded foam insulation
    vapour barrier
    400 mm reinforced concrete slab
112 Bracket
118 Column

Example 22

**Primary school and kindergarten, Amsterdam, The Netherlands**

1986

Architect:
Herman Hertzberger, Amsterdam
with Henk de Weijer

Structural engineers:
Evers Partners, Ijmuiden

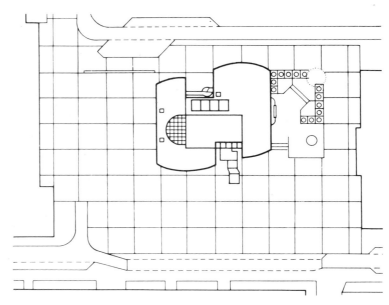

Site plan   scale 1:1000

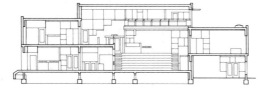

Section   scale 1:500

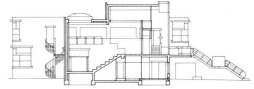

Elevation   scale 1:500

This new building replaces an old school building in the immediate vicinity. The primary school, integrated in a heavily built-up residential area, contains nine classrooms clustered in groups of three around a central core with hall and common rooms. The central hall is illuminated from above and forms a meeting place where all manner of activities merge. The classrooms, on the other hand, are located in the side wings to permit undisturbed teaching and learning. However, these wings are not isolated from the core but indeed linked to it by way of seating areas, stairs and partly glazed internal walls. The interleaving of the various levels, the exciting views into the urban surroundings and the largely transparent building envelope create a socially dynamic atmosphere. The classrooms are divided by sliding partitions, so the rooms can be enlarged as required. The principal, mainly glazed, elevation of this three-part structure, standing alone like a sculpture on the Ambonplein, faces west. The full effect of the building is not lost on the enclosed square, which forms the focal point of this district. In contrast to the horizontal accent of the side wings, the central section is characterised by the vertical lines of the fully glazed staircase. The facing concrete masonry gives scale to the building and contributes greatly to the friendly, inviting appearance. The two leaves of masonry with partial-fill cavity insulation enclose the reinforced concrete frame. The light-coloured concrete masonry and the untreated fair-face concrete form a harmonious ensemble with the white door and window frames. Precast concrete plank flooring units plus lintel and spandrel panel claddings of precast concrete elements round off the use of concrete in this structure.

Primary school and kindergarten, Amsterdam

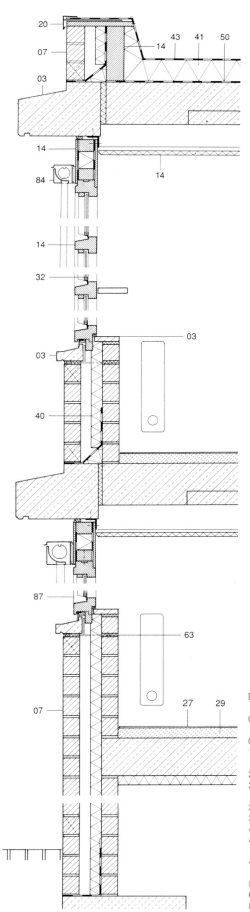

Detail  scale 1:20

03 Precast concrete element
07 Concrete masonry
14 Timber / timber derivative
20 Sheet metal flashing
27 Rubber and plastic floor coverings
29 Screed
32 Glass
40 Thermal insulation
41 Closed-cell thermal insulation
43 Bituminised roofing felt
50 Vapour barrier
63 Fixing strap
84 Sunblind
87 Window element

# Example 23

**Letter sorting centre, Cologne, Germany**

1991

Architects:
Joachim & Margot Schürmann, Cologne
with Wolf Dittmann,
Christian Becker,
Wilfried Euskirchen

Structural engineers:
Varwick-Horz-Ladewig, Cologne

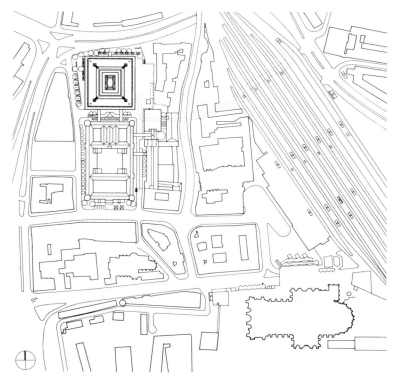

Location plan  scale 1:5000

The site of Letter Sorting Centre No. 3 is in the immediate vicinity of Cologne's main railway station and is connected to it via a tunnel. The advantage of the proximity of the station had to be weighed against the restrictions placed on urban developments, the confines of an inner-city site and the need to accommodate the many functions of a modern postal distribution centre.
The architects decided to arrange the operations in concentric "circles" around an inner courtyard. This specially designed structure contains 4000 m² of usable floor space in total, but without exceeding a building height appropriate to this neighbourhood. The first basement level constitutes the largest operations area; this is also where the tunnel to the station begins. The upper floors contain the public hall and offices, which are grouped around the landscaped inner courtyard. The restrictions of the plot made it necessary to construct two underground parking levels for goods vehicles and cars below the building. Large spans result in the entire building being largely free of intervening columns. A reinforced concrete frame in fair-face quality on a 7.5 x 15.0 m grid serves as the loadbearing structure. Columns and beams are used sparingly, but have been shaped very effectively. To match the natural stone masonry of the adjacent Cologne Post Office No. 1, the infill panels between the facade columns are of concrete masonry. The colour scheme is restrained: grey masonry and white structural steelwork. The use of lightweight concrete for the columns and thermal insulation at the parapet prevented the formation of thermal bridges. The characteristic four corner towers of the old post office building were also imitated in the new structure. These house tea kitchens and common facilities. Where functions permit, glass is used for the walls to enable passers-by a view into the building.

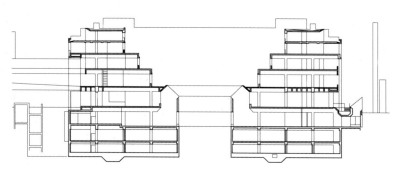

Section  scale 1:1000

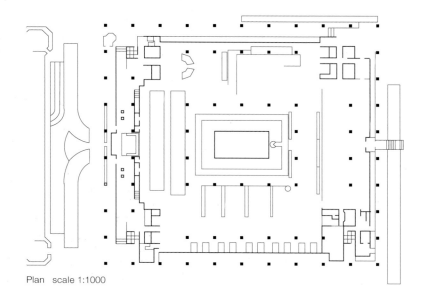

Plan  scale 1:1000

Letter sorting centre, Cologne

Example 23

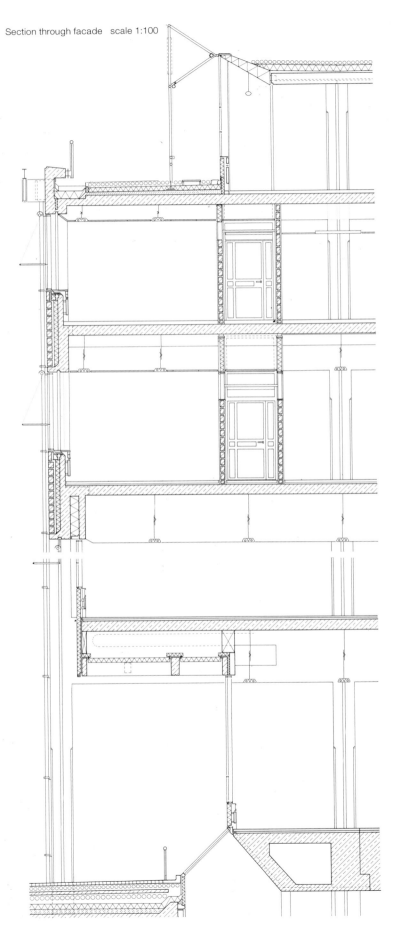

Section through facade   scale 1:100

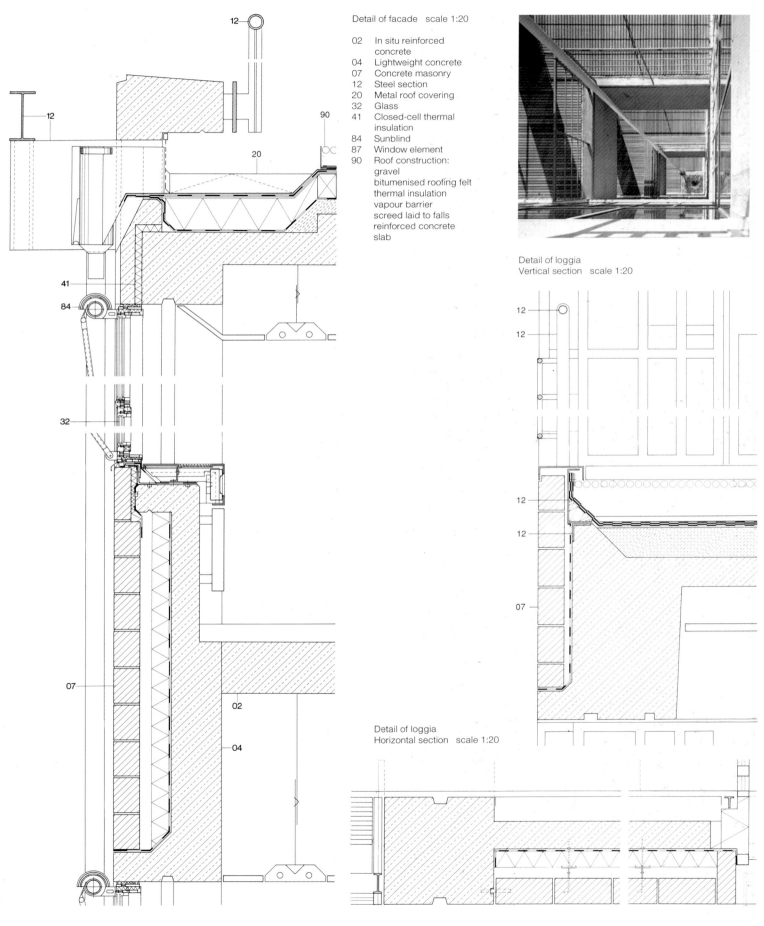

Detail of facade  scale 1:20

02  In situ reinforced concrete
04  Lightweight concrete
07  Concrete masonry
12  Steel section
20  Metal roof covering
32  Glass
41  Closed-cell thermal insulation
84  Sunblind
87  Window element
90  Roof construction:
    gravel
    bitumenised roofing felt
    thermal insulation
    vapour barrier
    screed laid to falls
    reinforced concrete slab

Detail of loggia
Vertical section  scale 1:20

Detail of loggia
Horizontal section  scale 1:20

253

## Railway station, Lyon, France

1994

Architect and structural engineer:
Santiago Calatrava, Zurich/Paris
Assistants: Alexis Bourrat, Dan Burr,
Sebastian Mémet, David Long

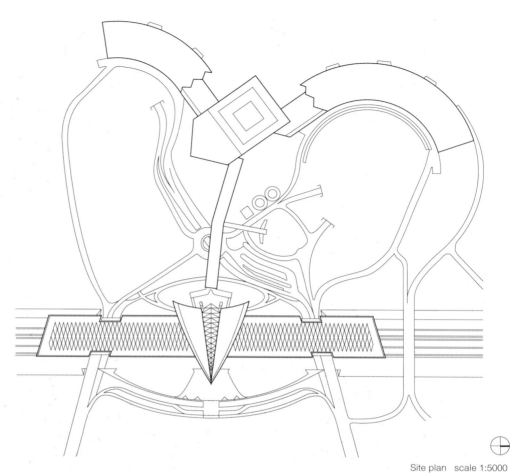

Site plan  scale 1:5000

Satolas, a suburb of Lyon, is the location of this TGV station. Together with Lyon Airport, it is being developed into an interchange for road, rail and air traffic.

The complex is divided into a long hall covering the platforms, a 36 m high main hall at 90° to it, and the elevated walkway forming the link between station and airport terminal.

The expressionist architecture of the main hall has an organic element, reminding the observer of a bird with outstretched wings rising above the railway tracks. White in situ concrete is employed for the supports and lateral abutments to the main hall. Their sculpted form complements and intensifies the dynamic architecture of the structural steelwork of the hall.

In contrast to this, the concrete structure of the roof over the tracks – inverted V-shaped wall elements carried on 53 m span, curving reinforced concrete trusses – have a repetitive, static effect. The reinforced concrete space frame of the 450 x 56 m hall is constructed in white in situ concrete and is supported on diamond-shaped concrete blocks. The cross-section of these blocks tapers towards the support, similarly to the concrete abutments of the open, lattice-type concrete loadbearing structure to the hall roof. This lends the roof a lightness, which is successful here owing to the skilled use of the material.

The edges of the delicate concrete lattice of the curving roof are infilled with precast concrete panels which carry the sheet metal roof covering. Two rows of diamond-shaped rooflights guarantee plenty of daylight in the interior. These rooflights and those in the side walls of the hall reinforce the weightlessness of the concrete construction thanks to the interesting play of light and shade created by the openings in the roof and the glass infills.

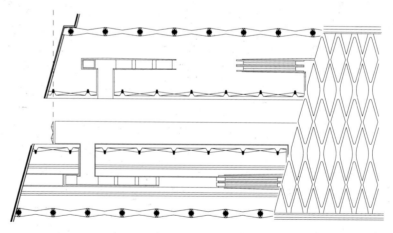

Plan  scale 1:1000

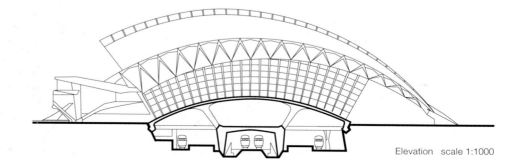

Elevation  scale 1:1000

Railway station, Lyon

Example 25

**University library, Mannheim, Germany**

1989

Architect:
Gottfried Böhm, Cologne
Assistants: Jürgen Minkus (project manager)
and Bertsch-Friedrich-Kalcher, Stuttgart

Structural engineers:
Züblin AG, Stuttgart

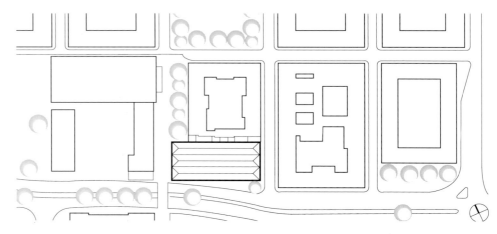

Location plan   scale 1:2500

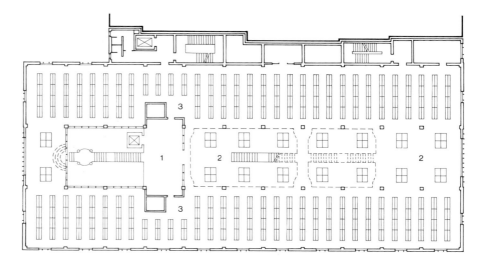

Plan of ground floor
scale 1:500

1   Lecture theatre
2   Foyer
3   Library entrance
4   Deliveries

Plan of upper floor
scale 1:500

1   Foyer
2   Library, reading room
3   Bookstacks

The library building for Mannheim University is an example of the use of precast concrete elements. The shape of the building conforms to the strict Baroque street grid of the city. However, displacing the southern part of the complex by half a bay creates a mediating factor between the neighbouring buildings. To the west the library projects two bays forwards. Vertically, the four-storey building is divided into a tall plinth zone, behind which are the lecture theatre and the overlying library storeys. The horizontal arrangement of these storeys is similar to that of a hall church (nave plus two aisles), with a glass-covered, interrupted central "nave".
The "concrete trees", decorated with architectural motifs, ranged along the palace elevation at ground level should be seen as a response to the adjoining, very busy road. The shape of the lecture theatre, which is repeated in the line of the facade, breaks away from the strict orthogonal grid. The round windows in the coloured precast concrete cladding panels are decorated with frames of glass blocks and ornamental glass. These allow daylight to enter the library storeys.
This building, characterised by its use of precast concrete elements, gains credibility among its neo-Baroque neighbours by virtue of its calm basic shape making use of new types of concrete details. The columns are given their dark colour through the use of porphyry chippings, and the facade cladding its light shade thanks to a yellow Jurassic stone aggregate. Thus, the constructional system of this precast concrete edifice manifests itself. The concrete surfaces were lightly brushed and washed directly after striking the formwork.

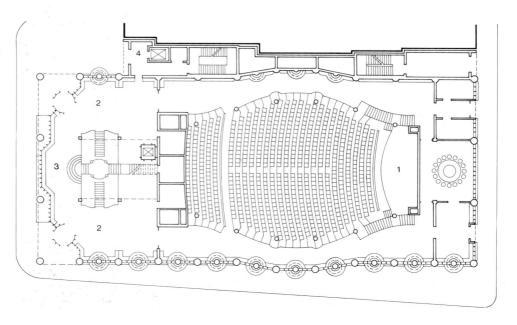

University library, Mannheim

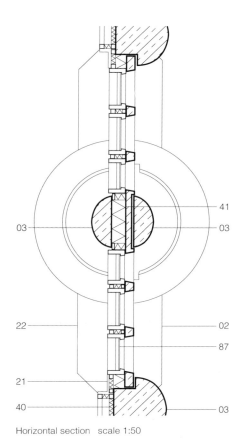
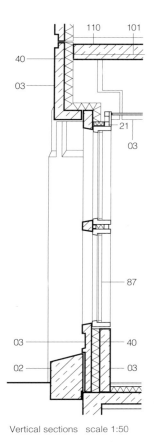
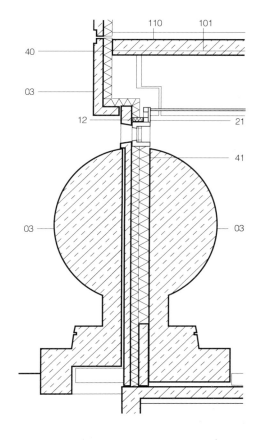

Horizontal section   scale 1:50

Vertical sections   scale 1:50

- 02  Reinforced concrete
- 03  Precast concrete element
- 12  Steel
- 21  Plasterboard
- 22  Timber derivative
- 40  Thermal insulation
- 41  Closed-cell thermal insulation
- 101 Precast concrete flooring unit
- 110 Concrete topping

Example 26

**Office building, Stuttgart, Germany**

1985

Architect:
Gottfried Böhm, Cologne
Project team: Dörte Gatermann (project leader),
Jürgen Minkus, Frederico Valda,
Klaus Beckmannshagen
Bertsch-Friedrich-Kalcher, Stuttgart

Structural engineer:
Karl Heinz Bökeler, Züblin AG, Stuttgart

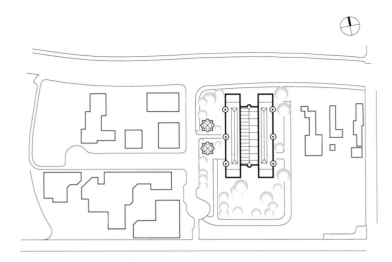

Location plan   scale 1:4000

The headquarters of this contractor is located on the southern outskirts of Stuttgart in an area between other office and commercial buildings. The architect has created an urban link between the two villages of Möhrigen and Vaihingen, now incorporated in Stuttgart's suburbs, in the form of two office wings, nearly 100 m long and six to eight storeys high, and an intermediate glass-covered atrium. The two office wings house some 700 employees. The elongated block form of the complex represents a particularly beneficial solution for an office design. Furthermore, the glass atrium is exploited for its passive solar energy gains. Interesting in this project is the use of precast concrete elements. All the loadbearing columns and all the beams of the framework were prefabricated in reinforced concrete and assembled on site. Sections of the floor slabs and shear walls are of in situ concrete. The concrete of the facade is coloured red with an iron oxide pigment, although a white cement was used for the spandrel panels. This led to the varying shades of red. The effect of the continuous, loadbearing columns was accentuated by incorporating a semicircular nose, also loadbearing, on the front of each column. This semicircular form also continues across the portal frame supporting the roof. Twin semicircular projections mark the vertical junction with the spandrel panels, required for structural and building science reasons. This provided the opportunity to include a groove for inserting a preformed plastic sealing strip. The spandrel panels themselves splay outwards like a small canopy over each window; this arrangement emphasises the individual windows, preventing them from merging into a ribbon of glass, and also serves to throw rainwater clear of the windows.

The perimeter and inner columns each support 50 mm thick precast concrete planks which, together with a layer of reinforced in situ concrete, form a continuous construction.

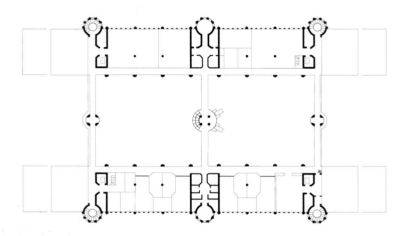

Plan of upper floor   scale 1:1000

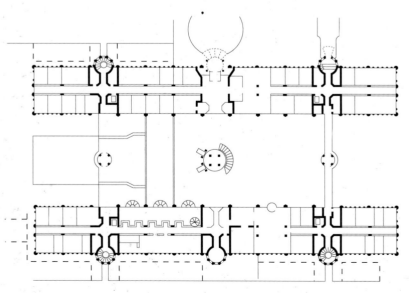

Plan of ground floor   scale 1:1000

258

Office building, Stuttgart

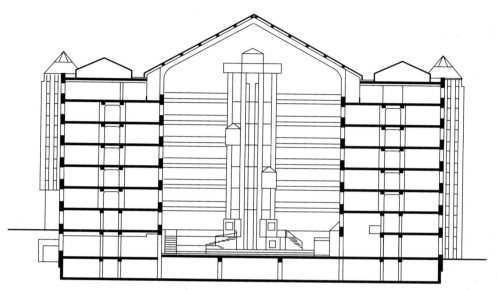

Section scale 1:500

Example 26

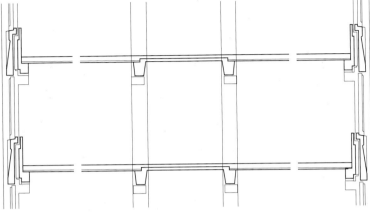

Section   scale 1:100

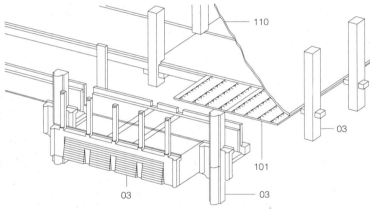

Isometric view of loadbearing structure

03  Precast concrete element
92  Cast-in, pre-curved joint baffle of hard PVC
101 Precast concrete flooring unit
110 Concrete topping

Elevation, section   scale 1:50

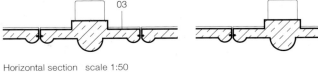

Horizontal section   scale 1:50

Office building, Stuttgart

Example 27

## Office building, Canberra, Australia

1974

Architect:
Harry Seidler, Sydney

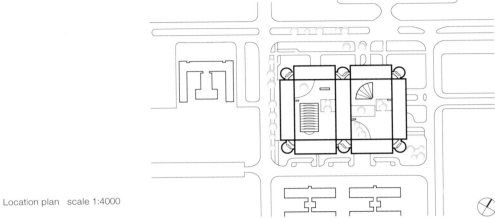

Location plan   scale 1:4000

The Trade Group Offices on Kings Avenue in Canberra's government district house three independent but related ministries employing a total of about 3,250 staff. Adjacent to the flexible office floor areas are a computer centre, a conference hall and a cafeteria. Each ministry has its own entrance. The windowless cylinders set back into the facade at the corners and on two sides house stairs and lifts. Conference hall and cafeteria are separate structures located in the two inner courtyards.
The interlinked office floors guarantee easy communication between departments. The brief for flexible utilisation plus possible extension resulted in a precast concrete construction with prestressed floor elements and beams, and precast concrete columns matching the height of the windows.
The 26 m long I-section facade beams are carried on the stocky precast concrete columns arranged in pairs. The 16 m long flooring units are attached along the neutral axis of the facade beams, whose depth is equal to half the storey height. The anchorages for the prestressing tendons, provided with stainless steel caps, are exposed externally in the centre of the web.
The frameless heat-absorbing glass fixed with simple neoprene gaskets is set back deep between the facade beams, whose soffits form the lintels and top surfaces the window sills. Services are concealed in the space between top flange and web of facade beam, and adjacent the floor beam, hidden behind a suspended ceiling.
The side walls, like the cylinders, comprise sandblasted precast concrete elements. The shape of the prestressed precast concrete components follows structural requirements. However, the advantage of the system used for the Trade Group Offices lies in the choice of primary span direction and the connection with the prestressed floor beam.

Office building, Canberra

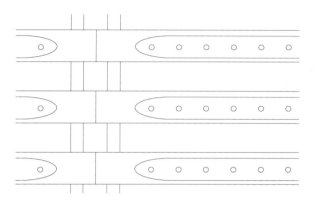

Part-elevation scale 1:200

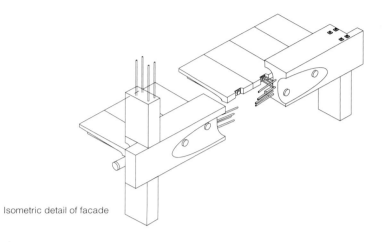

Isometric detail of facade

Details scale 1:50

- 02 Reinforced concrete
- 03 Precast concrete element
- 05 Prestressed concrete
- 08 Clay brickwork
- 20 Sheet metal
- 32 Glass
- 85 Lighting
- 86 Ventilation
- 110 Concrete topping
- 122 Prestressing tendon

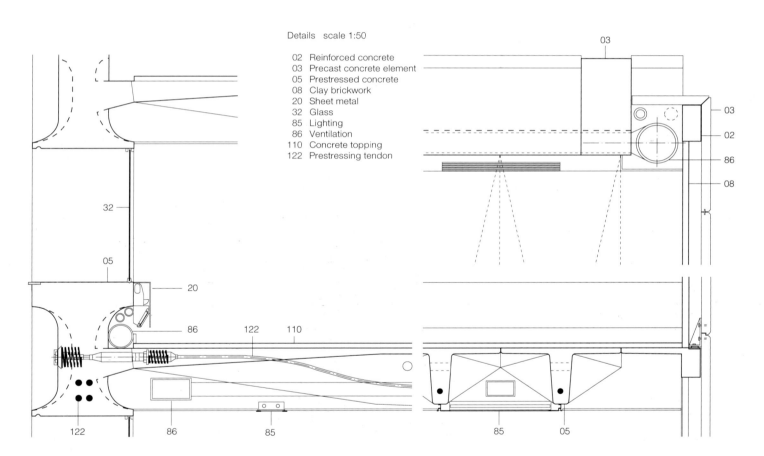

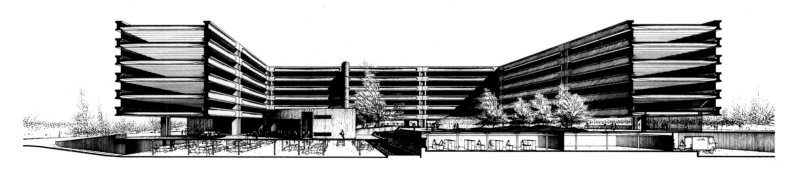

263

## Office building, Dortmund, Germany

1994

Architect:
Eckhard Gerber, Dortmund

Structural engineers:
Polónyi & Fink, Cologne
Hochtief, Essen

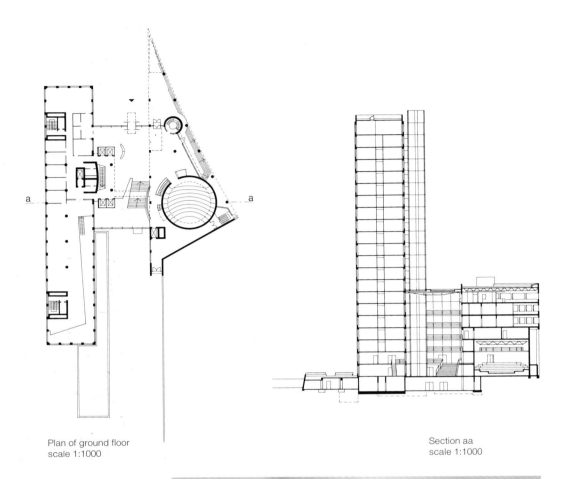

Plan of ground floor
scale 1:1000

Section aa
scale 1:1000

The new high-rise block for a publisher forms the western boundary to the broad railway station forecourt and closes a gap in the streetscape at this point. In a continuation of the adjoining block development, at the base there is a curving block following the line of Königswall and ending with an acute angle. A transparent steel-and-glass hall links this block with the actual tower. A glass access tower splits up and enhances one longitudinal side. The primary functions within the complex are clearly discernible from outside: offices in the tower, and cafeteria, conference facilities and auditorium in the low-rise block.

The choice of construction and materials reflects an attempt to achieve a "workshop" character, leaving the nature of the activities in the building with as much scope as possible. Reinforced concrete is used for all loadbearing members. The concrete surfaces left exposed internally correspond to the smooth, light-grey precast concrete elements of the external facade. They consist of concrete grade B 35 with aggregate grading 0-2 and 8-16 mm. The light colour was obtained without the need for special additives. The concrete cover is 25 mm internally and externally. Against the background of grey concrete, white steel stairs and balustrading dominate the interior of the link beneath the delicate steel-and-glass envelope, ensuring an agreeable visual harmony.

Office building, Dortmund

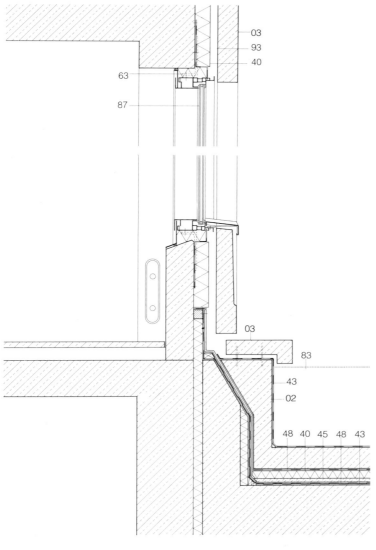

Details of facade scale 1:20
Vertical section
Horizontal section

02 Reinforced concrete
03 Precast concrete element
40 Thermal insulation
43 Bitumenised felt
45 Mastic asphalt
48 Separating layer
63 Retaining bracket
82 Panel
83 Water
87 Window element in aluminium frame
93 Air space

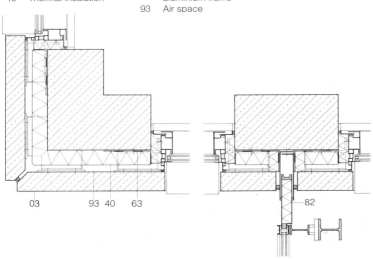

265

Example 29

**Museum, Houston, USA**

1987

Architect:
Renzo Piano, Paris
with Richard Fitzgerald

Structural engineers:
Ove Arup & Partners, London
Peter Rice

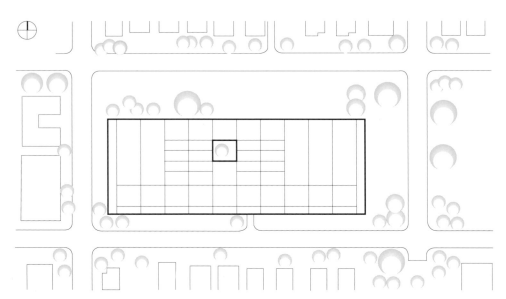

Site plan   scale 1:2000

This private museum in Houston contains important collections of modern and African art. It is located amid subtropical vegetation in a residential district. To fit neatly into its surroundings, the building has been designed like a long pavilion. A perimeter arcade surrounds the whole building. While the restoration workshops and photography darkrooms are located in the basement, the foundation's other facilities, e.g. auditorium, bookshop, restaurant, are housed in the existing nearby bungalows.
To admit plenty of daylight without glare, special louvres were developed. These long, blade-like overhead louvres are used not only in the exhibition rooms but also over the garden courtyards, the workrooms and the internal and external circulation zones.
The shape of the (adjustable) louvres was determined in numerous model tests and computer simulations. The outcome is a highly natural, organic louvre form – the dominant element in the museum's architecture. They consist of a combination of ferrocement panels beneath cast steel supporting members.
The upper part of the louvre is structural and serves as the bottom chord of the overlying cast iron girder construction supporting the solar control glass and exhaust-air ducts. The lower part is curved in such a way that it conceals the services, shields against direct sunlight and also scatters indirect light.
Various types of sand and cement were investigated to establish the right sort of white, reflective material for the surface. Best results were obtained with a white marble aggregate.
After the cross-section had been checked, a prototype was produced and its loadbearing behaviour tested in England. Then several of these complicated elements were manufactured in the USA according to the final design and erected near the site as a full-scale mock-up measuring 6 x 12 m.

# Museum, Houston

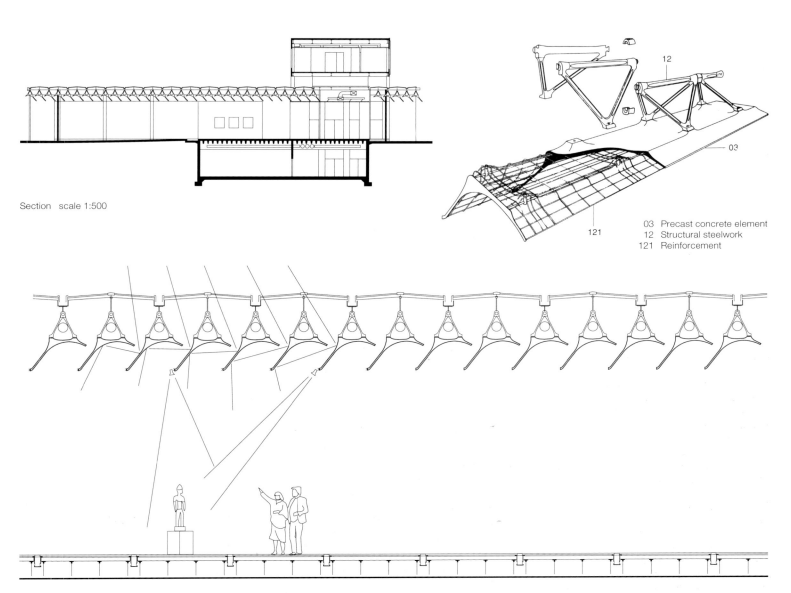

Section   scale 1:500

03  Precast concrete element
12  Structural steelwork
121 Reinforcement

Part-section

## Administration centre, Nottingham, UK

1994

Architects:
Michael Hopkins & Partners, London

Structural engineers:
Ove Arup & Partners, London

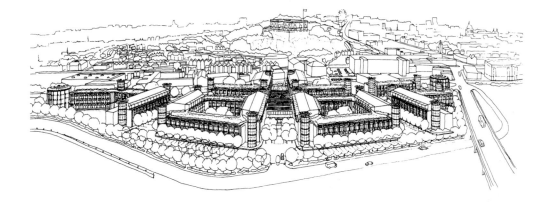

This office development covers a large area of a former industrial site on the edge of Nottingham city centre. The site is bounded to the north by the Nottingham Canal and to the south by a railway line. The complex of buildings is made up of blocks with large glazed, circular towers at the corners. These house stairs but also act as giant air ducts whose stack effect promotes the natural ventilation of the buildings. A gently curving boulevard running east-west is intersected by radial transverse roads with their focus at Nottingham Castle. The three- and four-storey blocks are grouped around landscaped inner courtyards. Public amenities such as a central reception hall, kindergarten, sports hall and restaurant complement the offices.

The width of the office floors was limited to 13.6 m in order to achieve natural lighting and ventilation. A slightly offset central access corridor renders possible a combination of individual and open-plan offices. The key feature of the energy concept is the avoidance of artificial climate control. The heat generated by artificial lighting and solar radiation is reduced satisfactorily by the use of daylight and effective sunshading.

The precast concrete vaulted flooring units guarantee the thermal storage capacity necessary for passive climate control. These span the entire width of the wing (13.6 m). They are supported on – likewise – prefabricated tapering columns of industrially produced Nottingham clay bricks.

Administration centre, Nottingham

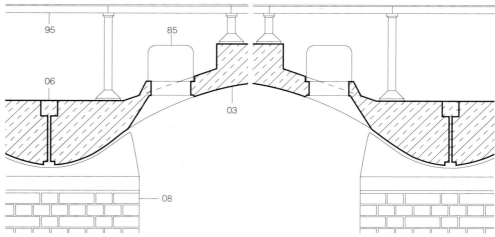

Detail of upper floor slab   scale 1:20

03  Precast concrete element
06  Grout
08  Prefabricated facing masonry column
85  Downlight
95  Cavity floor

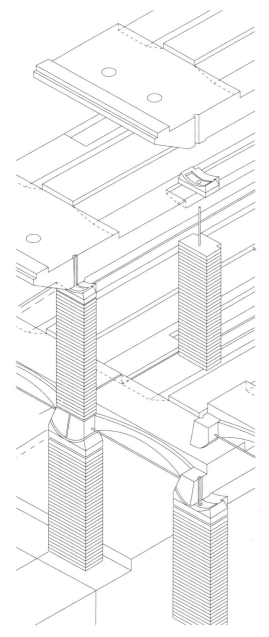

Example 31

**Industrial building, Bussolengo Barese, Italy**

1982

Architect:
Angelo Mangiarotti, Milan

Structural engineers:
BVC STL, Milan

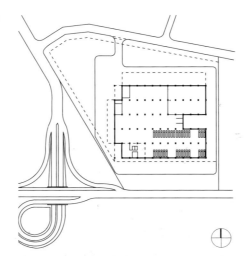

Site plan   scale 1:5000

This building in Bussolengo is a good example of a single-storey shed employing precast concrete elements.
This system was developed for various industrial uses that require large uninterrupted spans and good lighting. The construction in this case is based on a rectangular structural grid measuring 20.0 x 10.6 m. Capable of extension lengthways and sideways, this system could also be employed for other industrial and commercial applications.
Three principal elements determine the construction: H-section columns (with concealed integral rainwater downpipe), prestressed inverted Y-beams and ribbed roofing units. The tops of the columns are shaped to fit into the fork of the upturned Y. Plastic-coated plywood forms were employed for the columns; this results in a completely smooth surface finish. Both bottom "flanges" (in which the prestressing tendons are located) of the beams rest on neoprene pads on the columns. The roofing units have four ribs, spaced 800 mm apart, notched at the ends where they are supported on the prestressed beams. These were cast in steel forms. The depth of the roof slab is just 35 mm, and the units are 2.5 m wide. The openings between the columns can be filled with precast concrete panels, with trapezoidal profile sheeting, or with a post-and-rail construction with glazing. The columns are 5.0 m high and the beams 1300 mm deep, so a two-storey construction with intermediate mezzanine floor is also feasible.

270

Industrial building, Bussolengo Barese

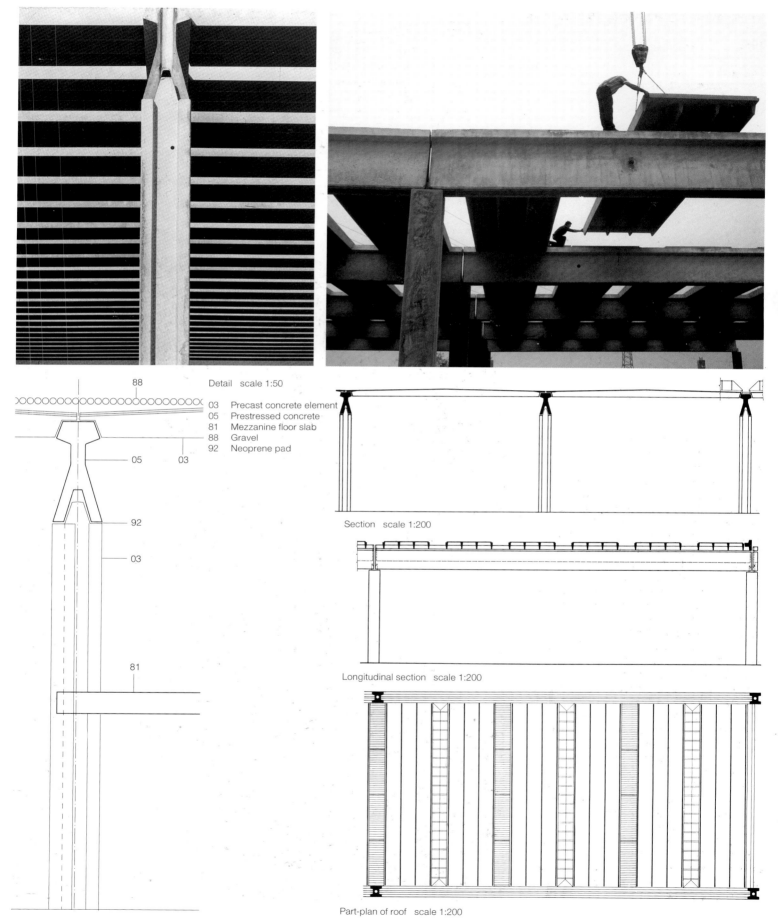

Detail scale 1:50

03 Precast concrete element
05 Prestressed concrete
81 Mezzanine floor slab
88 Gravel
92 Neoprene pad

Section scale 1:200

Longitudinal section scale 1:200

Part-plan of roof scale 1:200

Example 32

**Sports stadium, Bari, Italy**

1989

Architect:
Renzo Piano, Genoa
with S. Ishida, F. Marano,
O. di Blasi (project manager), L. Pellini

Structural engineers:
Ove Arup & Partners, London
Peter Rice, T. Carfrae, R. Kinch,
A. Lenczner

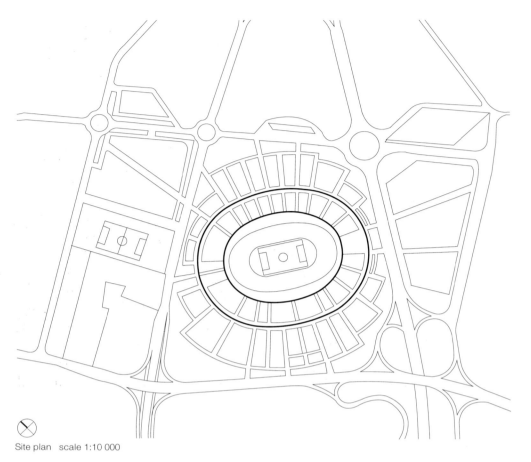

Site plan   scale 1:10 000

The new football stadium in Bari is designed to hold 60,000 spectators and is located in a suburb away from the city centre. In the flat Apulian landscape the stadium structure is readily visible from afar. The stadium is surrounded by a great expanse of park-like landscaping, which forms a counterweight to the monumental architecture. To rule out possible antagonism, rival fans are kept separate on the way to and from the parking areas.
The stadium consists of a radial system with 26 axes, corresponding to the entrances. The lower part of the grandstand is sunk into the ground like an arena. The main circulation zone – between the high- and low-level parts of the grandstand – appears to be a continuation of the encircling ground level, resulting in a transparency between the football pitch itself and the stadium's surroundings. The upper part of the stadium is elevated above ground level and comprises 312 crescent-shaped precast concrete elements. These were cast on site and lifted onto the concrete columns. Ancillary rooms, offices and service facilities are accommodated beneath the cantilevering high-level grandstand. Below these are the players' rooms and warm-up areas, with their access corridors serving as escape routes in an emergency. A translucent roof of Teflon-coated glass fibre fabric is stretched over a steel frame spanning between crescent-shaped precast concrete ribs – also spanning over the gaps between the high-level grandstand sections.
The unpleasant wind eddies associated with many stadiums are avoided here because the distance of the front edge of the roof from the edge of the pitch is greater than the height of the roof above the pitch. Small openings are included in the rear wall behind the final row of seats in order to guarantee a continuous flow of air.

Sports stadium, Bari

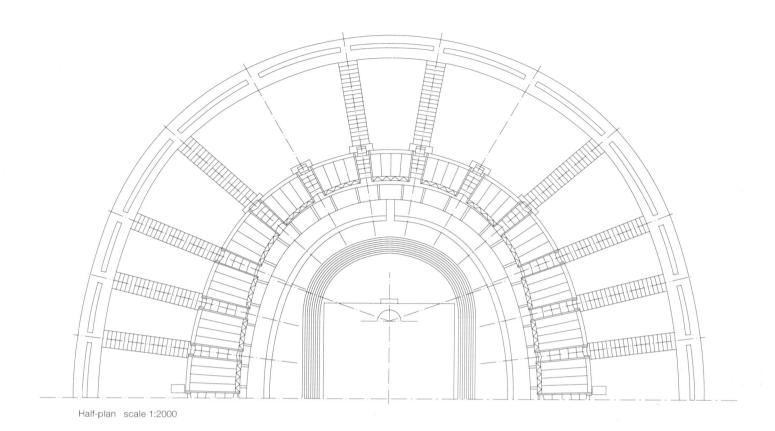

Half-plan scale 1:2000

Example 32

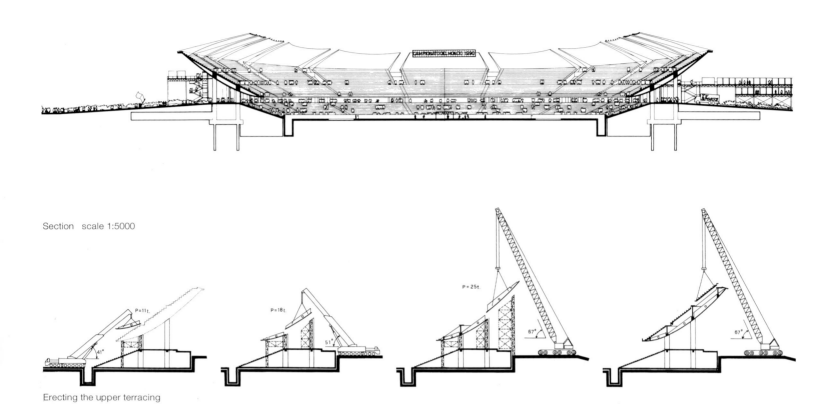

Section scale 1:5000

Erecting the upper terracing

Sports stadium, Bari

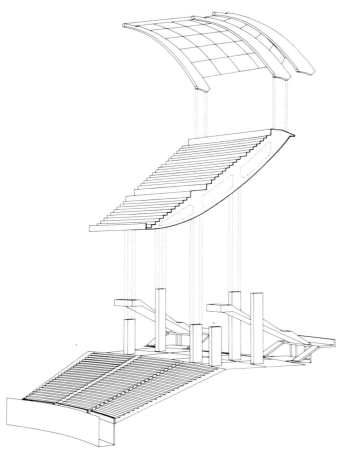

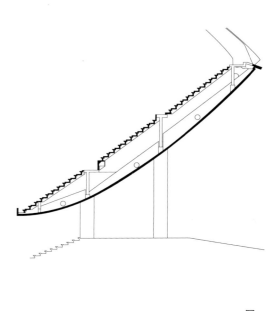

Section   scale 1:500

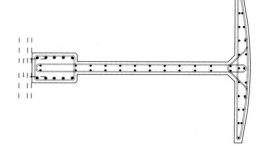

Detail   scale 1:50

**Underground station, Canary Wharf, London, UK**

1999

Architects:
Foster & Partners, London

Structural engineers:
Ove Arup & Partners, London

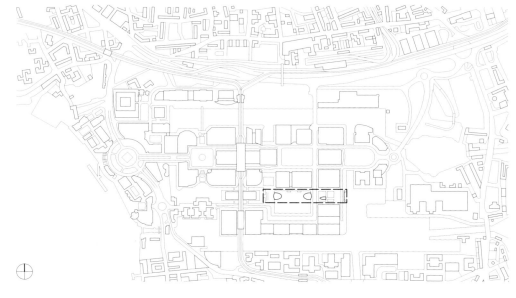

The extension to the Jubilee Underground Line was opened in 1999. Its 11 new stations, four tunnels beneath the River Thames and length of 12.2 km made it the most complicated structure ever to be built below the British capital. Today, it connects the City of London with the vast urban development area on the former London Docklands. The need to provide an adequate link to the business district established on the Isle of Dogs during the 1980s generated the main impetus for embarking on the extension to the existing underground line. "Canary Wharf" is the largest of the new stations and is in the immediate vicinity of its namesake – London's tallest building. The station is sunk two storeys deep into the former West India Dock and at peak times can handle up to 40,000 passengers. A park laid out on the roof of the station is a welcome tranquil tract for office workers in this district dominated by office blocks. Merely three shell-shaped glass roofs at surface level divulge the existence of the station. During the day these roofs permit daylight to penetrate through to the station, while at night they become shimmering illuminations. From inside, the glass roofs provide visible points of orientation, allowing the number of direction signs to be minimised. Some 20 escalators serve the station, which is only interrupted by one central row of columns and so is easy for passengers to navigate despite its large size. Offices, kiosks and ancillary rooms flank the ticket concourse, thus allowing it to remain as an unobstructed open area. The choice of material – fair-face concrete, stainless steel, glass – was influenced by issues of durability and maintenance. The transition between fair-face concrete roof and slender concrete columns was accentuated by cast iron bearings.

Underground station, Canary Wharf, London

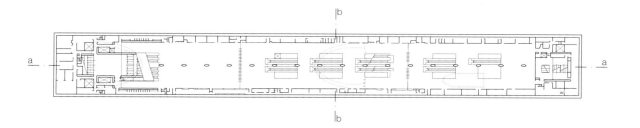

Location plan
scale 1:12 000
Section aa
Ticket concourse
scale 1:2000

277

Example 33

Section bb
scale 1:200

a  Machine room
b  Escape tunnel
c  Services
d  Left-luggage lockers
e  Platform
f  Telephones, ticket machines
g  Ticket concourse
h  Elevated walkway

Axonometric views
not to scale
Section through balustrading
scale 1:10

02  In situ reinforced concrete
12  Steel
32  Glass
62  Cast-in anchor rail
91  Floor covering: paving slabs mortar bed screed
118 Column
127 Tie
137 Hinge

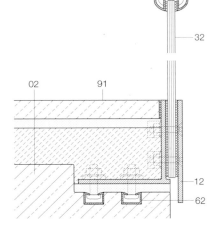

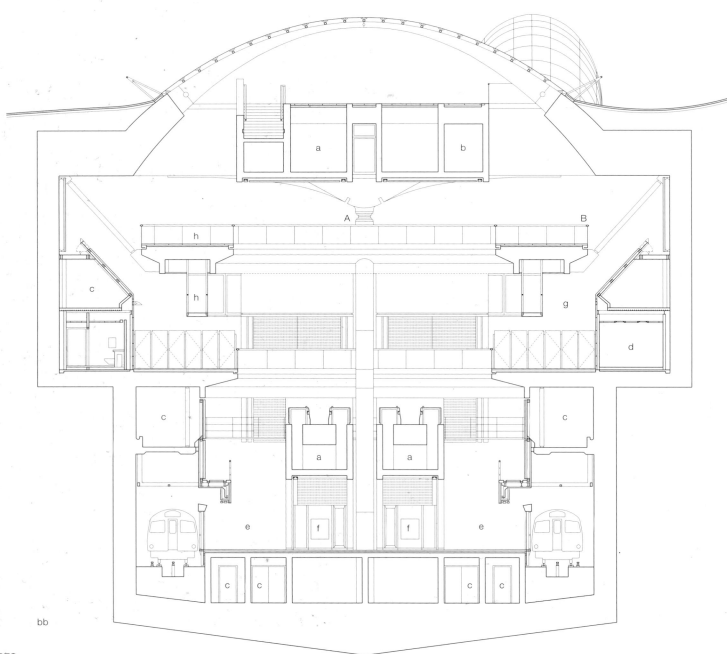

Underground station, Canary Wharf, London

A

B

# Index and glossary

Acid etching → 73
Additional reinforcement → 109, 114, 117, 127, 149
Additive → 11, 50, 52, 53, 65, 134, 170, 264
  Added to cement during production
Admixture → 47, 49, 50
  Added to concrete during mixing
Admixture/additive
  Admixture added to concrete mix during mixing
  Additive added to cement during manufacture
Aggregate → 10, 12, 27, 47, 48, 49, 50
  The standard defines this as a granular material for use in building. Aggregate may be a natural, industrial or recycled material.
Airborne pollutant → 85
Air-entrained concrete → 52
Air pore → 52
Amount of reinforcement
Anchor, dowel → 10, 99, 102, 117, 120, 134, 135, 136, 137, 138, 149, 150, 154, 165, 166, 173, 174, 181, 189, 222, 262, 278
Anhydrite → 59, 65, 224
Anti-crack reinforcement
Application directives
Arch-type folded plate structure
As-struck fair-face concrete
As-struck finish
Autoclaved aerated concrete → 61, 64, 93
  Type of concrete whose microstructure contains air pores also after setting. It is produced in an autoclave by adding a propellant (e.g. aluminium powder, calcium carbide or hydrogen superoxide) to mortar (finely-ground or fine-grained aggregate containing silicic acid, cement and/or lime, water). The gases thereby produced swell the concrete, which then hardens rapidly through the use of steam curing (DIN 4164/4165/4166). It is characterised by low weight and very good thermal insulation properties. It is readily cut and drilled and accepts nails easily.
Axial/longitudinal deformation
Axial/longitudinal force
Axial/longitudinal stress

Beam and slab, T-beam slab
Beam grid slab → 118
Beam-type folded plate structure → 151
Behaviour in fire → 46, 94, 95
Bending load/action
Bending moment diagram
Bending reinforcement → 117, 157
Bending strength → 58
Bending stress → 136, 157, 166
Bentonite slurry
Blastfurnace cement → 47, 48
  Standardised cement containing cement clinker and gypsum and/or anhydrite as well as, if necessary, ground inorganic minerals, plus rapid-cooled and consequently vitrified blastfurnace slag (cinder sand 36-80% by wt).
Bleeding
  A tendency for water to separate out from concrete mixes. Cement has a density about three times that of water and so tends to settle within the cement paste. This can lead to a clear layer of water on the surface of the concrete. This tendency for the water to separate out increases considerably with the water/cement ratio; it is greater in coarsely ground than in finely ground cements. In concrete the water does not separate out quite so much as in pure cement paste because some of the mixing water is used for wetting the fine aggregate.
Blemishes → 68, 73
Blending/mixing (of cements)
  Each cement is optimised with regard to its setting properties. The mixing of cements should therefore be avoided if possible. If mixing cannot be avoided, the suitability of the mix must be verified.
Blowhole → 129
Board formwork → 157
Bolster chisel → 69
Bond → 12, 13, 14, 35, 37, 42, 74, 80, 96, 102, 112, 114, 120, 121, 127, 134, 167, 168, 188, 198, 201, 217
  An "adhesive effect", based on adhesion or capillary forces, exists between the steel and the hydrated cement. This "adhesive effect" or adhesion depends on several factors, including the roughness and cleanness of the surface of the steel; this alone is insufficient for a good bond and is disrupted after just minor displacements. Without adhesion, a friction resistance is generated between the steel and the concrete when pressures acting transverse to the steel are present. (Such transverse stresses can arise from transverse compressive stresses due to loads or shrinkage or a swelling of the concrete. The coefficient of friction is high, $\mu = 0.3$-$0.6$, owing to the surface roughness of the steel. In the case of a mechanical interlock between the surface of the steel and the concrete, the "concrete corbels" interlocking with the irregularities must first be sheared off before the bar can slide within the concrete.) Shear resistance is the most effective and most reliable type of bond and is necessary to exploit high steel strengths. As a rule, it is achieved through the use of rolled ribs but can also be present with highly twisted bars with a suitable profile due to the "corkscrew" effect. With ribbed reinforcing bars the magnitude of the shear resistance depends on the shape and inclination of the ribs, their height and clear spacing.
Bonding coat/course
  A layer which improves the bond between an existing concrete substrate and a new bed of mortar, e.g. during the refurbishment of reinforced concrete surfaces, or layers of concrete, e.g. in the production of composite cement screeds.
Bored pile → 171, 172
Bracing, stiffening → 129, 143, 150, 230, 242
Brick dust
Brick, masonry unit
Bridge beam → 140
Brushing and washing → 72
Building materials class
Building services → 93, 94, 106, 109, 110, 118, 119, 124, 134, 138, 202
Bulk density
  Quotient of mass and volume including any voids. The bulk density can depend heavily on pretreatment, type of storage and tipping.
  Examples:
  Gravelly sand, B32 grading curve
  · dry: approx. 1900 kg/m$^3$
  · 3% surface moisture: approx. 1650 kg/m$^3$
  Cement
  · loosely tipped: 900-1200 kg/m$^3$
  · vibrated: 1600-1900 kg/m$^3$
Bundled reinforcement → 146
Bush hammering → 69
Buttress → 10

Calcium alumino ferrite → 65
Cantilever (floor) slab
Cantilevers with central beam
Capillary (pore)
  Cement is able to bond approx. 40% of its mass to water (hydration), which corresponds to a water/cement ratio of 0.40. If a cement paste exhibits a higher w/c ratio, the water which cannot be bonded by the cement is known as excess water. The space this occupies in the hydrated cement represents a system of fine, often interlinked, pores (> 100 nm) known as capillary pores. The quality of the hydrated cement and hence the concrete decreases as the volume of capillary pores increases.
Capillary water
  Water in the capillary pores. It can defy the force of gravity and rise as a result of surface tension in the pores.
Carbonation → 52, 53, 54
  The formation of calcium carbonate from the hydrated lime of the hydrated cement as a result of the effect of carbonic acid: $Ca(OH)_2 + CO_2 \rightarrow CaCO_3 + H_2O$. The carbonic acid can come from the surrounding air or from water containing carbonic acid. Carbonation is a crucial factor in the corrosion of reinforcement in reinforced concrete. The concrete cover must always be thick enough to prevent the carbonated layer from reaching the reinforcement.

Cast-in steel (component)
Cast-in-situ (floor) slab
Cast-in-situ pile
Cast iron → 266, 276
Cavity insulation → 82, 92, 99, 102, 103, 220, 248
Cavity, air gap
Cellular raft
Cement-bound building board
Cement clinker → 65, 66
  This ensues in various clinker phases during the manufacture of cement in rotary kilns by heating the raw materials until they are sintered.
Cement content → 52, 80
Cement gel
  This ensues from the cement paste during hydration of the cement. The products of hydration (essentially calcium silicate hydrate and calcium hydroxide) are designated cement gel. The cement particles are initially surrounded by a thin coating. The products of hydration form in the water-filled void which encloses each individual cement particle. It occupies slightly more than twice as much space as the cement from which it ensues but has a lower volume than that taken by the original cement with the unbonded water. The remaining intermediate spaces between the products of hydration in lamellar, stadia, foil and fibre form are known as gel pores.
Cement mortar → 29, 59, 99, 166
Cement paste → 68, 72, 242
Cement paste requirement
  Cement (blended if applicable) and water, admixtures and additives if required, form the cement paste in the fresh concrete. The amount required depends on the aggregate and the consistency of the concrete. As they have smaller surface areas and can be worked and compacted more easily, stumpy (spherical, cubic), smooth aggregates have a lower requirement than roughly broken, plate-like, elongated and splintery materials. Laboratory tests determine the cement paste requirement. To do this, the cement paste is produced with the necessary water/cement ratio and added to a weighed quantity of aggregate with a dry surface until the desired consistency is achieved. This method should be used for gap-graded aggregate or for a high fines content.
Cement requirement
  Concretes with a particular quality require a cement and water content in a certain ratio. If, for example, the aggregate grading or the consistency of the concrete mix changes, the water or cement paste requirement changes, too, i.e. when the water/cement ratio remains constant, the cement requirement must also change.

# Index and glossary

Cement screed → 59, 217, 219
  Made from cement, aggregate and water plus, if applicable, admixtures and additives, to DIN 18560. Divided into compressive strength classes according to tests on 28-day-old prisms. Strength classes ZE 55 and ZE 65 are generally produced as hard-aggregate screeds.
Cement slurry/laitance, surface l.
Cement strength class → 47, 48, 53
Cement-trass mortar → 59
Centring screw → 170
Ceramic (tile) finish
Chalking
  The detachment of pigments and fillers liberated as a result of degradation of the binder at the surface of a coating.
Chamotte concrete
  A type of concrete to which chamotte aggregate has been added to improve its heat resistance.
Change in length → 63
Channel rail
Channel section floor unit → 121, 127
Chip concrete
  Broken natural stone material is used as the aggregate in the production of this type of concrete. Its workability is usually as good as gravel concrete. The compacting factor test is preferred for assessing consistency. The tensile bending and splitting tensile strengths of this type of concrete are approx. 10-20% greater than those of gravel concrete for the same compressive strength owing to the irregular cubic to splinter-like grain shape and owing to the angular and rough surface of the broken aggregate. The "green" concrete strength is also better. Used in all fields of concrete construction.
Cinder sand → 48, 65
Circular foundation → 168
Circumferential force → 157
Circumferential prestress → 156
Circumferential reinforcement → 168, 169
Clamp/formwork vibrator
Clay brick → 9, 11, 25, 84, 86, 87, 200, 201, 263, 268
Climbing formwork → 134, 244
  Formwork for a tower-like structure with a more or less constant plan shape that is raised at regular intervals. The main element is a large formwork element similar to that used for walls. The remainder consists of corbels anchored at the base of the structure. This serves as a working platform for aligning and supporting the formwork components. A platform for subsequent work can be suspended underneath if required. Typical applications for climbing formwork are bridge piers, lift shafts, chimneys, cooling towers, silos, etc.
Closed porosity
  This is created in concrete by using foaming agents or propellants. The resulting enclosed, spherical expansion pores mean that water can only be transported in the form of vapour under normal ambient conditions.
Coal tar epoxy resin → 73
Coating → 28, 46, 56, 66, 68, 69, 73, 74, 75, 81, 120, 127, 149, 222
Coefficient of thermal expansion → 62, 63
Cohesion → 64
Collar, tie, tension member
Coloration → 68, 73
Colour coat(ing) → 66
Colour scheme → 73, 250

Coloured pigment → 66, 68, 69, 72
Column
Column grid → 18, 117, 124, 125, 127, 151, 224, 242
Column head (enlarged)
Column-mounted crane → 142
Comb chiselling → 69
Compactability, ease of compaction
  The property of fresh concrete which reveals how much work needs to be done in order to compact a certain quantity of concrete. This property is part of the workability.
Compacting → 9, 50, 51, 60, 64, 68, 129, 134, 137, 157, 170
  This is the step following placing of the concrete in which a low-voids, dense concrete microstructure is attained by forcing out the air bubbles. Screeds are generally compacted with lightweight vibration tampers. Concrete members are compacted using various methods depending on type of construction and consistency: tamping, rodding, "shock treatment", compression, centrifugal force, vacuum dewatering, rolling.
Compacting, compaction → 15, 47, 51, 57, 129, 134, 137, 164, 202
  DIBt approval is required for shear-resistant adhesive joints between steel plates and reinforced concrete components. Reinforced concrete components can be strengthened by attaching steel plates, or segments in precast construction and unsupported cantilever work can be glued together. This method should not be used for dynamic loads, and fire resistance must be proved in every single case. DIN 4227 part 3 covers the bonding of segments. Cold-curing synthetic resins or cements with synthetic resin additives can be used as binders.
  (Injected) joints in areas in which shear forces must be transferred are provided with fine serrations – similar to a finger joint in laminated timber construction.
Compaction by vibration → 17
Composite cement → 47, 65
  Standardised cement with main type of cement being CEM V with three main constituents. The, in terms of quality, irreplaceable main constituent is Portland cement clinker; CEM V/A must contain at least 40%, CEM V/B at least 20%. Other main constituents can be cinder sand, natural trass or naturally tempered pozzolanas or coal tar pulverised fuel ash. DIN EN 197 part 1 covers the properties and composition of such cements, DIN EN 206 part 1 and DIN 1045 part 2 their use in concrete.
Composite column → 129, 134, 175, 230
Composite construction → 15, 17, 238
  A type of construction in which steel beams and steel sheeting are connected to the concrete to carry the loads jointly.
Composite floor (slab)
Compressive strength (of concrete) → 10, 17, 18, 48, 50, 51, 52, 53, 54, 55, 56, 57, 58, 59, 60, 64, 120
Compressive strength of aggregate
  Naturally occurring sands and gravels and the aggregates obtained from these are normally so strong owing to the preceding natural exposure that they can be used for producing concrete of normal grades. The compressive strength of customary aggregates is 150-300 N/mm².
Compressive stresses in concrete → 62
Concrete admixture/additive → 47, 49, 50
Concrete cover → 53, 60
Concrete cover allowance
  The values given in DIN 1045 for concrete cover to reinforcement are minimum values. In order to be able to maintain this at every point of a component under site conditions, the structural engineer must add a value ($\Delta c$) to these minimum values (min. c). The nominal concrete cover (c) is then: c = min. c + $\Delta c$. According to the recommendations of the German Committee for Reinforced Concrete (DAStB), the concrete cover allowance is generally 10 mm for external components but only 5 mm if special measures are employed.
Concrete for radiation shielding
Concrete for retaining aqueous liquids → 56
Concrete with high impermeability
Concrete for water-retaining purposes
  In the test to DIN 1048 water penetration should not exceed 50 mm (average of three tests on samples). This type of concrete is used for components permanently subjected to water on one side (e.g. tanks, locks, swimming pools, pipelines).
Concrete grade → 116, 129, 130, 132, 230, 236, 264
Concrete microstructure → 50
Concrete mix → 13, 46, 51, 56, 57, 63, 65, 73, 113, 167, 173, 196, 202, 242
  The mix ratio is specified in terms of proportions by weight of the individual constituents. The moisture content of the aggregate must be taken into account. The concrete must contain an amount of cement that guarantees the required compressive strength and, in the case of reinforced concrete, protection of the steel against corrosion. The aggregates, their make-up according to nominal sizes and the grading of the aggregate must correspond to the test for suitability during production of the concrete and enable proper working of the concrete. Fair-face concrete in particular requires concrete which is easily worked and has a good water-retaining ability. A consistent concrete mix is particularly important.
Concrete provision
  That part of the concrete production chain that includes all the processes from unloading, storage and batching of the raw materials to the mixing of the concrete and transportation of the fresh concrete to the building site. This work takes place in batching plants. We distinguish therefore between ready-mixed and site-mixed concrete, depending on where it is produced. Concrete batching plants include storage facilities for cement, aggregates and, if necessary, admixtures, a water supply, dosage equipment and mixers. The most important aspect of this part of the production is the mixing of the concrete.
Concrete strength → 47, 60, 138
  Concrete strength at an age of a few hours or days can be improved with a suitable concrete mix and/or heat treatment. Important for faster occupation of the structure, earlier striking of the formwork for precast components, better freezing resistance of newly

### Concrete grades and mechanical properties [N/mm²] to EC 2

| Grade | | C12/15 | C16/20 | C20/25 | C25/30 | C30/37 | C35/45 | C40/50 | C45/55 | C50/60 |
|---|---|---|---|---|---|---|---|---|---|---|
| Compressive strength | $f_{ck}$ | 12 | 16 | 20 | 25 | 30 | 35 | 40 | 45 | 50 |
|  | $f_{cm}$ | 20 | 24 | 28 | 33 | 38 | 43 | 48 | 53 | 58 |
| Tensile strength | $f_{ctm}$ | 1.6 | 1.9 | 2.2 | 2.6 | 2.9 | 3.2 | 3.5 | 3.8 | 4.1 |
|  | $f_{ctk; 0.05}$ | 1.1 | 1.3 | 1.5 | 1.8 | 2.0 | 2.2 | 2.5 | 2.7 | 2.9 |
|  | $f_{ctk; 0.95}$ | 2.0 | 2.5 | 2.9 | 3.3 | 3.8 | 4.2 | 4.6 | 4.9 | 5.3 |
| Modulus of elastic. | $E_{cm}$ | 26000 | 27500 | 29000 | 30500 | 32000 | 33500 | 35000 | 36000 | 37000 |
| Ultimate strain | $\varepsilon_{cu} \times 10^{-3}$ [1] | -3.6 | -3.5 | -3.4 | -3.3 | -3.2 | -3.1 | -3.0 | -2.9 | -2.8 |
|  | $\varepsilon_{cu} \times 10^{-3}$ [2] | -3.5 | -3.5 | -3.5 | -3.5 | -3.5 | -3.5 | -3.5 | -3.5 | -3.5 |

[1] σ–ε line for determining internal forces

[2] σ–ε line for determining dimensions of cross-section

Preferred grades are printed in **bold** type.

The figures in the table are linked by the following relationships:
Compressive strength
  characteristic value $f_{ck}$
  average value $f_{cm} = f_{ck} + 8$
Tensile strength
  average value $f_{ctm} = 0.3 \, f_{ck}^{2/3}$
  lower fractile $f_{ck,0.05} = 0.7 \, f_{ctm}$
  upper fractile $f_{ck,0.95} = 1.3 \, f_{ctm}$
Modulus of elasticity
  average value $E_{cm} = 9500 \, f_{cm}^{1/3}$

The characteristic value of the resistance (the load-carrying capacity) calculated from the characteristic strength values $f_{ck}$ for concrete, $f_{yk}$ for steel and $f_{pk}$ for prestressing steel is determined by dividing it by the partial safety factors.

# Index and glossary

placed concrete.
Concrete surface finish
Concrete with exposed aggregate finish
Concreting → 23
That part of the concrete production chain that includes conveying, placing, spreading and compacting as well as finishing surfaces not in contact with formwork and which must be worked while the concrete is still fresh. Operations must be arranged such that the specified pours can be completed with the shortest possible transportation and conveying times. The best solution is to place the concrete immediately after mixing. If, in exceptional cases, this is not possible, then the concrete must be protected against the weather (sun, wind, rain). Generally, site-mixed concrete should be placed and compacted within 30 minutes during dry, warm weather and within 1 h during cool, wet weather.
Concreting opening → 60

### Concrete cover to reinforcement

The reinforcement requires concrete cover in order to ensure
- protection against corrosion,
- maintenance of the bond,
- fire protection (p. 96).

Minimum dimensions min. c for concrete cover to protect the reinforcement against corrosion [mm] to EC 2

| Environment category | Examples of environment cat. | Reinforcing steel | | | Prestressing steel | | |
|---|---|---|---|---|---|---|---|
| | | gen. | stressed skin struct. | ≥ C 40/50 | gen. | stressed skin struct. | ≥ C 40/50 |
| 1 | Interiors of residential or office buildings (only applies if no more severe conditions are present over a longer period during construction). | 15 | 15 | 15 | 25 | 25 | 25 |
| 2, 2a | • Rooms with a high humidity (e.g. laundries)<br>• External components<br>• Components in non-aggressive soil and/or water | 20 | 15 | 15 | 30 | 25 | 25 |
| 2b | • External components subjected to attack by frost<br>• Components in non-aggressive soil/water, with frost<br>• Internal components in high humidity conditions and subjected to frost | 25 | 20 | 20 | 35 | 30 | 30 |
| 3 | External components subjected to frost and de-icing salts | 40 | 35 | 35 | 50 | 45 | 45 |
| 4 (also with frost) | • Components in splashing water zones or components immersed in seawater, one surface of which is exposed to the air<br>• Components in salt-laden air (directly on the coast) | 40 | 35 | 35 | 50 | 45 | 45 |
| 5, 5a | Mildly aggressive chemical atmosphere (gaseous, liquid, solid) Aggressive industrial atmosphere | 25 | 20 | 20 | 35 | 30 | 30 |
| 5b | Moderately aggressive chemical atmosphere (gaseous, liquid, solid) | 30 | 25 | 25 | 40 | 35 | 35 |
| 5c | Highly aggressive chemical atmosphere (gaseous, liquid, solid) | 40 | 35 | 40 | 50 | 45 | 50 |

Conformity, compliance
Agreement between the properties of a concrete as produced and its prescribed (specified) properties. Conformity is verified by means of tests on random samples and evaluation of the results according to established statistical methods. As 100% conformity is not normally possible, the standard specifies conformity criteria in which the permitted deviations are laid down. If these criteria are met, the production may be assumed to be "in conformity". Proof of conformity is always an obligation of the concrete producer, irrespective of whether production takes place in a ready-mixed concrete works or an on-site batching plant.
Conformity/compliance criteria
Criteria which must be met to prove the conformity of a concrete production. The standard specifies criteria, above all for the compressive strength, but also for consistency, water/cement ratio and other relevant properties of the fresh and hardened concrete. The criteria can be requirements for an absolute maximum or minimum value, ranges of values or other statistical parameters based on an overall test value.
Consistency/viscosity
Measure of the workability and compactability of fresh concrete. Classified as very stiff, stiff, plastic, soft, very soft, fluid and very fluid according to the definitions of viscosity ranges.
Consistency/viscosity range
A range defines the limits within which the viscosity may vary. Determined by means of the viscosity test. DIN 1045 part 2 distinguishes between seven viscosity ranges (C0, C1/F1, C2/F2, C3/F3, F4, F5, F5, F6) and includes the associated measures of viscosity (mound size, compacting factor). EN 206 parts 1-6 specify four classes: slump class, Vebe class, compacting factor class, mound size class, with the associated measures of viscosity.
Construction joint → 68, 112, 120, 180, 183, 187, 189
Continuity effect → 112
Continuous beam → 119, 151
Continuous drum/barrel mixer, continuously rotating drum m.
Continuous piling
Continuous slab → 117
Continuous trough/tub mixer, cont. open-top/pan m.
Conventional reinforcement → 17, 120, 147
Cooling, lowering the temperature (of concrete)
Measures to reduce the temperature of the concrete in mass concrete components or during hot weather. Achieved, for example, by providing shade or sprinkling the coarse aggregate with cooling water from, for example, a deep well, or by adding ice or providing a supply of liquid nitrogen.
Corbel → 40, 124, 125, 127, 142, 149
Corrosion → 51, 52, 53, 54, 57, 60, 120, 137, 150, 167, 173, 242
Steel corrodes considerably in the presence of water and carbonic acid. Iron oxide "rust" forms: approximate composition $2Fe_2O_3 = Fe(OH)_3$. This flakes off easily. Critical for the rusting process is the presence of carbonic acid, which initially forms the readily soluble iron(II) carbonate. Absorption of water and oxidation in the presence of atmospheric oxygen gives rise to iron(III) hydroxide. This splits off some of the water and becomes rust. The formation of rust entails an increase in volume.
Cover to bar nearest concrete surface
Crack → 15, 16, 17, 60, 110, 111, 112, 113, 120
We distinguish between surface and separating/splitting cracks. Surface cracks only affect a small part of the cross-section and frequently resemble a network or mesh. Separating cracks affect large parts of the cross-section (e.g. tension zone, web) or the whole cross-section. The risk of cracking can be reduced or prevented by means of constructional measures, the concrete mix, the method of placing, careful curing and the provision of joints.
Crack formation
A natural property of concrete. Fine, usually invisible cracks occur in reinforced concrete components subjected to tension or bending even under service loads. This is taken into account in reinforced concrete design. The most important constructional measures taken to prevent undesirable cracking are the inclusion of reinforcement, which does not prevent cracks but distributes them in such a way that many narrow and hence harmless cracks ensue, and the provision of joints. Depending on the ambient conditions, crack widths of 0.1-0.4 mm are regarded as totally acceptable. In most cases it is sufficient in reinforced concrete construction simply to conceal the cracks; paint or plaster is usually adequate. When applying a paint finish, it is usually necessary to ensure that the crack is closed first, at least at the surface.
Crane beam → 143
Creep → 17, 62, 63, 110, 111, 140, 150, 156, 163, 185
Curing
Concrete requires sufficient moisture for the setting process. Furthermore, when new it must be protected against the influences of heat, cold, rain, snow, wind (drying out), running water, chemical attack, soiling, also against vibration and shock insofar as these could weaken the concrete microstructure. Curing activities depend on the extent and nature of possible influences. Dry curing (without water) has proved to be good at protecting against drying out, e.g. covering with plastic sheeting (which in the case of fair-face concrete should not touch the surface). Water, high humidity and fluctuating temperatures promote efflorescence, particularly in spring and autumn.

**D**eep foundation → 106, 168, 171, 172, 173
Deflection → 110, 113, 114, 116, 127, 143, 144, 163, 185
Deformation index
Degree of hardening
Dense concrete → 50, 81
Dense microstructure → 50, 198
Cohesion between hydrated cement and aggregate with a minimum proportion of voids. Prerequisite for durability in normal-weight concrete. It is the

result of a favourable fresh concrete mix using clean aggregate, but also intensive mixing, compaction and curing.

Density of concrete
Depends primarily on the densities of the raw materials used in the concrete but also on the compaction and moisture content of the hardened concrete. In the air-dried state, the density fluctuates, for example, for normal-weight concrete, between 2100 and 2400 kg/m$^3$.

Designed (concrete) mix
A concrete for which the required properties and additional requirements are specified to the manufacturer, who is responsible for producing a concrete which complies with this specification. This is normally the case in the ordering and supply of ready-mixed concrete. The site staff inform the ready-mixed concrete works of the properties required to suit operations on site when placing an order. The ready-mixed concrete supplier guarantees to provide the specified properties and carries out the conformity analysis for the standardised properties.

Detection of (steel) reinforcement
The detection of layers of reinforcement in finished reinforced concrete components. Reinforcement detectors are electromagnetic devices which can also be used to detect services.

Development of strength, strength gain → 51
The development of strength of the cement and the concrete depends on age, water/cement ratio, type of cement, cement strength class, admixtures and ambient conditions (temperature and humidity). At the start it is faster, then later becomes slower and slower until full hydration is achieved. The influencing factors have a particularly noticeable effect during the initial few days. A low water/cement ratio and high cement strength result in the strength developing faster; likewise, a higher temperature. An undisturbed development of strength requires adequate moisture in the concrete.

Diaphragm wall → 43, 173
Directrix → 156
Displacement → 110, 140, 159, 172
Displacement pile → 172
Distribution of shear stresses
Dome → 10, 11, 17, 23, 30, 31, 41, 44, 78, 93, 104, 152, 157, 158, 159, 163, 210
Double-tee floor unit → 121, 122, 124, 125, 126
Dowel bar → 51, 57, 183, 189
Downstand beam → 109, 151, 185
Driven pile → 171, 172, 175
Driving rain load → 81, 82
Dry batched materials
A consistent mix of materials consisting of cement, dried aggregate and, if applicable, admixtures produced in the works and suitably packaged for storage. Adding the correct amount of water and subsequent mixing is required, in accordance with the directive for producing dry packaged materials and the dry spraying method. The method for producing sprayed concrete in which the dry packaged materials, comprising binder, aggregate and, if applicable, concrete admixtures, are conveyed to the spraying nozzle pneumatically, where the water, and liquid concrete admixtures if required, is mixed in.

Drying (out) behaviour
Duct, sheath
Conduit through which the tendons are passed. Grout is subsequently injected to provide corrosion protection.
Duration of fire resistance
Dust trap → 76
Dusting → 68
The process of fine grains becoming detached from a concrete surface as a result of inadequate microstructural cohesion. The surfaces affected appear rough. Possible causes are a low cement content, premature drying of the concrete due to inadequate curing, or highly absorbent formwork.

Early setting/hardening
An accelerated change in consistency, towards a stiffer consistency, in contrast to the normal behaviour of fresh concrete and cement mortar.

Early shrinkage cracks
Often incorrectly called "contraction cracks". These occur mainly on exposed surfaces of fresh or newly placed concrete as a result of too rapid drying out. This process is also known as "plastic shrinkage". As long as the concrete is still workable, these cracks can be closed by revibrating the concrete. Such cracks can sometimes be very deep. However, "crazing" is of only very limited depth.

Early strength → 48, 51
Edge beam → 31, 164, 165, 166
Edge disturbance → 156, 163
Edge strip → 166
Effective water content
Total of mixing water plus surface moisture on aggregate or already present in additives and admixtures.

Effective water/cement (w/c) ratio
1. The quantity of water which determines the water/cement ratio is the sum of the surface moisture on the aggregate plus the mixing water. If part of water in the cement paste is absorbed by the aggregate (particularly lightweight aggregate), the effective w/c ratio is calculated from the remaining water available for hydration divided by the cement content. A prescribed w/c ratio is kept within relatively tight tolerances when the required amount of mixing water is increased by the amount absorbed by the dry aggregate (in particular lightweight aggregate) within the first 30 min.
2. If entrained air contents higher than 1.5% by vol. are achieved, e.g. through the use of additives, the entrained air content p exceeding 15 l/m$^3$ should be added to the water content w in kg/m$^3$. The effective w/c ratio is thus (w+p)/z.

Effects of wind
Wind, temperature and relative humidity define the ambient conditions under which new concrete should harden. Strong winds bring with them the risk of premature drying of the concrete. This must be avoided by means of suitable curing measures.

Efflorescence → 73
A light, veil-like discoloration of the concrete surface which impairs the appearance but does not normally have any affect on the quality of the concrete. Caused by water enriched with hydrated lime evaporating from the surface of the concrete. Upon contact with the air the hydrated lime (calcium hydroxide) is converted to calcium carbonate, which is practically insoluble in water. Efflorescence is generally weakened by the effects of the weather and often disappears altogether over the course of time.

Elastic theory → 128, 157
Elastomeric bearing → 111, 120, 136, 191
Electric heating (of concrete)
Elliptical shell → 163
Embossing → 69
Energy-saving(s), energy economy → 78, 87, 88, 90, 104, 194

Entrained air
Total quantity of air remaining in the concrete even after careful compaction. Made up of the compaction pores and, if applicable, the air pores generated by air entrainers. Specified in % by vol.

Entrapped air
Air inclusions which remain in the concrete even after careful compaction.

Epoxy resin → 73, 173
Equilibrium condition → 157, 159
Equivalent diameter → 51, 57
Equivalent load → 61, 62
Expanded clay → 48, 49, 50, 66, 87, 93, 94, 247
Expanded polystyrene bead concrete → 49
Contains aggregates made from foamed plastic beads. Owing to the low density of the aggregate, segregation occurs easily. The raw material for producing polystyrene (new designation: polyphenylethene) is styrene (vinyl benzene, phenylethene), which is produced from benzole and ethylene.

Expanded shale → 48, 49, 50, 66
Expanding cement, sulphoaluminate c.
Non-standardised cement which, during hydration, does not shrink like all other cements, but instead increases its volume somewhat. This swelling, normally increased ettringite expansion, is, however, controlled so that expansion cracks do not occur. This type of cement is usually produced by grinding and mixing ordinary Portland cement with high-alumina cement (as the propellant), and gypsum or calcium alumina sulphate and free lime. It is not produced in Germany.

Exposure class
Classification of the chemical and physical ambient conditions to which the concrete can be subjected during its service life. Also the conditions that can have a corrosive effect on the concrete, the reinforcement or metal inserts but are not considered as loads in the sense of the structural analysis.

External tanking → 81, 129, 170, 180, 187, 189
External vibrator → 64, 164

Facade cladding → 225, 256
Facade panel → 135, 136, 137, 179, 189, 212
Facing concrete → 248
A layer of concrete with a different mix placed in front of the loadbearing concrete of panels and other elements for architectural, building science or also acoustic reasons. This layer must be firmly bonded to the underlying concrete and should be at least 10 mm thick. The cement content can fluctuate between 350 and 450 kg/m$^3$ depending on the mix. Pulverised fuel ash (PFA) should not be used in such concrete in order not to impair the resistance to freezing and de-icing salts.

Facing skin/leaf
Fair-face concrete → 20, 21, 22, 33, 34, 36, 37, 41, 42, 43, 66, 68, 69, 72, 73, 77, 137, 138, 195, 196, 200, 202, 206, 212, 216, 218, 220, 228, 232, 234, 244, 248, 276
Ferrocement → 118, 157, 158, 195, 266
Composite material made from cement mortar and wire reinforcement for thin-walled precast concrete elements. Also used for boat-building.

Fibre-reinforced concrete → 224, 225
A concrete to which fibres are added to the cement paste or fresh concrete in order to improve the tensile strength, impact strength and deformability of the hardened concrete. Asbestos, plastic, glass and steel fibres are used in practice.

Fibre-reinforced sprayed concrete
General term for sprayed concrete to which fibres in the form of snips of various materials have been added as reinforcement in order to improve certain properties. The main types are steel, glass, carbon and various synthetic fibres, e.g. acrylic and polyaramide, which are intended to reinforce the hydrated cement matrix. They must, however, exhibit increased alkali resistance.

Filler(s) → 117, 150, 154
Finely dispersed inorganic, natural mineral or synthetic material with a primarily physical effect. These should not have any negative effects on concrete properties such as durability.

Final creep coefficient → 62, 63
Final shrinkage strain → 62, 63
Fine grain
Size of grain which can just pass through the narrow spaces between the spheres of the original nominal grain size when these are packed tightly together. In theory the fine grain is ≤ 0.155 x the next largest grain.

Fine grinding → 70
Fire (compartment) wall → 97, 99, 103, 150, 230
Fire protection → 46, 78, 94, 95, 96, 97, 98, 99, 101, 102, 103, 106, 113, 114, 121, 127, 134, 142, 183, 186
Fire resistance → 14, 61, 94, 95, 96, 99, 101, 102, 103, 113, 122, 123, 127, 134, 144, 146, 186
Fire resistance class → 94, 95, 96, 99, 101, 102, 103, 113, 122, 123, 144, 146, 186
Flame cleaning → 72
Flash/quick set
Non-standardised cements that solidify as soon as they are mixed with water, at the same time giving off a noticeable amount of heat. Employed, for example, for sealing work in hydraulic engineering.

Flat slab → 16, 109, 117, 118, 119, 120, 128, 129, 134, 174, 202, 240, 242
Flat slab with flared column heads → 18
Flexural disturbance → 156, 161, 163

283

# Index and glossary

Floor slab formwork
Floor slab suitable for vehicular traffic
Floor slab with ducts/ducting → 126
Flooring system → 106, 109, 114
Flowable concrete → 64
Fresh concrete whose mound size following the consistency test measures 490-600 mm and exhibits good flowability and adequate cohesion. It is produced from a fresh concrete of plastic to soft consistency range (KP/KR) by subsequently adding a superplasticiser. As superplasticisers are only effective for a limited time, e.g. 30 min, this type of concrete is often produced on site by adding the plasticiser to the truck mixer. There are guidelines to cover this procedure. Contrary to earlier assumptions, this type of concrete must still be compacted when being placed, but minimum compaction is adequate. As a rule, the other properties of this concrete – strength, shrinkage, creep, etc. – do not differ from those of the original concrete without superplasticiser with identical compaction.

Flying buttress → 23
Foamed concrete → 50
Foamed slag → 48, 49, 66, 94
Foaming agent
Used for producing foamed concrete and very lightweight mortar. This creates a large number of small, discrete air bubbles directly in the mortar or concrete. It is added during mixing (longer mixing time required) or first mixed with water to create a foam which is then mixed into the mortar or concrete. Stabilising additives ensure that the high air content is retained while the concrete is worked.

Folded plate structure → 23, 35, 151, 152, 164, 165, 167
Force due to change in direction
Fork support → 147, 150
Formation of plastic hinge → 113
Formwork panels → 33, 62, 66, 68, 108, 134, 196, 198, 220
Formwork texture → 66
Formwork tie/tie-bolt
This connects together formwork for walls etc. in such a way that their position is not altered by the pressure of the concrete as it is poured into the form. The simplest type is the twisted wire tie. After striking, the ends protruding above the surface of the concrete are cut off. However, rust causes unsightly blemishes on the surface. Many systems have been developed and patented to avoid this problem. Ties can be either completely extracted for reuse, or remain within the concrete (ends unscrewed or broken off). The recess in the concrete is filled with repair mortar. Ties often serve as spacers at the same time. Ties for basement walls, channels, tanks, etc. must be formed in such a way that water under pressure cannot seep through.

Formwork, shuttering, falsework
Temporary "mould" for virtually all concrete works and components. High dimensional accuracy is required and in many cases this must be coupled with adequate loadbearing capacity. Surface finish and appearance of the concrete very much depend on the nature of the formwork, and so to a large extent can be determined beforehand. Deformation of the formwork must be avoided in order to guarantee the dimensional accuracy of the concrete. Formwork panels and supports are subjected to the pressure of the wet concrete, its self-weight and additional loads such as wind, imposed loads, stored materials, etc. The formwork must be secured against toppling, collapse, buckling, bulging, etc, by strengthening it with struts, rails, posts and anchors. In addition, it must be clean and impervious. Timber, steel and plastic or combinations of these are ideal materials for formwork. Most types of formwork must be treated with a release agent before concreting in order to prevent the concrete adhering to the formwork.

The most common types of formwork are:
· fixed forms, e.g. for foundations, walls, columns, beams, slabs
· movable forms, e.g. climbing formwork, sliding formwork
· special forms, e.g. vacuum formwork, inflatable forms
· battery moulds
· formwork for slipforming
· timber formwork
· pneumatic formwork
· permanent formwork

Frame-type folded plate structure
Free-form shell → 151, 164
Freezing (of the concrete)
During cold weather solidification and development of strength are delayed. In freezing conditions the development of strength almost stops. If water in newly placed concrete freezes, the concrete microstructure can be loosened by the ensuing ice pressure or can even burst. The concrete mix chosen should therefore be such that, with protection, it withstands a one-off freezing condition without damage. Experience has shown that concrete with a compressive strength ≥ 5 N/mm$^2$ provides the necessary freezing resistance.

Freezing/frost resistance
A concrete is resistant to freezing when it can withstand a single freeze-thaw cycle without damage. This is the case with a compressive strength ≥ 5 N/mm$^2$. Prerequisite for this is protecting the concrete against excessive ingress of moisture.

Frost apron → 149, 168
Frost heave
Frost/freezing resistance → 137
The property of concrete and aggregate to withstand frost and freezing conditions without damage.

Fround freezing
Method of making the soil stable and impervious to water. The advantages depend on the application but include: preventing lowering of the water table, preventing settlement of neighbouring buildings, avoiding the need for support to excavations.

Full prestress
The tensile stresses are completely neutralised by the prestress in the concrete cross-section.

Full-displacement bored pile

Gap-graded aggregate
Graded aggregate in which one or more nominal sizes of aggregate between the finest and coarsest is omitted. The grading curve is irregular, running horizontally at the points where sizes are missing, if there are no corresponding oversize or undersize grains. Gap-grading can be advantageous; however, a check to ensure adequate workability is necessary.

Gasket → 262
Gas-silicate concrete
This is produced either on the basis of artificial raw materials using steam curing and lime or cement, or from a mixture of lime and ground (silicate-reactive) natural sand. Compressive strengths exceeding 100 N/mm$^2$ can therefore be achieved. The low free lime content means that the reinforcement must be protected against corrosion.

Gaussian curve(s)
German Building Technology Institute → 129, 138
Girder → 107, 109, 114, 138, 140, 142, 147, 149, 150, 151, 156, 230, 238, 266
Glasscrete
Type of construction using glass and intermediate reinforced concrete ribs in which the interaction of these materials is necessary to accommodate the loads. Design and workmanship is covered by DIN 1045 section 20.3. Used for producing lightly loaded walls, floors and roofs.

Glass fibre-reinforced concrete → 224, 225
A concrete to which alkali-resistant glass fibres in the form of snips are added in order to improve the tensile strength, impact strength and elasticity of the hardened concrete. The proportion of fibres is relatively low: 1-5% by vol. Glass fibres are easily integrated in a cement-bound matrix by various means.

Gluing/bonding (to) concrete
DIBt approval is required for shear-resistant adhesive joints between steel plates and reinforced concrete components. Reinforced concrete components can be strengthened by attaching steel plates, or segments in precast construction and unsupported cantilever work can be glued together. This method should not be used for dynamic loads, and fire resistance must be proved in every single case. DIN 4227 part 3 covers the bonding of segments. Cold-curing synthetic resins or cements with synthetic resin additives can be used as binders. (Injected) joints in areas in which shear forces must be transferred are provided with fine serrations – similar to a finger joint in laminated timber construction.

Grab set
A process in which the water necessary for hydration of the cement is removed from the hardened concrete (e.g. by heat, wind). Setting comes to a stop.

Grade → 48, 49, 50, 51, 52, 54, 55, 57, 58, 60, 61, 66, 87, 91, 96, 113
All the grain sizes between two adjacent grading limits; designated by the upper and lower grading limits.

Graded aggregate
Aggregate consisting of a mixture of coarse and fine grains (sand). A graded aggregate can be produced without prior separation into aggregate sizes but also by mixing together coarse and fine aggregates.

Grading coefficient
Aggregate grading and water requirement of aggregate in concrete, determined from the sum of the residues on the sieves of a standard sieve set in % (weight-volume proportion).

Grading curve
Grain/particle size
Granolithic concrete → 217
Type of concrete whose aggregate comprises synthetic hard materials or particularly hard natural stone, e.g. Zechite. It is employed for concrete surfaces requiring a high resistance to wear and abrasion, e.g. industrial floors.

Gravel → 10, 18, 21, 43, 48, 49, 66, 67, 75, 102, 112, 143, 149, 168, 172, 181, 182, 183, 185, 187, 222, 235, 237, 253, 271
Gravel concrete
Designation for normal-weight concrete made from sand and gravel. This term is usually used when we wish to distinguish this type of concrete from another type, e.g. chip concrete.

Grinding
Grinding additive
Substance added to cement during grinding.

Grinding fineness → 65
The grinding fineness of a cement is assessed according to its specific surface and calculated by means of air permeability measurements in cm$^2$/g to DIN 1164 part 4. This standard requires a grinding fineness of at least 2200 cm$^2$/g; this may be reduced to 2000 cm$^2$/g in special cases. There is no upper limit because there is no pertinent engineering reason. The range 2800-4000 cm$^2$/g is regarded as average. Cements below 2800 cm$^2$/g are classified as coarse, those above 4000 cm$^2$/g as fine. Very fine cements have values between 5000 and 7000 cm$^2$/g. Finely ground Portland cements exhibit fast hydration and high initial strength.

Grip aid → 61
Grit → 48, 49, 66, 70
Gross/bulk density
Bulk density is a suitable expression for tipped, loose materials, e.g. sand, gravel.

Gross/bulk density class
Ground anchor → 172, 173, 174
(Ground granulated) blastfurnace (slag) cement
In addition to Portland cement clinker, the main constituent is cinder sand. The following types are available: blastfurnace cement with 36-80% by wt cinder sand, Portland blastfurnace cement with 6-35% by wt cinder sand, Portland pulverised fuel ash blastfurnace cement with 0-20% by wt cinder sand.

Ground slab → 51, 85, 111, 140, 149, 168, 170, 172, 174, 180, 181, 182, 183, 187, 189
Grout → 49, 69, 95, 111, 120, 121, 127, 137, 147, 150, 166, 167, 168, 170, 172, 173, 174, 180, 186, 269
A type of concrete or mortar with a fluid consistency that is used for filling recesses, erection openings etc., but also for the bed of grout beneath bearings which, owing to their expansion or

position, are not suitable for packing. Most contain expanding agents to compensate for shrinkage. Their flowability is much improved by the addition of bead-like additives (e.g. fly ash from electrostatic precipitator).
Guying, tying → 165
Gypsum → 11, 43, 59, 65, 84, 93, 99

Haematite → 51
Hairline crack, microcrack → 16
Hairpin bar → 164
Handling (of concrete)
The handling of fresh concrete begins with the transfer of the ready-mixed concrete from the truck to the building site or, in the case of site-mixed concrete, with the emptying of the mixer. It ends at the point where the concrete is to be placed. Method of handling and concrete mix should be coordinated such that segregation is reliably prevented. Furthermore, the method of handling (crane skip, pump, conveyor belt, etc.) depends on the circumstances of the particular building site, the quantity of concrete to be placed, the distance to be covered (horizontal and vertical), the dimensions of the component and the plant available. Skips are preferred primarily for soft (KR) or plastic (KP) concrete. Segregation is unlikely with this method of handling provided the discharge flaps close tightly and hence prevent cement paste from leaking out. Skips travelling long distances on rails over uneven ground can lead to segregation of the concrete, especially concrete of soft consistency (KR). Only stiff concrete (KS) should be transported in open trucks. In doing so, it should be covered with tarpaulins or plastic sheets in order to prevent it drying out or becoming saturated by rainwater. Only plastic concrete (KP) is suitable for use with conveyor belts; transporting stiff (KS) or soft (KR) concrete on a conveyor belt leads to a risk of segregation. When pumping fresh concrete through pipelines, it should be ensured that segregation does not occur because this can cause a blockage.
Hardened concrete → 50, 51, 53, 56, 57, 58, 63, 64, 65, 72, 80, 154
Hardening
The progressive change in consistency, towards a stiffer consistency, of fresh concrete or cement mortar after its production up until it is no longer workable.
Harsh mix
Property of concrete mixes of low compactability and with a serious tendency to segregate. Mixes with a low fines content and those with angular aggregates tend to have this property. Air pores counteract this effect.
Headed stud → 117, 127, 128
Hearting concrete
In concrete technology we distinguish between concrete in the centre and concrete near the edge.
Heat of hydration → 48, 59, 170
Heat treatment
High temperatures accelerate the hardening of the concrete. The concrete already worked is heated to achieve a high early strength. Besides heating individual or several concrete constituents and heating the concrete during mixing, the compacted concrete can be heated by the heat of hydration of the cement, by steam at atmospheric pressure (steam curing), by hot air, by heating moulds and formwork, by heat radiation, by electric heaters or by pressurised steam (autoclaving). Heat treatment can be divided into four phases: acclimatisation, heating time, holding time, cooling time. The majority of methods are only suitable for producing precast concrete elements. Proper use of heat treatment requires certain rules to be followed. The 28-day compressive strength of concrete hardened at higher temperatures is generally somewhat lower than that of concrete hardened initially at lower or normal temperatures.
Heating (the concrete)
Measures carried out during concreting in cold weather:
1. Heating of the mixing water and/or aggregate in order to guarantee a minimum temperature for the fresh concrete during placing.
2. Heating of the air surrounding the concrete or the formwork in order to influence the hardening of the concrete or prevent freezing of the newly placed concrete.
Heavy spar (earth), baryte
High-strength concrete → 9, 17, 50, 52, 64
Normal-weight or heavy concrete exceeding grade C 50/60, and lightweight concrete exceeding grade LC 50/55. High compressive strengths are generally achieved by using very low water/cement ratios in conjunction with highly effective superplasticisers and silicate dust.
Also: Hardened concrete with extraordinary resistance to external influences.
Holding, delay
Hollow core floor unit → 122
Holorib sheeting
Hot cement
Cement with a temperature of 70-80 °C as delivered. These high temperatures can occur in cement during periods of high demand, particularly in summer, if insufficient time is available for cooling. The influence of the temperature of the cement on the temperature of the fresh concrete is usually overestimated. With a cement content of 300 kg/m³, a 10 °C rise in the temperature of the cement only raises the temperature of the fresh concrete by 1 °C. Hot cement can be safely used in concrete.
Hydrated cement → 47, 50, 53, 61, 63, 65, 68, 70
Hydrated/slaked lime, calcium hydroxide
Hydration → 48, 59, 80, 111, 170
1. Generally: The reaction of a substance with water in which the water becomes bonded to the substance concerned. This is a chemico-physical process.
2. In cement: The bonding of the water and the cement during setting transforms the cement paste into hydrated cement. After complete hydration the cement bonds approx. 25% of its original weight to water chemically and approx. 10-15% physically. The chemically bonded water cannot evaporate. The total content of bonded water (water of hydration) is about 40% by wt, corresponding to a water/cement ratio of 0.40. A higher w/c ratio in hydrated cement always leads to capillary pores. Hydration depends on temperature and is accelerated by higher temperatures. It progresses faster in the initial hours and days and becomes slower over the course of time (maturing). It stops when there is not sufficient water available. Early and adequate curing of the concrete is therefore necessary.
Hydraulic binder → 47
Hydraulicity → 9, 12
Water-bonding capacity: the property of a binder to set and remain solid hydraulically, i.e. with water both in the air and underwater.
Hydrophobic cement
A water-repellent Portland cement to DIN 1164 to which small amounts of hydrophobic substances are added during production. It is not harmed by moisture (rain) and reacts with water only upon mixing by machine after disintegration of the cement grains through friction with the aggregate or the ground. This type of cement is produced in strength class 32.5 R and is primarily employed for ground consolidation.
Hydrophobic impregnation → 74
Hyperbolic paraboloid

Identity test
Test to determine whether a selected batch or loading originates from a conforming total amount. It is carried out by the purchaser of the ready-mixed concrete at the building site. DIN 1045 part 3 Annex A covers the object, scope and frequency of such tests. The test depends on whether a designed or a prescribed mix is being supplied. The standard also describes acceptance criteria.
Impact analysis → 149
Impact load → 142, 149
Impermeable concrete → 81, 129, 174, 179, 180, 185, 187, 189
In situ concrete, cast-in-place c., site-mixed c.
Injection anchor → 173, 174
Injection grout
Employed to fill the tendon ducts of prestressed post-tensioned concrete components. DIN 4227 part 5 covers its production, properties and testing.
Injection method
Epoxy resin is compressed under high-pressure by means of special apparatur when injecting closed-off cracks in concrete in order to seal components and interlock the sides of the crack.
Insulating material → 88, 91, 94, 95, 100, 182, 185
Intermediate grain
Grain size that is smaller than the coarse grain (usually the largest grain) but larger than the fine grain and so prevents mutual contact between the coarse grain of particulate media. The intermediate grain size is theoretically ≤ 0.225 x the next larger grain.
Internal force → 113, 157, 159, 160, 161, 162, 163, 164
Internal tanking → 170
Iron oxide content → 47, 65
Iron oxide yellow/red/black → 66, 67

Joint → 10, 14, 15, 18, 43, 61, 65, 66

Kaolin, china clay

Latent hydraulic
A latent-hydraulic binder is only triggered, or produces strengths interesting in engineering terms, upon the addition of an activator. Cinder sand is the most significant substance of this kind in concrete technology.
Lateral buckling
Lateral restraint → 147, 148, 150, 172
Lava cement
Non-standardised cement made from Portland cement clinker and lava dust.
Leaching
Water seeping through a concrete body can absorb soluble components from the cement, sometimes also from the aggregate. If this water reaches the surface and evaporates, the dissolved components remain as a residue on the surface and form blemishes. Caused by poor workmanship, porous concrete, badly formed construction joints or damage within the concrete component. Low-lime cements are less at risk. The addition of trass or similar products can limit this problem and maybe even prevent it.
Lean concrete
Concrete with a low cement content and hence low strength. Used for blinding layers, to compensate for unevenness in the subsoil and as a protective layer above/below damp-proof membranes, and for consolidating and filling pockets in the subsoil.
Levelling screed → 113, 114
Lift (of concrete)
Light brushing and washing → 72
Light(weight) sand
Lightweight concrete → 49, 50, 51, 52, 57, 58, 59, 60, 61, 62, 63, 64, 66, 82, 83, 87, 93, 94, 100, 101, 175, 180, 184, 195, 198, 199, 205, 238, 239, 250, 253
Lightweight concrete block → 49, 58
Lightweight prestressed concrete → 167
Density ≤ 2000 kg/m³; suitable aggregates are expanded shale/clay, foamed slag and pumice.
Lightweight reinforced concrete → 184
Lightweight concrete for the production of loadbearing, reinforced and, occasionally, prestressed components and structures is a concrete with a closed microstructure, i.e. without voids, and is produced wholly or partly using lightweight aggregates. In contrast to normal-weight concrete, this type of concrete is characterised by a lower self-weight (max. 2000 kg/m³). The aggregates used are mainly expanded clay/shale, foamed slag and pumice, which also improves the thermal conductivity. DIN 4219 covers this type of concrete. It corresponds to concrete grade LB 15 at least. The mixes required for a certain strength and density can only be determined by means of a suitability test. In doing so, the cement content must be at least 300 kg/m³.
Lime mortar → 8, 9, 10, 59
Lime popping/blowing
This can be caused by liberated lime (CaO) in cement when the lime is present in large quantities in a coarse crystalline form. This is because the reaction with water progresses very

slowly and is not yet completed when the cement starts to set. Cements to German standards do not contain lime.
Limestone → 6, 9, 11, 12, 47, 48, 65, 66, 67, 69, 72
Limestone concrete
Concrete containing crushed limestone aggregate exhibits a lower thermal expansion than concrete with sandstone aggregate and allows dimensions to be reduced by 10% while still guaranteeing fire resistance.
Lime-trass mortar → 59
Limited prestress
Tensile stresses may also occur within the concrete cross-section as a result of imposed loads.
Loadbearing behaviour → 151, 152, 156, 157, 159, 161, 164, 166, 170, 266
Loadbearing structure
Loadbearing wall → 18, 110, 117, 135, 180, 181
Loading class → 112, 116
Loss of prestress → 17
Reduction of the stress in the cross-section of a component, e.g. due to creep of the concrete.
Low-shrinkage concrete
The effect of shrinkage of the concrete can be reduced by using a low hydrated cement content, low water/cement ratio, cement with a low grinding fineness, in reinforced concrete by including shrinkage reinforcement, by delaying the onset of drying and by using suitable types of construction (effective thickness).
Low-slump concrete → 52, 114, 134

Machine mixing, mixing by machine
Generally, concrete must be mixed in a suitable mixer. The batched raw materials for the concrete are mixed together until an even consistency is achieved. At least 30 s is regarded as sufficient mixing time for good quality mixing plant, other mixers require at least 60 s. It is recommended to increase the mixing time considerably when including concrete admixtures.
Magnetite → 51
Manual surface treatment
Masonry infill panel → 20
Masonry mortar → 49, 59, 61, 99
Mass concrete → 47, 52
Concrete for components with thicknesses exceeding about 1 m. The compressive strength is usually less important than the heat liberated as a result of hydration of the cement. The slow loss of heat from the interior forms a temperature gradient from the middle to zones near the surface. The concrete in the middle wants to expand more than that near the edges and this is not possible without differential deformations. Restraint stresses ensue – compression in the middle, tension near the edges. Early thermal contraction cracking occurs once the tensile strength is exceeded. Concrete technology measures involve the use of:
• low-heat cements,
• cement paste requirement,
• a large maximum aggregate size,
• a low fresh concrete temperature, if necessary by cooling the water and the aggregate (e.g. by adding ice),
• admixtures to reduce the amount of water (e.g. plasticiser),
• and cooling.

Constructional measures are:
• the provision of expansion or dummy joints,
• the placing of concrete in smaller pours,
• the embedding of pipes for cooling water,
• the provision of thermally insulating formwork (timber).
Maturing, afterhardening
Concrete continues to harden even after the 28th day and becomes even more solid, provided it does not dry out completely. The degree of this maturing varies depending on the cement, concrete mix and other variables. Over a long time the strength can reach five times the 28-day figure in special cases. Generally, maturing over decades is of little practical significance. However, the development of strength up to an age of 3, 6 or even 12 months can be relevant in certain cases. The higher the water/cement ratio and the lower the ambient temperature, the greater is the increase over the 28-day strength. With a w/c ratio in the normal range of 0.50-0.70 and an ambient temperature of the order of magnitude of +20 °C, we can expect the strength of concrete at 180 days to lie between 105 and 160% of the 28-day figure, depending on the strength of the cement.
Maturity → 63
Expression for the current state of the compressive strength of concrete still undergoing hardening, expressed in °Cd or °Ch. If we compare compressive strength and maturity, we obtain a diagram with a scatter that only applies to that particular cement. However, this can be taken as a rough guide for other cements as well.
Measure of consistency/viscosity
The physical variable measured by means of the viscosity test. EN 206 parts 1-6 recognises the following: slump test (in mm), Vebe degrees (in s), compacting factor (as a ratio), mound size (in mm).
Mechanical surface treatment
Membrane forces → 156, 157
Membrane mortar → 35, 107, 151, 175
Membrane theory → 160
Membrane-type shell → 151, 152, 161, 162, 163, 164
Meridian curve → 156, 157, 162
Meridional force
Mesh reinforcement → 161, 162, 164
Two layers of cold-formed steel bars intersecting to form a right-angled mesh. Normally connected by spot welding. Bar diameters are 2.5-12 mm, but for structural purposes not less than 4 mm. Applications include reinforced concrete floors/walls and as shrinkage/anti-crack reinforcement.
Method for the heat treatment of concrete. We distinguish between two systems:
1. Heat applied from outside by way of electric heaters. This method includes electrically heated formwork or blankets and, in precast concrete construction, the heating of mortar at joints by way of heat conductors included in the proximity of the joints between precast components.
2. Heating internally by way of heating wires (up to approx. 40 V) laid in the formwork prior to placing the concrete.
Method of resolution
Microclimate → 88
Microstructure of wet concrete
Cohesion between cement paste and aggregate. A poor microstructure leads to, for example, segregation, the water separating out and poor workability.
A good microstructure is achieved through a favourable blend of aggregates and, if necessary, the use of admixtures/additives.
Minimum cement content
Minimum concrete cover → 57, 60
Minimum grain size
Lower grading limit for a nominal size of aggregate or graded aggregate.
Mixing (of concrete)
Fresh concrete is produced by mixing the raw materials. Normally, good quality concrete has to be mixed by machine. The mixers used are either of the continuous drum or trough type. DIN 459 parts 1 and 2 cover the size, output and assessment of mixing. Concrete mixed by hand is only permitted in exceptional cases for small quantities of very low grade concrete. The raw materials are mixed together until an even consistency is achieved. Mixing may only be carried out by experienced personnel. For site-mixed concrete the mixer operator must have a copy of the mixing instructions; for ready-mixed concrete a delivery note is necessary. The mixing ratio of binder to superficially dry aggregate to water (z/g/w) must be specified by weight because this specifies the proportions of the individual components accurately. DIN 52170 can be used to determine the mixing ratio of fresh concrete and concrete already hardened with an accuracy approaching about ±10%. The time required to mix the materials to obtain an even consistency depends on the mixer and the concrete mix. 30 s is recommended for pan mixers and up to 60 s and longer for drum mixers. Flowable concrete requires several minutes.
Mixing water → 80
Any water occurring naturally is suitable for use as mixing water, provided it does not contain constituents which have an unfavourable effect on the hardening or other properties of concrete, or impair the protection of the concrete against corrosion.
Modulus of elasticity, Young's modulus → 61, 63, 111, 157
Moisture control → 78, 96
Moisture due to the building process
Moisture in aggregate
Moisture in the pores of the aggregate in the concrete. This is only significant in the case of porous aggregate.
Moment of inertia → 140
Monitoring class
Division of concrete into classes according to strength, ambient conditions and special properties with varying monitoring requirements.
Mortar → 278
A mixture of binder, sand and water which solidifies and hardens after a certain time. Mortar is employed to bond together masonry units (masonry mortar), to render surfaces of components (rendering) or to make good blemishes (repair mortar). Mortars of low density for improving the thermal insulation are known as lightweight masonry or insulating mortars.
Mortar group
Masonry mortars are divided into three groups depending on their composition. Group I (hardly used these days) is only approved for walls ≥ 240 mm thick and for buildings with max. 4 full storeys, also for all non-loadbearing walls. It may not be used for vaulting, reinforced masonry and basement walls. Groups II and IIa are suitable for all loadbearing walls, also in basements, with the exception of reinforced masonry and vaulting. These mortars are always necessary when the masonry is to be loaded at an early stage. Group III is for all walls apart from external skins and collar joints in double-leaf masonry. It is primarily used in heavily loaded components such as piers or in reinforced masonry. Group IIIa is for masonry according to suitability test (EM) in accordance with DIN 1053 sheet 2. Mortars of groups II and IIa may not be used simultaneously on the same building site.
Mould growth → 80

Natural stone/rock
No-fines concrete → 50
A concrete which does not include aggregate below 4 mm or consists of only one nominal size of aggregate. The proportion of cement paste is reduced such that this just coats the pieces of aggregate and does not fill the voids between these. An even coating and hence cementing of the grains is therefore more important for the compressive strength than the water/cement ratio. This type of structure is produced, for example, for materials with good thermal insulation properties (hollow blocks).
No-fines pore
Voids between the grains of graded aggregate, which in concrete must be filled by the hydrated cement. Engineering and economic considerations usually demand the lowest possible voids content in concrete. One way of achieving this is by choosing a favourable graded aggregate.
Non-destructive testing (NDT) of concrete
A method of testing to establish the compressive strength of concrete in a structure without having to remove test cores. The most important methods are:
• Measuring the rebound (R) with a Schmidt or rebound hammer,
• Measuring maturity with a computer-controlled instrument,
• Measuring the indentation (d) of a hard steel ball (Brinell hardness),
• Measuring the sound transmission with ultrasonic equipment.
Non-loadbearing facade, cladding → 178, 184, 185
Normal-weight concrete → 48, 49, 50, 59, 61, 62, 63, 64, 66, 80, 83, 87, 102, 103, 183, 206

Oil release agent
This is usually an aqueous emulsion of chemically inert minerals; however,

versions on a biological basis (e.g. oil-seed rape) have recently been introduced. This is applied to the formwork so that the concrete does not adhere to it upon striking. Only oils which do not leave blemishes on the surface of the concrete should be used. Care should be taken to ensure that oil does not come into contact with the reinforcement because this impairs the bond between steel and concrete.

Oil shale combustion residue
On-site casting plant
Opening → 29, 41, 42, 43, 60, 61, 82, 83, 110, 111, 115, 116, 117, 119, 128, 134, 140, 149, 150, 173, 174, 185, 198, 206, 236, 254, 270, 272
Opus caementitium
Ordinary Portland cement → 47, 48, 65
  Standardised cement produced from finely ground Portland cement clinker to which gypsum and/or anhydrite as well as, if applicable, inorganic mineral substances have been added.
Outer leaves of concrete
  Method for building walls in which hollow blocks (of lightweight concrete, polystyrene or other lightweight material) are built up dry in courses so that a double-leaf "shell" is produced that is then filled with concrete. This method is often proposed for DIY builders and is useful for detached houses, garages, etc. The blocks act as permanent formwork, as a substrate for plaster/rendering and provide thermal insulation. Services are easily incorporated in the cavity (prior to concreting) or in the soft shell (after concreting).
Oven-dry density → 48, 50, 51, 61, 66
Overlap → 69, 112, 136, 137, 140, 161, 162, 167

Pad foundation → 168
Pan mixer, pan-mill m.
Partial prestress
  Permitted by EC 2, whereby a continuous transition between reinforced and prestressed concrete exists.
Passive corrosion protection
  This is guaranteed by the concrete cover to the reinforcement.
Patent wall formwork panel
PCC component, facing layer
PCC component, loadbearing layer
Penetration → 58, 74, 126, 181, 183, 184, 185
Perforated facade → 138, 140
pH value
  (potentia hydrogenil = hydrogen concentration) Unit of measurement for hydrogen concentration and hence acidity or alkalinity, i.e. the concentration of acidic or alkaline reactions in a solution. This value is equal to the negative of the common logarithm of the concentration of hydrogen ions. A pH value < 7 indicates an alkaline, 7 a neutral and > 7 an acidic reaction of the solution. Shortly after mixing the concrete the mixing water assumes a very high pH value exceeding 12.5 because alkalis from the cement dissolve. The water in the pores of the hydrated cement also exhibits the high pH value of a saturated calcium hydroxide solution. Corrosion protection for the reinforcement is essentially determined by the pH value of the surrounding medium. At a value > 10 a passive layer forms around the steel,

## Partial safety factors for material properties

| Loading combination | Concrete $\gamma_c$ | Reinforcing/Prestressing steel $\gamma_s$ |
| --- | --- | --- |
| Basic loading | 1.5 | 1.15 |
| Accidental loading | 1.0 | 1.00 |

which protects it even if moisture and oxygen manage to penetrate. If the hardened concrete dries out, the carbonic acid in the air can diffuse into the very fine pores of the hydrated cement and react there with the calcium hydroxide to form calcium carbonate. This reduces the pH value of the solution in the pores of the hydrated cement, i.e. the hitherto highly alkaline medium is slowly neutralised from outside. The pH value drops to a little under 9 with the natural $CO_2$ content of the air (0.03% by vol.). This continual process from outside to inside is known as carbonation. With a pH value < 9 the passive layer on the steel is no longer stable, i.e. a normal degree of corrosion of the reinforcement can take place as soon as moisture and oxygen are available.

Phonolite/clinkstone cement
  A Portland pozzolanic cement made from Portland cement clinker, 20-35% by wt tempered phonolite plus gypsum and/or anhydrite. In comparison to ordinary Portland cement, it has a longer setting time and a lower early strength. At low temperatures there is a greater delay in the setting and so it remains workable for longer and requires a longer period of curing.
Piled foundation → 172
Placing (concrete) → 17, 48, 62, 114, 140
  The fresh concrete is poured into the formwork to give it its final shape prior to setting. In doing so it should not be allowed to segregate. All loose debris should be removed from inside the formwork and, if necessary, the formwork prewetted prior to placing the concrete. The concrete should not be allowed to fall freely over a height > 2 m, otherwise a tremie pipe should be used. Speed of placing – above all for columns and walls – must match the loadbearing capacity of formwork. The concrete pour should not be interrupted if possible, particularly for fair-face concrete. The concrete should be placed in equally thick horizontal layers (lifts) whenever possible; 500 mm is a good guide.
Plaster and masonry binder
  Finely ground hydraulic binder for plasters and masonry mortars to DIN 4211. Its main constituents are cement, stone dust and admixtures (air entrainer, plasticiser, retarder) plus sometimes also hydrated lime. Mixed with water, this binder sets both in the air and underwater and remains solid underwater. It must exhibit volume stability and a compressive strength of min. 5 N/mm² after 28 days.
Plastic theory → 157
Plate → 14, 15, 23, 35, 107, 110, 111, 112, 114, 115, 117, 121, 127, 134, 135, 143, 147, 149, 151, 152, 154, 155, 156, 157, 164, 165, 167, 168, 170, 228
Plate effect/action
Pocket foundation → 168, 170

Point tooling → 69
Poker vibrator → 60
Poker/internal vibrator
Polishing → 70
Polonceau truss → 147
Polymer concrete
  A mixture of reactive plastics and dry aggregate. Duromers like epoxy resin (EP), unsaturated polyester (UP) and polyurethane (PUR) but also thermoplastics like polymethyl methacrylate (PMMA) are the most popular binders (proportion approx. 5-15%). The properties of this type of concrete can be varied over a wide range by varying the type and quantity of the aggregate (gravel, furnace-dried quartz sand, fillers) and the resin. It is characterised by high chemical resistance, rapid hardening and high mechanical strength, and complements cement-based concrete for special applications. Current fire resistance requirements and high costs restrict its use in loadbearing components.
Polystyrene concrete
  An autoclaved aerated concrete in which the aggregate consists partly of foamed polystyrene. Density lies between 300 and 1600 kg/m³. This material is used mainly for thermal insulation plaster.
Pore water
  Water contained either in gel and/or capillary pores. Most of the water remaining in the gel pores is adsorbed chemically on the walls of the pores as a monomolecular layer of water. This bonding force almost reaches the value of chemically bonded water. The water remaining in the capillary pores is held together by the hydrogen bridge bond within the water molecules, and the layer adjacent to the walls of the pores adheres as a result of capillary forces. This bond is relatively weak and so the water molecules can evaporate.
Porous microstructure
  1. In normal-weight concrete a porous microstructure is the result of a poor mix plus inadequate mixing, working and curing. It reduces strength, impermeability and durability.
  2. Lightweight concrete normally exhibits such a porous microstructure. We distinguish between no-fines and grain pores. Both types of pores are present when producing single-sized concrete with a porous aggregate.
Porphyry → 66, 75, 256
Portland blastfurnace cement → 47, 48
  Standardised cement which, apart from cement clinker and gypsum and/or anhydrite as well as, if applicable, ground inorganic materials, contains rapid-cooled and, as a result, vitrified blastfurnace slag, the cinder sand. The amount of cinder sand is 6-35% by wt.

Portland burnt shale cement → 47, 65, 67, 75
  Apart from cement clinker, this contains 6-35% by wt fired oil shale.
Portland cement clinker → 65
Portland composite cement → 47, 65
  Cement according to EC standard, consisting of min. 65% by weight Portland cement clinker plus specified proportions of cinder sand, natural pozzolana, pulverised fuel ash (PFA) and/or fillers depending on their effectiveness, and up to 5% by wt other constituents.
Portland limestone cement → 47
  Besides Portland cement clinker, this contains 6-20% by wt limestone.
Portland pozzolanic cement → 47
  According to the European cement standard, this must contain min. 65% by wt Portland cement clinker, max. 35% by weight natural pozzolana, pulverised fuel ash (PFA) and/or fillers.
Portland pulverised fuel ash (PFA) cement
  Apart from cement clinker, this contains up to 20% by wt pulverised fuel ash (PFA).
Post-tensioning → 154
  (post-tensioned prestressed concrete) Post-tensioned prestressing involves the tendons, as usual, being housed in narrow conduits (ducts) which are left exposed during concreting. After the concrete has reached the necessary strength, the tendon is stressed by means of a special hydraulic jack and, upon reaching the specified stress, is anchored at the end of the component. After stressing, grout is forced into the duct under high pressure in order to protect the steel against corrosion and improve the load-carrying capacity (DIN 4427 part 5).
Pour
  Structures and components of in situ concrete to DIN 1045 are divided into separate segments to suit production processes or working hours. The construction joints thus formed must be capable of accommodating all the stresses that occur.
Pozz(u)olana → 12
  Substances named after the town of Pozzuoli near Naples, e.g. pozzuolana, santorin and trass as a natural plus brick dust and pulverised fuel ash (PFA) as artificial versions. In chemical terms these substances consist mainly of reactive silicic acid, in concrete technology terms they are classified as additives. They form insoluble compounds with hydrated lime; however, the reactions are very slow and of low stability.
Precast concrete (PCC) c./e.
Precast concrete external wall → 135, 136
Precast concrete floor plank → 114, 127
Precast concrete (floor) slab
Precast concrete panel → 39, 40, 121, 128, 135, 136, 149, 181, 254, 270
Precast concrete pile → 172
Precast concrete plank/panel → 127, 225, 248, 258
Precast concrete wall panel → 129, 149
Prefab(ricated) component/element, (of concrete:) → 15, 70
Prefabricated/proprietary formwork unit
Prefabrication, pre-assembly
Preparations for winter working, p. f. working in cold weather
  Special precautions need to be taken

Index and glossary

during cold weather (below +5 °C). During production of the concrete these are: increasing the cement content, using higher cement strength classes, increasing the temperature of the fresh concrete. Measures to protect the hardening concrete are: thermally insulating blankets, heating, longer striking times. DIN 1045 calls for minimum temperatures for the fresh concrete. Also: A concrete for which the required composition (mix) and the raw materials to be used are specified to the manufacturer, who is responsible for supplying a concrete which complies with this specification. This is a possible option in the ordering and supply of ready-mixed concrete. The site staff are responsible for ensuring that all the necessary concrete properties are reached by the mix ordered and carry out the necessary quality control. This requires a concrete testing laboratory equipped with the necessary apparatus and staffed by experienced personnel.

Prescribed (concrete) mix
The mix depends on the application of the concrete and is described in the relevant standard. The desired properties do not have to be proved first using trial mixes; they are assumed to be met when the given concrete mix is adhered to. The use of this type of concrete is only permitted for grades ≤ C 16/20 and exposure classes X0, XC1 and XC2. It should only be used in exceptional cases because the prescribed mixes lie very much on the safe side in terms of intended properties. Trial mixes nearly always result in more economic and usually also better engineered concrete mixes.

Pressed concrete
Concrete of a stiff (damp soil) consistency that is compacted by pressure, if necessary combined with vibration. This method is mainly used in concrete works. The particular advantage of this concrete is its high strength in the "green" state, which permits the formwork/mould to be struck immediately after compaction. Employed for producing paving stones, smaller concrete components and concrete roof tiles.

Pressure grouting
Injection of cement mortar under pressure into the tendon ducts of prestressed concrete, or injection of cement slurry for ground consolidation.

Pressure injection
1. In post-tensioned prestressed concrete, grout is injected under pressure into the voids in the ducts.
2. Cracks in reinforced concrete external components can impair the corrosion protection of the reinforcement. With crack widths > 0.3 mm it is advisable to inject, for example, epoxy resin, into the cracks. Higher tensile forces can then also be transferred in the immediate vicinity of the injection zone.

Pressure of wet concrete
Pressure of wet concrete on formwork
Prestressed concrete → 8, 9, 16, 17, 63, 64, 77, 112, 120, 121, 122, 134, 143, 144, 145, 146, 147, 148, 149, 150, 154, 164, 166, 175, 185, 195, 242, 243, 263, 271

In prestressed concrete the concrete is precompressed by prestretching the tendons in order to eliminate the (theoretical) tensile stress. The prestressing force is transferred to the concrete either by anchorages or bond. The tendons are grouted into ducts in the case of in situ concrete components. After the concrete has reached the necessary strength, the tendons are stressed with the help of hydraulic jacks, with these being braced against the concrete. Stress and elongation are measured during the process. The tendons are fixed in the anchorages by way of wedges, clamps or bolts. Afterwards, the ducts are filled with cement mortar, which produces a composite action. The tendons are protected against corrosion by applying grease and by placing them in polyester sheaths. This method is known as post-tensioned prestressed concrete (unbonded tendons). In the case of precast concrete components, the tendons, usually stranded wires, are stressed before placing the concrete. The jacks are braced either against special abutments or the formwork. After the concrete has reached the necessary strength, the prestress is transferred to the concrete by the bond between tendons and concrete (pretensioned prestressed concrete). Bonded tendons also possible with in situ concrete. Flared-out tendons (fan/cage anchorages) can be used to form an internal anchorage at one end of the tendons. With smaller tendons it is also possible to remove the anchorages after the prestressing force is transferred through the grout to the wall of the duct and the concrete (injection anchorage). In cast-in-place concrete the tendons are incorporated with a varying height matching the area of the bending moment diagram. The forces due to the change in direction counteract the load. In precast components changes of direction are not used because they are difficult to implement in practice. The dimensions of concrete components and amount of steel required is less than with normal reinforced concrete. In Germany prestressing methods also require approval certificates.

Prestressed concrete (floor) plank
Prestressed edge member → 35
Prestressed shell → 152, 162
Prestressing bed
The formwork in a precast concrete works in which the prestressing tendons are pretensioned prior to concreting the component.

Prestressing steel → 173
Type of steel employed for prestressed concrete components. Must be approved by the DIBt; such approval is only issued for a limited period. Pretensioning wires must be min. 5.0 mm dia. or, in the case of a non-circular cross-section, min. 30 mm$^2$, whereby individual wires must be min. 3.0 mm dia. However, wires of min. 3.0 mm dia. or, in the case of a non-circular cross-section, min. 20 mm$^2$ are permitted for pipes of prestressed concrete or temporary reinforcement.

Prestressing tendon
Tension member made from prestressing steel and used to generate a pretension. This includes individual wires, bars and stranded wires prefabricated in a factory.

Prestressing wire → 121
Prestressing, pretensioning
Concrete has a high compressive strength but low tensile strength (about 1/10 of the compressive strength). In order to be able to load a concrete component in compression across its whole cross-section when subjected to bending, it is prestressed using tendons. The compressive strength of the entire concrete cross-section is therefore fully exploited (stressed condition). And this means that considerably higher loads and longer spans (e.g. in bridge-building) are possible.

Pretensioning
(pretensioned prestressed concrete)
In this case the tendons in a prestressed concrete member are stressed prior to concreting. After the concrete has reached a strength of at least 0.8 $\beta_{w28}$, the tendon is released from its temporary anchorages and the prestressing force transferred into the component via the adhesion between steel and concrete.

Proof of conformity/compliance
Proportion of filler(s)
Mix ratio 1:X (by weight) between a binder (e.g. 1 part cold-curing synthetic resin) and the non-reactive component (e.g. X parts furnace-dried quartz sand) added to it. "Highly filled" or "lean" mortar mixes can hence be identified by their relatively low proportion of binder.

Proportion of fines, p. o. fine aggregate
The proportion of aggregate which passes through the 0.063 mm sieve. Takes the form of stone dust in broken aggregates, otherwise mainly cohesive constituents (clay, silt). Higher contents have a negative effect on the properties of the concrete (increased water requirement and shrinkage, reduced bond between cement matrix and aggregate). DIN 4226 limits the maximum content of such fine particles. Sieves (mesh size 0.063 mm) are used to determine the proportion of fines; the settling test may be used to estimate the amount.

Proportion of voids
The proportion of pores in graded aggregate related to the volume of aggregate.

Pulverised fuel ash (PFA) → 50, 65
Fine combustion residue consisting of coal dust which accumulates during the scrubbing of exhaust gases from steam generators in power stations. It is added to cement and concrete. It consists partly of bead-like particles with pozzolanic properties. The composition depends heavily on the type and origin of the coal and the combustion conditions.

Pumice → 48, 49, 50, 66, 85, 87, 99, 118, 128, 166
Pump(ing) mix
Fresh concrete which is pumped through pipes or hoses to the point at which it is to be placed. It therefore requires a higher water/cement ratio and a longer setting time.

Punching → 117
Punching reinforcement → 117

Purlin → 143, 144, 145, 146, 147, 149, 150

Quartz → 49, 51, 52, 66, 72, 75, 167
Quintling method → 114

Raft foundation → 168, 169, 170
Rapid-hardening binder/cement
A hydraulic binder that sets particularly quickly after mixing owing to the addition of chemicals (accelerators). Preferred for sealing leaks, for grouting anchors and similar applications. However, the final strength is usually well below that of normal binders. These binders can also be produced by adding high-alumina cement (min. 20% by wt) to binders with a lime content. Those containing chloride may not come into contact with the reinforcement owing to the risk of corrosion.

Rapid-hardening binder/cement, rapid-setting b./c.
A hydraulic binder that exhibits high initial strength, i.e. hardens particularly quickly. In contrast to rapid-hardening binders, the point at which setting begins is no different from that of normal cements.

Rapid-hardening cement
This is distinguished by a short solidification time and high initial strength compared to normal cements. These materials are proved for use in building and are produced from a raw dust with a high alumina content. The main applications are the joining of precast components or grouting of anchors.

Rate of pouring
The max. permissible speed at which fresh concrete can be poured into the formwork limits the pressure of the wet concrete on the formwork and should be adapted to the strength of the formwork. The pressure that occurs depends on the fresh concrete properties such as gross density, consistency, temperature, setting time and method of compaction.

Rate of setting/hardening
Chronological progression of the development of strength, which in the case of cement paste depends on many influences, e.g. grinding fineness of cement, type of cement, temperature of cement paste, water/cement ratio, admixtures and method of curing.

Ready-mixed concrete → 57, 58
Concrete whose constituents are batched off the building site and which is conveyed in vehicles to the building site ready to be incorporated in the structure. We distinguish between works-mixed and truck-mixed concrete. Today, about 85% of all concrete used on building sites comprises ready-mixed concrete. It should be used as soon as possible after being delivered. The inclusion of a retarder (VZ) means that the concrete remains workable for longer. However, it should be remembered that the effect of the retarder depends on the temperature. It must always be guaranteed that the concrete is used before it stiffens.

Rebound → 51, 167
Recess, cut-out → 110, 147, 188, 234
Reconstituted stone, cast stone → 59, 68, 70, 182, 247
Redistribution of loads/forces
Refractory concrete

This type of concrete is essential for service temperatures ≥ 250 °C. With Portland cement it can withstand constant temperatures up to approx. 1200 °C and with high-alumina cement up to 1700 °C. This makes it suitable for use in combustion chambers and blastfurnaces. The thermal expansion of the aggregate must correspond to that of the hydrated cement. Limestone should not be used here. Quartz transforms at approx. 600 °C and increases in volume. Proven aggregates for refractory concrete include chrome ore, blastfurnace slag, corundum, magnesite, chamotte, silicon carbide and brick chips.

Reinforced concrete (RC)
Reinforced concrete truss → 147, 254
Reinforced precast concrete
Manufactured in factories or on-site plants and transported from there to the building site where they are erected or installed. In Germany the quality of such components is monitored by the quality control agreement of the concrete and precast concrete industries and the official materials testing centres.

Reinforcement, reinforcing steel/bars
→ 6, 9, 13, 14, 15, 17, 19, 24, 46, 51, 52, 53, 54, 57, 60, 62, 68, 69, 77, 87, 95, 96, 103, 106, 109, 111, 112, 113, 114, 115, 116, 117, 118, 119, 120, 121, 122, 127, 129, 130, 132, 134, 136, 137, 140, 146, 147, 149, 150, 154, 156, 157, 158, 159, 161, 162, 164, 165, 167, 168, 169, 170, 171, 172, 173, 181, 185, 187, 189, 202, 225, 267

In reinforced concrete the tensile forces are accommodated by steel reinforcing bars which, for this purpose, are placed in the areas subjected to tension (tension reinforcement). However, they may also be used to accommodate shear forces (shear reinforcement) and to strengthen the compression zone (compression reinforcement).

Reinforcing cage → 134
Release agent
The pretreatment of the formwork, above all in fair-face work, should achieve an even colouring, permit easy striking of the formwork without damaging the concrete and help preserve the material of the formwork. Different substances have different effects. In Germany delivery, use and testing is covered by the directive of the German Concrete Association.

Removal of soiling/blemishes from concrete
Mechanical or chemical methods, less often also thermal methods (e.g. for oil damage), are employed. The choice of method depends on the type of soiling, the depth it has penetrated into the substrate and the nature (or sensitivity) of the concrete surface to be cleaned:
1. Cleaning with water, the effectiveness of which can be considerably improved by using superheated water (steam phase) and the addition of surfactants (wetting agents). The development of modern high-pressure jet equipment (with pressures up to 3000 bar) permits optimum cleaning performance with low water consumption, assuming the right choice and arrangement of nozzle.
2. Blasting with solid particles (quartz sand, silicon carbide, slag dust) is also an excellent cleaner but the great amount of dust generated places restrictions on its use. In addition, the surface of exposed aggregate is roughened by angular particles and is hence vulnerable to new soiling. Similar effects can be expected with wet sandblasting, albeit less severe.
3. Wet cleaning with chemical agents assumes a good deal of experience and detailed knowledge of the product in order to avoid damaging the surface of adjacent, sensitive components (e.g. of glass or aluminium). Both alkaline and acidic wet cleaning agents as well as organic solvent mixtures, also in the form of pastes, are employed. Official stipulations must be complied with when disposing of excess cleaning substances and water used for rinsing.
4. Graffiti is difficult to remove with solvent mixtures alone and often requires additional, abrasive blasting methods. Off-the-shelf "anti-graffiti" products often have a hydrophobic effect as well as acting as oil and grease repellents and this prevents paint sprays from penetrating deeply into the substrate, thus easing subsequent cleaning. Consequently, they are ideal for preventive measures.

Resin pocket → 69
Resistance to chemical attack
Concrete can be destroyed through the prolonged effects of waters, soils and gases that contain chemically aggressive substances. DIN 4030 is used to assess the degree of attack. The resistance of a concrete to chemical attack essentially depends on its impermeability. It must be so impermeable that the greatest depth of water penetration in the test to DIN 1048 for a "weak" attack does not exceed 50 mm and for a "severe" attack 30 mm. Concrete with a high resistance to "weak" chemical attack may have a water/cement ratio of max. 0.60 and for a "severe" attack max. 0.50. Concrete that is subjected to "very severe" chemical attack over a prolonged period must be protected against direct ingress of aggressive substances. It must also comply with the requirements for a "severe" attack.

Resistance to mechanical/physical damage
Generally, concrete is characterised by a high resistance to the mechanical effects of loading, restraint, impact and abrasion. The older the concrete, the lower the water/cement ratio, the better the compaction, the lower the volume of voids, the better the curing, the better is its resistance to mechanical damage. Resistance to freezing conditions, in particular frost and de-icing salts, is improved mainly by the introduction of artificial air pores (aerated concrete). The thermal stability of the hydrated cement and the aggregate determines the resistance of refractory concrete. There are currently no uniform test procedures for determining resistance to mechanical damage.

**Reinforcing steel to EC 2**
*DIN figures for comparison*

| Type of product | Bars | | Rings | | Welded mesh reinforcement | |
|---|---|---|---|---|---|---|
| Steel grade | B500 H | B500 N[1] | B500 N | B500 N | B500 H | B500 N |
| Nominal diameter $d_s$ [mm] | 6 – 40<br>6 – 28 | | 6 – 16 | 4 – 16 | 6 – 16 | 4 – 16<br>4 – 12 |
| Yield strength $R_e$ [N/mm²] | 500 | | | | | |
| Tensile strength $R_m$ [N/mm²] | 540<br><br>550 | 515[2]<br>525 | 540 | 515[2]<br>525 | 540 | 515<br>525<br>550 |
| Total elongation [%] at ultimate tensile force | 5.0<br><br>1.0 | 2.0 | 5.0 | 2.0 | 5.0 | 2.0<br><br>1.0 |
| Elongation at rupture fatigue strength [N/mm²] (amplitude $2\sigma_A$) | $d_s$ 6 – 20<br>200<br>$d_s$ > 20<br>150 | | 200 | | 100 | |
| *Suitable for welding using:* | E, MAG, GP, RA, RP | | | | E, MAG<br>RP | |

H ductility high
N normal

[1] Cut from rings and straightened
[2] Not yet finally stipulated

*Methods of welding:*
E manual metal-arc welding, $d_s \geq 8$ mm
MAG active gas metal-arc welding, $d_s \geq 6$ mm
GP pressure gas welding
RA flash-butt welding
RP resistance fusion spot welding

Restraint
Restraint stress → 60, 111, 167, 170
Restraint/internal/residual stress
Occurs in concrete primarily as a result of shrinkage of the hydrated cement; this dries out as the water evaporates. Uneven thermal expansion of cement and aggregate can also lead to such stresses in concrete.
Retarder → 49, 72, 114, 236
Revibration
Supplementary measure to increase the quality. Even without a retarder, concrete can be revibrated between two and seven hours after mixing. This helps to close up shrinkage and settlement cracks as well as voids beneath horizontal reinforcing bars and pieces of coarse aggregate.
Rhine gravel → 67, 75
Rhine sand → 75
Rib (element)
Ribbed shell → 159
Ribbed slab → 103, 117, 118
Rigid polyurethane foam
Rigidity (in bending)
Rigidity, stiffness → 165, 166
1. In mechanics: the reciprocal of the deformation of a component caused by a unit force.
2. Generally: the ability of a construction to accommodate static and dynamic loads.
3. In building science: the strength of layers of the construction, e.g. for sound insulation.
4. In concrete technology: degree of consistency of a concrete or mortar mix.
Ring beam → 23, 150, 158, 166
Risk of settlement → 174
Roman cement → 12
Roof truss
Roof-slung gantry crane → 142
Rotational shell → 151, 156, 157, 162
Routing of (building) services
Ruled surface → 152, 154, 156, 159

Sand bed technique
Sand requirement
Angular, irregularly shaped aggregate and, in particular, broken material, requires a sand-rich mix in order to achieve workable concrete, whereas stocky, rounded aggregate obtained from sand and gravel pits requires less sand. Cement-rich concretes, concretes of soft consistency and concretes with entrained air have a lower sand requirement.
Sandblasting → 69, 72
Sandstone → 72, 73, 206
Sandwich panel → 95, 128, 129, 136, 137, 138, 148, 150, 183, 185, 189, 191
Scaffold(ing)
Temporary works to enable work to be carried out at higher levels on a structure. Steel or lightweight metals have in most cases replaced timber these days, and fixed scaffolding has given way to mobile or movable scaffolds, which depend on the ground conditions. Scaffolds are covered by DIN 4420. Long-span formwork and scaf-

folds must be analysed structurally, in doing so, the requirements of DIN 1052, 1050, 1074 and 1054 must be followed. Deformations in the structure are to be taken into account by suitable cambering. The scaffold tubes used today are all 48.3 mm O.D.; different loadbearing requirements are taken into account by using different wall thicknesses and materials. All components are therefore fully interchangeable, are assembled using simple couplings and are hence economic. Besides full scaffolding, movable scaffold towers and adjustable floor centres (girders) made from steel circular hollow sections are used as fixed scaffolds.

Seal, gasket → 68, 95, 188

Segregation → 60
The separation of the coarser and finer constituents of the concrete. This process can occur during transportation and placing of the fresh concrete. Suitable measures must be instigated to prevent this.

Self-compacting concrete → 9, 50, 51
Fresh concrete with a fluid consistency. This technology is based on the use of new types of additives/admixtures and a higher quantity of ultra-fine material. The fresh concrete flows "like honey" and measurements of consistency depend on time. It is possible to allow the concrete to flow without segregation within large forms, even with heavy reinforcement and closely spaced bars. Deaeration of the fresh concrete takes place automatically without the need for any additional compaction.

Self-weight → 49, 51, 116, 117, 120, 122, 123, 136, 143, 144, 145, 146, 147, 154, 156, 161, 162, 163

Serrations → 170

Service core → 106, 111, 129, 134, 138, 140

Setting → 12, 16, 47, 50, 59, 68, 134, 154
Setting of the cement paste to form hydrated cement. The development of strength in ordinary Portland and Portland blastfurnace cement essentially depends on the hydration of the fast-reacting tricalcium silicate and the slower reacting dicalcium silicate. In Portland pozzolanic cement the trass reacts as a natural pozzolana with calcium hydroxide liberated during the hydration of the Portland cement clinker, forming calcium silicate hydrate. The hydration products of Portland burnt shale cement correspond to those of ordinary Portland cement.

Shear force → 117, 120, 128, 155, 156, 160, 161, 164

Shear force component → 156

Sheath → 120, 166, 173

Shrinkage → 16, 17, 62, 63, 110, 111

Shrinkage crack → 236

Shrinkage reinforcement → 111

Shrinkage strain → 111

Shrinkage stress → 111
Essentially caused by cooling of the heat of hydration in the concrete. This stress occurs at the surface of a reinforced concrete component also as a result of the hydrated cement drying out and can lead to the formation of cracks.

Sieve oversize
Grain above grading limit

Sieve undersize
Grain below grading limit

Silberkühl shell

Silicic acid → 12

Single strand → 242

Single-size aggregate
Grains of virtually the same size (dia. of largest grain = 2 x dia. of smallest grain) with approx. 35% by vol. voids for no-fines concrete.

Single-sized concrete
No-fines concrete which uses aggregate of only one nominal size.

Single-span beam → 127, 154, 242

Single-storey shed → 17, 23, 31, 39, 106, 112, 142, 143, 144, 145, 146, 147, 148, 149, 150, 151, 152, 154, 155, 156, 158, 160, 162, 164, 166, 175, 270

Single-tee floor unit → 121

Site-batched concrete

Sizing of components → 95

Slab supported on beams/walls

Slenderness ratio → 95, 114, 149

Solidification/setting time
Period of time between the viscosity states of a concrete. After adding the mixing water the fluid cement paste changes to a solid, hydrated cement. This change of state from liquid to solid does not take place abruptly. We distinguish between three phases (initial set, solidifying, final set). Starting with an initial viscosity (Vo), the standard viscosity, the three phases are assigned to viscosity ranges to DIN 1164. The setting time is defined by the viscosity VA at the onset of setting and the viscosity VE at the end. The viscosity is measured by means of the Vicat needle apparatus. According to DIN 1164 setting should not start before 1 h after mixing and must be completed within 12 h.

Solidifying/setting
Defined increase in viscosity of cement paste within fixed time limits.

Spacer → 137, 189, 225
Variously shaped small components made from various materials, e.g. concrete, plastic, for attaching to the reinforcement. Their task is to create a space between the reinforcement and the formwork so that the newly compacted concrete can fully enclose the steel bars and ensure that the prescribed concrete cover to DIN 1045 is maintained.

Span moment → 119

Span ratio → 119, 127

Spandrel panel, (roof:) parapet → 28, 70, 94, 95, 119, 135, 136, 140, 225, 248, 258

Spherical half-dome → 157

Splitting, cleaving

Sprayed concrete, gunite, shotcrete
Concrete that is conveyed in an enclosed high-pressure pipe or hose to the point of application, where it is sprayed out and thereby highly compacted. DIN 18551 distinguishes between the dry (water added at the spraying nozzle) and the wet (pressurised delivery of a wet mix) methods. The former is used mainly for making good damaged components, the latter for producing components.

Sprayed mortar
Cement mortar with aggregate of max. 4 mm grain size produced like sprayed concrete.

Spraying aid
Liquid or powder concrete admixtures: accelerators and – less often – stabilisers. Their composition is specially matched to the needs of sprayed concrete work. The most important requirements of such an admixture are rapid setting and minimal negative effects on the final strength of concrete. These admixtures reduce rebound and, in particular, improve adhesion of even thicker layers on vertical and overhanging surfaces. Furthermore, they enable sprayed concrete to be applied to wet substrates, even those carrying some water.

Spreading, expanding, splaying

Spun concrete → 60, 129, 134, 172, 195, 206
A type of concrete that is compacted by spinning it in rotating hollow moulds. The heavy components of the plastic to soft concrete used are forced outwards by the centrifugal force and the water inwards. This results in an effective water/cement ratio of approx. 0.30. This method can be used to produce components with a centrosymmetric cross-section, e.g. spun concrete pipes, masts, piles, posts, columns and linings for steel and cast iron pipes (for corrosion protection). The reinforcement is placed in a horizontal steel mould, the concrete poured and the mould closed. This is enclosed with steel rings mounted on neoprene-covered rollers. An electric motor rotates the mould about its own axis. The concrete is compacted by the centrifugal force at 300-400 rpm, reaching an acceleration of 25-27 g in doing so. The degree of compaction is 90%. This means that with a full mould the compacted concrete surface equals 90% of the full cross-section. The void has a circular cross-section irrespective of the external shape. Components with a circular cross-section therefore have a void equal to 1/3 x the outer radius. Strong spacers hold the reinforcement securely in position. A helix is sufficient for the outer reinforcement. The longitudinal reinforcement is prevented from buckling inwards by arch action. This method allows concrete grades of C 90/100 (B 100) to be achieved with a suitable mix.

Standing time (for mortar)
The time that mortar to be mixed with water must be left to stand before being used after initially mixing the binder (building lime) with the sand. This time is specified in the instructions supplied by the manufacturer.

Starter bars

State Building Code → 138

Steam curing
Method for the heat treatment of concrete in order to increase the early strength. Primarily used in precast concrete works, the compacted fresh concrete is usually exposed to saturated steam at atmospheric pressure at less than 100 °C in special facilities. The reduction in final strength is negligible with a long treatment time (approx. 1 d). With shorter times a decrease in final strength of 20-40% must be reckoned with.

Steel beam grid → 121, 137

Steel fibre-reinforced concrete
The addition of steel fibres to the fresh concrete increases the compressive, tensile bending, splitting tensile and impact strengths as well as deformability and resistance to cracking of the finished concrete.

Steel fibre-reinforced sprayed concrete
Sprayed concrete to DIN 18551 to which steel fibres are added in order to attain certain properties. The fibres benefit the:
• tensile strength,
• shear strength,
• impact strength,
• wear resistance.
The steel fibres used should be about 0.3-0.5 mm dia. and 20-30 mm long. They are added to the original mix (wet or dry) using various techniques. A proportion of fibres of approx. 5-8% (related to the dry weight of the concrete) is common. The fibres are distributed in all directions in the hardened concrete but most lie parallel to the application surface. The corrosion behaviour of exposed steel fibres near the surface of the concrete is only important from an aesthetic point of view.

Steel grade (for reinforcement) → 60, 96
Steel bars with a practically circular cross-section. According to DIN 1045 the bars must exhibit certain strength properties. DIN 488 distinguishes between various grades.

Steel mesh/fabric (reinforcement)

Steel shot → 48, 49, 50, 51, 72

Stiffening ribs → 23, 156, 163

Stone dust → 50

Stone rift saw → 70

Strain → 14, 18, 28, 29, 35, 60, 62, 63, 65, 109, 111, 112, 114, 116, 128, 136, 142, 147, 148, 150, 151, 154, 156, 163, 167, 170, 172

Streaking → 68

Strength in tension, tensile strength

Strength of fresh concrete
Strength of concrete prior to the onset of hydration. It depends mainly on the water content and the compaction energy, but also on quantity of cement and grinding fineness of cement, concrete admixtures and aggregate. The addition of fibres can increase this value two-fold. It normally lies between 0.1 and 0.5 N/mm².

Stress cracking corrosion (SCC)
If a prestressing tendon is not adequately protected against damaging influences, corrosion progresses from the surface to the interior and forms a notch in the steel which can cause failure without warning (collapse of part of the canopy of the Berlin Congress Hall in 1980, see p. 159).

Stress(ed) condition

Stressed skin structure → 23, 106, 107, 143, 149, 151, 153, 154, 155, 156, 157, 159, 161, 163, 165, 167

Stressing steel
Reinforcement incorporated in components made from prestressed concrete. Workmanship should comply with DIN 1045 sections 13 and 18. DIN 4227 applies to prestressed concrete components with limited or full pretension.

Striking/stripping the formwork

Strip foundation/footing

Strut, compression member → 14, 107, 120, 150

Studrail → 117, 120

Substance detrimental to concrete
Waters and soils which contain free

# Index and glossary

acids, sulphides, sulphates, certain magnesium and ammonium salts or certain organic compounds, also gases containing hydrogen sulphide or sulphur dioxide, can attack concrete. We distinguish between two types of chemical attack: sulphates have an expansive effect, acidic and soft waters as well as exchangeable salts and vegetable or animal oils and fats have a solvent effect. DIN 4030 specifies criteria for assessing the degree of attack of natural waters and soils.

Suevit-trass cement
   A Portland pozzolanic cement with trass from the Nördlinger Ries area.

Sulphur concrete
   A type of concrete in which the customary gravel-sand mix is processed with liquid sulphur (melting point 120 °C) as a binder. It has a high resistance to acids and salt solutions but is highly combustible.

Support conditions → 113, 122, 124, 151
Support reaction → 149, 157, 164, 166
Surface of the concrete, concrete face → 59, 68, 72, 73, 74, 113, 114, 137, 185, 244
Surface retarder → 72
Surface roughness (of aggregate)
   Besides the shape, the surface properties of an aggregate have a significant effect on the compactability and hence on the water requirement of the concrete. Splinter-like and rough pieces require a greater amount of sand for compaction. However, aggregate with a rough surface exhibits greater tensile bending strength owing to its better interlock within the microstructure.

Surface texture/pattern
Suspended roof → 106, 151, 165, 166, 167, 175
Synthetic resin mortar → 154
   The difference between this and polymer concrete is usually only the higher binder content and the aggregate size limited to max. 4 mm. The main use of this is for making good surface damage and the production of self-levelling courses.

Synthetic resin plaster → 82, 83
Synthetic resin skim (plaster) coat
Synthetic roofing felt → 214

Table formwork → 31, 69, 114
Tamped concrete → 9, 13
   Designation of concrete according to type of compaction. Tamping means compacting the concrete by means of compressive impacts. Generally, tamping is only advisable for concrete of consistency KS. Owing to the limited depth penetration, the finished, tamped layer should not be thicker than 150 mm. Manual tamping – or better still machine tamping – of a layer should be carried out until the concrete is soft and the surface closed and consistent.

T-beam → 14, 118
Technical surface treatment
Tee unit → 122, 127, 155
Teflon fibre
Temperature during hydration process
Temperature of the concrete → 56
   Low temperatures delay and high temperatures accelerate the setting process. Temperature also affects the solidification behaviour of fresh concrete. The temperature of the concrete must therefore be taken into account when concreting in extremely cold and extremely hot weather. As a rule, concrete should only be placed at temperatures between +5 and +30 °C.

Tensile strength → 9, 17, 59, 111, 120
Tension load/action
Tension reinforcement → 118, 121
Test for compliance, conformity
Tests on concrete
   Tests whose results permit statements to be made about the expected properties of the hardened concrete. The sample of fresh concrete must be representative of the quantity of concrete to be assessed. The following tests are common: fresh concrete density, consistency, cement content, water/cement ratio.

Tests on trial mixes → 51
   Tests carried out prior to starting production of a concrete in order to establish the mix ratios for a new concrete mix or range of mixes so that all the specified requirements for the fresh or hardened concrete are satisfied. These tests serve to establish the concrete mix, prior to its use, using the intended raw materials and consistency under the given conditions of the building site. They ensure reliable workability and that the required properties are reliably attained.

Textile-reinforced concrete
   A composite material with cement as the main binder in which the tensile forces are accommodated by textile inserts. The textile inserts are formed from single fibres (filaments) spun to form yarns; the textile reinforcement is produced from these yarns. This production process allows a much more precise alignment of the reinforcement to suit the application than is the case with conventional steel reinforcement. The textile reinforcement can be much thinner and the main force trajectories three-dimensional and prefabricated to suit the respective loading case. This type of reinforcement is lighter and, in comparison, more efficient than conventional steel reinforcement. It does not suffer from the corrosion risks of reinforced concrete. This type of concrete permits the production of very lightweight and small components with a high loadbearing capacity. As the production of textile reinforcement is comparatively expensive and this method of construction has not yet been adequately covered by standards owing to the need for further research, it is practically insignificant in concrete construction at the moment and reserved for special applications.

Textured fair-face concrete → 69
Threaded bar → 134, 140, 150, 172
Three-pin truss design
Tie, tension member
Tilt-up form(work)
Top-down method → 174
Trailing tamping vibrator → 154
Translation surface → 159, 160
Transverse bending → 156
Transverse/shear load
Trapezoidal profile sheeting → 143, 147, 149, 150, 155, 270
Trass → 11, 12, 19, 39, 40, 50, 59, 230
Trass blastfurnace cement
   A type of cement covered by building authority approvals. It consists of cement clinker, gypsum and/or anhydrite, up to 25% by wt trass and up to 50% by wt cinder sand.

Travelling formwork
   Formwork which is moved continually or incrementally; used, for example, in cut-and-cover tunnelling work.

Travelling overhead crane → 11, 12, 19, 39, 40, 50, 59, 230
Tremie pipe → 60, 173
   This prevents segregation of the concrete when being placed from a great height. This is a gently tapering sheet steel pipe made up of segments to suit the height required. Openings in the reinforcement for inserting a tremie

**Types of cement** and composition to EN 197, mass proportions in %[1]

| Type of cement | Designation | Abbreviation | Main constituents | | | | | | Minor constituents[2] |
|---|---|---|---|---|---|---|---|---|---|
| | | | Portland cement clinker | Cinder sand | Natural pozzolana | Pulverised fuel ash (PFA) rich in silicic acid | Fired oil shale | Limestone | |
| | | | K | S | P | V | T | L | |
| CEM I | Portland cement | CEM I | 95 – 100 | – | – | – | – | – | 0 – 5 |
| CEM II | Portland blastfurnace cement | CEM II/A–S<br>CEM II/B–S | 80 – 94<br>65 – 79 | 6 – 20<br>21 – 35 | –<br>– | –<br>– | –<br>– | –<br>– | 0 – 5<br>0 – 5 |
| | Portland pozzolanic cement | CEM II/A–P<br>CEM II/B–P | 80 – 94<br>65 – 79 | –<br>– | 6 – 20<br>21 – 35 | –<br>– | –<br>– | –<br>– | 0 – 5<br>0 – 5 |
| | Portland pulverised fuel ash (PFA) cement | CEM II/A–V | 80 – 94 | – | – | 6 – 20 | – | – | 0 – 5 |
| | Portland special pozzolanic cement | CEM II/A–T<br>CEM II/B–T | 80 – 94<br>65 – 79 | –<br>– | –<br>– | –<br>– | 6 – 20<br>21 – 38 | –<br>– | 0 – 5<br>0 – 5 |
| | Portland limestone cement | CEM II/A–L | 80 – 94 | – | – | – | – | 6 – 20 | 0 – 5 |
| | Portland pulverised fuel ash (PFA) blastfurnace cement | CEM II/B–SV | 65 – 79 | 10 – 20 | – | 10 – 20 | – | – | 0 – 5 |
| CEM III | Blastfurnace cement | CEM III/A<br>CEM III/B | 35 – 64<br>20 – 34 | 36 – 65<br>66 – 80 | –<br>– | –<br>– | –<br>– | –<br>– | 0 – 5<br>0 – 5 |

[1] The figures given in the table refer to the main and minor constituents of the cement as listed without calcium sulphate and cement additives.
[2] Minor constituents may be fillers or one or more main constituents provided these are not main constituents of the cement.

pipe must be allowed for during planning. Heaps of concrete are avoided by positioning the pipe at close intervals.
Trussed floor slab → 120
Tuff, tufa
Type of bolster chisel → 69

**U**ltra-fine particle/grain (of aggregate, cement, admixture, additive)
Undermining → 73
Underwater concrete → 174
Unsupported edge of shell
Upstand beam → 119, 240

**V**acuum-dewatered concrete
Concrete already placed and compacted and given a special type of curing. Excess water is removed from the surface of the concrete with the help of a vacuum generated by special equipment, e.g. vacuum mats. If there is simultaneous vibration to compact the voids previously filled with water, this vacuum treatment increases the strength and durability of the concrete. The water/cement ratio falls (e.g. by 0.10), early strength and final strength rise and shrinkage decreases. The effect of the vacuum method does not penetrate very deep. But it improves the outer layers of the concrete, those subjected to the hardest conditions. Formwork to components that have been vacuum treated can be struck early, often immediately after vacuum treatment. This method has proved worthwhile with hard concrete. It can be used on vertical and horizontal surfaces.
Vapour diffusion → 82, 85, 86
Vented/ventilated facing concrete
Vibrated concrete
Designation of concrete according to type of compaction. The shape and dimensions of the components determine the type and size of suitable compaction plant (vibrators). We distinguish between poker vibrators, vibration tampers and clamp vibrators. Vibration is the customary method of compaction for plastic concrete.
Vibration/vibrating tamper
Vibrator opening → 60
Vibratory tamper → 167
Vierendeel girder → 107, 138, 140, 151
Void → 69, 87, 117, 129, 138, 172

**W**affle (floor) slab
Water content → 80
Total of mixing water plus surface moisture on aggregate in a concrete mix.
Water permeability
1. If common, normal aggregates are used, the permeability of the hydrated cement determines the permeability of the concrete. The permeability of the hydrated cement depends on the capillary pore volume. Up to about 20% by vol., capillary pores are discrete, meaning that the permeability is practically zero. This is the case after complete hydration up to a water/cement ratio of about 0.50. Hydrated cement remains impermeable even after complete hydration for w/c ratios ≥ 0.70. If the concrete is required to be impermeable, this means that water should not be allowed to pass through the concrete and the side of the component not in contact with the water should not exhibit any moist patches.
2. The permeability of concrete in a component or structure is also partly determined by influences during production, e.g. temperature, curing, reinforcement. Special workmanship and constructional measures are called for when building impermeable concrete components so that such components do not exhibit any defects, i.e. leaks, cracks and permeable joints.

Water requirement
The quantity of water (l/m$^3$) required to achieve a certain consistency of the fresh concrete. With some experience this can be estimated from the grading curve. This depends on the consistency and is determined by the type, largest grain size and grading curve of the aggregate and the fines content. Empirical tables, formulas and groups of curves are available for estimating the quantity of water per cubic metre of compacted fresh concrete. This is given in kg/m$^3$ and depending on the consistency and that aggregate mix (e.g. expressed by the grading coefficient). The figures generally apply to gravel concrete without admixtures/additives. They should be increased when broken aggregate and/or admixtures are added, or decreased when using plasticisers or air entrainers. The actual quantity of water required to achieve a certain consistency can only be determined from a trial mix.
Water table → 87, 106, 170, 173, 174
Water/cement (w/c) ratio
Water-retention capacity
A property of the fines (binder and fine aggregate) which retains the mixing water added during production of mortar or concrete.
Waterstop/water bar → 187, 189
Wax release agent
This is applied to concrete formwork and prevents a bond forming between this and the fresh concrete. Used similarly to oil release agents.
Wet storage
A method of curing concrete and mortar test specimens. After striking, test specimens are to be stored on a grid underwater or in a moisture chamber ready for the quality control and suitability tests. Test specimens of lightweight concrete should be prevented from absorbing moisture while being stored.
Wet/moist curing
A method of curing concrete. The zones of the concrete near the surface are protected against drying out until sufficient setting has taken place. Methods include: not striking the formwork, covering with plastic sheeting, use of water-retaining covers, spraying of fluid curing agents. Spraying with water, as was used in the past, is damaging because it causes temperature stresses and hence cracks.
White cement → 47, 66, 67, 69, 71, 75, 258
A Portland cement produced from raw materials (limestone and china clay) essentially free from iron. In addition, reduced combustion conditions can prevent the formation of dark-coloured constituents in the Portland cement clinker (alumino ferrite). White cement is produced as Portland cement 42.5 R to DIN 1164 and may be used without restrictions just like ordinary Portland cement. Its primary uses are in fair-face concrete, coloured concrete or reconstituted stone as well as for other purposes in which the colour and its hydraulic properties are important. In Germany it is delivered in white sacks with black printing.
Wind girder/bracing
Wind load → 64, 95, 138, 140, 154, 163, 167
Wood particle concrete
Special type of mixed-pore concrete whose aggregate consists of wood chips. The wood should be pretreated with water glass (mineralisation).
Wood-wool lightweight building board → 50, 82, 83, 95

**Y**ield stress → 111, 117

# Index of names

**Architects, structural engineers**

**A**grippa → 10
Alder, Michael → 200
Alvarez Ordoñez, Fernando & Joaquín → 30
Anderson → 17
Ando, Tadao → 33, 42, 196, 220
Andreu, Paul → 226
Anthemios von Tralles → 11
Arup, Ove + Partners → 34, 141, 266, 268, 272
Ascoral Engineering Associates → 196
Aspdin, Joseph → 12
Atelier 5 → 32, 35, 218
Auer + Weber → 226

**B**aller, Hinrich & Inken → 230
Baudot, Anatole de → 20
Bauersfeld, Walter → 17, 23, 24
Becker, Christian → 250
Beckmannshagen, Klaus → 258
Berg, Max → 22, 23
Berlage, Hendrik Petrus → 18
Bertsch-Friedrich-Kalcher → 256, 258
Böbel + Frey → 238
Böhm, Gottfried → 35, 40, 256, 258
Bökeler, Karl Heinz → 258
Bofill, Ricardo → 37, 40
Bole & Partner → 141
Bomhard, Helmut → 39, 141
Botta, Mario → 36
Bourrat, Alexis → 254
Brauer, Karsten → 240
Broek, J. H. van den & J. B. Bakema → 141
Büchler, Walter → 234
Busch-Berger → 138
BVC STL → 270

**C**alatrava, Santiago → 254
Caligula → 10
Candela, Felix → 30, 159
Cantor, E. → 138
Carfrae, T. → 272
Castiglioni, E. → 164
Cato, Marcus Porcius → 10
Ceresa, Piero → 234
Cesba → 212
Ciriani, Henri → 212
Coignet, Edmond → 15, 16
Coignet, François → 13

**D**avid → 9
Derrida, Jacques → 41
Dischinger, Franz → 17, 24
Dittmann, Wolf → 250
Doehring → 15, 17
Domenig, Günter → 41
Domostatik → 242
Dorn, Roland → 242
Drexler, Gisela → 198
Dunster, Bill → 202
Dyckerhoff & Widmann → 24, 141, 165

**E**isenman, Peter → 41
Elzner → 17
Emperger, Fritz Edler von → 16
Endl-Storek, Stefan → 236
Erikson, Arthur → 33, 34
Esquillan, Nicolas → 155
Euskirchen, Wilfried → 250

Evers Partners → 248
Eyck, Aldo van → 38

**F**asanya, Ernest → 202
Finger & Fuchs → 218
Finsterwalder, Ulrich → 24, 165
Fitzgerald, Richard → 266
Forest de Bélidor, Bernard → 11
Foster & Partners → 276
Freyssinet, Eugène → 17, 23, 156
Freytag, Conrad → 15
Fuchssteiner, W. → 155
Fuses, Josep & Viader, Joan María → 228

**G**alfetti, Aurelio → 36, 210, 234
Garnier, Tony → 25
Gatermann, Dörte → 258
Gehry, Frank O. → 41
Gerber, Eckhard → 40, 138, 264
Gerkan, Marg & Partner → 240
Gieselmann, Berhard → 242
Gnädig, Miklós → 147
Goldberg, Bertrand → 138
Graves, Michael → 41
Gregotti, Vittorio → 39
Greschick & Falk → 200
Günther, H. → 155
Gwathmey, Charles → 41

**H**adid, Zaha → 42
Happold, E. → 202
Häußler → 154
Heidschuh, Carl → 15
Heinle Wischer & Partner → 141
Hejduck, John → 41
Hennebique, François → 15, 16, 17, 18, 44
Hentrich-Petschnigg + Partner → 141
Herrschmann, Dieter → 236
Hertzberger, Herman → 38, 248
Höltje, W. → 165
Hopkins, Michael → 202, 268
Hossdorf, Heinz → 155
Hovestadt, Klaus → 242
Hyatt, Thaddeus → 14

**I**shida, Shunji → 272
Isidoros von Milet → 11
Isler, Heinz → 164
Ito, Toyo → 43

**J**ackson → 15, 17
Jahn, Helmut → 138
John, Johann Friedrich → 12
Johnson, Charles Isaac → 12
Justinian → 11

**K**ahn, Albert → 19
Kahn, Louis → 33, 34, 36
Kezic, Neno → 202
Kinch, R. → 272
Kochta + Lechner → 236
Koenen, Mathias → 15, 16, 17
Koolhaas, Rem → 42
Küster, Plüdermann → 19

**L**ambot, Josef Louis → 13, 14
Lavers, Lucy → 202
Le Corbusier → 25, 26, 28, 32, 33, 34, 36, 37
Lehmbrock, J. → 158

Lenczner, A. → 272
Libeskind, Daniel → 41
Liedvogel, Heinrich → 24
Littmann, Max → 21
Long, David → 254
Loos, Adolf → 26
Lund → 17

**M**aillart, Robert → 16, 17, 21, 30
Mangiarotti, Angelo → 270
Matern, Eva → 242
Meier, Richard → 41
Mèmet, Sebastian → 254
Mendelsohn, Erich → 24
Menyhárd, J. → 155
Meyer, Rudolf → 198
Mies van der Rohe, Ludwig → 25, 27
Minkus, Jürgen → 256, 258
Miralles, Enric → 39
Mörsch, Emil → 16, 17
Monier, Joseph → 14, 15
Mora, Enrique de la → 30
Morger & Degelo → 43
Moser, Karl → 21
Müller, Hanspeter → 200

**N**ägele, Reinhard → 236
Nägeli, Walter → 41
Naumann & Strabag → 138
Nervi, Pier Luigi → 31, 118, 157-159
Neufert, E. → 155
Neufert, Peter → 35
New York Five → 41
Nötzold, F. → 138

**O**IKOS, Peter Herrle & Werner Stoll → 232
Olgiati, Valerio → 216
Otto, Frei → 39, 165
Otto, Ulrich → 238

**P**arker, James → 12
Perret, Auguste → 19, 20, 21, 28, 155
Pettenkofer, Max von → 13
Piano, Renzo → 266, 272
Pichler, Gerhard → 230
Plečnik, Josef → 21
Pleuser, Jürgen → 206
Plüdermann & Küster → 19
Polónyi, Stefan → 35, 106, 138, 155, 158, 165
Polónyi & Fink → 206, 264
Pringle, John → 202

**R**abitz, Carl → 15
Rainer, R. → 165
Ransome, Ernest Leslie → 15, 19
Reichel, Alexander → 224
Reitzel, Erik → 226
Rice, Peter → 266, 272
Rietveld, Gerrit Thomas → 27
Rudolph, Paul → 33, 34

**S**aarinen, Eero → 33, 165
Sacrez → 17
Scarpa, Carlo → 36
Schachner, Richard → 19
Schaller, Fritz → 165
Schelling, E. → 165
Schindler, Rudolf → 26, 27
Schinkel, Karl Friedrich → 15
Schipporeit, George → 138

Schlaich Bergermann & Partner → 226
Schnebli, Dolf → 36
Schönberg, Arnold → 41
Schürmann, Joachim & Margot → 250
Schuhmacher, Johannes C. → 220
Schulten, H. → 155
Schultes, Axel → 206, 244
Schumann → 16
Schwanzer, K. → 141
Schwarz + Weber → 240
Scobat BET → 212
Séailles, J. → 17
Seidler, Harry → 262
Sert, Josep Lluis → 37
Silberkuhl → 154
Sinan → 11
Siza, Alvaro → 42
Skidmore Owings & Merrill → 138
Smeaton, John → 12
Snozzi, Luigi → 36, 37
Spitzlei & Jossen → 206
Spreckelsen, Johan Otto von → 226
Steidle, Otto → 39, 40
Steiner, Rudolf → 17, 24
Stettner & Wald → 206
Stildors (Planungsgruppe) → 141
Stirling, James → 41
Striffler, Helmut → 35
Stubbins, Hugh A. → 166
Szrog, G. → 140

**T**ange, Kenzo → 33, 34
Thormälen & Peuckert → 242
Tilch, Axel → 198
Torroja, Eduardo → 30
Tschumi, Bernard → 41
Tsubui, Y. → 34
Tuch, H. → 155
Tyerman, T. E. → 13

**U**ngers, Oswald M. → 138
Utzon, Jørn → 33, 34

**V**aessen, Franz & Ehrhardt, H. J. → 141
Valda, Federico → 258
Van Berkel, Ben & Bos, Caroline → 41
Vanetta, Enzo → 210
Varwick-Horz-Ladewig → 250
Vitruvius → 10

**W**allot, Paul → 15
Wayss, Adolf Gustav → 15
Wayss & Freytag → 17
Weijer, Henk de → 248
Wilford, Michael → 41
Wilkinson, William Boutland → 13
Wilson → 17
Wolf, Harry C. → 138
Wright, Frank Lloyd → 20, 21, 26, 27, 118

**Z**üblin AG → 256
Züblin, Eduard → 19
Zumthor, Peter → 43

# Picture credits

## Picture credits

The authors and publishers would like to express their sincere thanks to all those who provided visual material, gave us permission to reproduce details or provided information and so contributed to the production of this book. All the drawings and diagrams in this book were specially commissioned. Photographs not specifically acknowledged were supplied by the architects and engineers named in the "Index of persons" or from the archives of the journal DETAIL. The numbers refer to the page or figure numbers.

### From photographers and picture archives

Beretta, S., Giubiasco, pp. 234-35
Bleyl, M., Berlin, pp. 232-33
Burkhard, B., Bern, p. 35 (1.81)
Casali, Milan, p. 270
Charles, M., Isleworth, pp. 203-05
Cook, P., London, pp. 272-75
Dupain, M., Artarmon, p. 262
Dyckerhoff AG, Wiesbaden, p. 67 (2.2.2, 2.2.8), p. 71 (2.2.16, 2.2.17, 2.2.18, 2.2.20, 2.2.21), p. 75 (2.2.22, 2.2.23, 2.2.26, 2.2.27)
Dyckerhoff & Widmann, Munich, p. 22 (1.36, 1.37), p. 23 (1.41, 1.42), p. 24 (1.44)
Ege, H., Lucerne, p. 255
Foto Marburg, Marburg, p. 21 (1.34), p. 31 (1.69)
Fotoprofis, Bremen, p. 167 (3.2.89)
Friedrich, R., Berlin, pp. 230-31
Gilbert, D./View, London, pp. 276-79
Habermann, K.J., Munich, pp. 268-69 top
Hamilton Knight, M., Nottingham, p. 269 bottom
Heidersberger, H., Wolfsburg, p. 236
Helfenstein, H., Zurich, p. 217
Hester, P., Houston, pp. 266-67
Huthmacher, W., Berlin, pp. 244-47
Institut Français d'Architecture, Paris, p. 7, p. 15 (1.14, 1.15)
Kinold, K., Munich, p. 26 (1.51), p. 28 (1.57, 1.59), p. 29 (1.63, 1.64), p. 36 (1.85), p. 37 (1.88), p. 38 (1.90), p. 39 (1.96), p. 40 (1.98, 1.99), pp. 45, 105, 177, 218-19, 234-35, 248-49
Klimek, S., Wrocław, p. 19 (1.26)
Knabben, T., Cologne, p. 139 (3.1.87), p. 141 (3.1.98)
Leiska, H., Hamburg, pp. 240-41
Leistner, D., Mainz, p. 257
Malagamba, D., Barcelona, p. 42 (1.103)
Monthiers, J.-M., Paris, pp. 213-15
Neubert, S., Munich, p. 19 (1.26)
Praßer, G., Cologne, p. 67 (2.2.6)
Richters, C., Münster, p. 41 (1.101), pp. 224-25, 264-65
Rohrbach Zement, Dotternhausen, p. 67 (2.2.5), p. 71 (2.2.15), p. 75 (2.2.25)
Rosendaal, J., Munich, p. 26 (1.52)
Roth, L., Cologne, pp. 242, 243 l.
Rühle, T., Cologne, pp. 251-53
Schindler Archive, University of California, Santa Barbara, Architectural Drawing Collection, p. 26 (1.53, 1.54)
Schittich, C., Munich, p. 11 (1.4), p. 21 (1.33), p. 32 (1.72), p. 33 (1.76), p. 36 (1.84), p. 37 (1.86, 1.87), p. 40 (1.97), p. 41 (1.102), p. 42 (1.106), p. 43 (1.107), pp. 193, 200-01, 221-23, 237
Schlesinger, H., Karlsruhe, p. 165 (3.2.82)
Schmölz, K.H., Cologne, pp. 260 top, 261
Shinkenchinku-Sha, Tokyo, p. 42 (1.104), p. 43 (1.105)
Shiratori, Y., Tokyo, p. 196
Suspa, Fa., Langenfeld, p. 120 (3.1.36, 3.1.37)
Suzuki, H., Barcelona, pp. 228-29
Dortmund University, Chair of Emer. Prof. A. Weißenbach, p. 173 (3.3.18, 3.3.19)
Verlag Bau + Technik, Düsseldorf, p. 10 (1.2), p. 12 (1.7), p. 14 (1.13), p. 24 (1.45), p. 28 (1.58), p. 33 (1.77, 1.78), p. 34 (1.79, 1.80), p. 35 (1.83), p. 38 (1.89), p. 66 (2.2.1), p. 67 (2.2.7, 2.2.9), p. 68 (2.2.10), p. 71(2.2.14, 2.2.19), p. 75 (2.2.24), pp. 198-99, 206-09, 226-27
Walti, R., Basel, p. 43 (1.106)
Wietzorek, U., Munich, p. 33 (1.75), p. 39 (1.95)
Züblin AG, Stuttgart, pp. 259, 260 bottom

### From books and journals

p. 10 (1.1), 11 (1 3) Lamprecht, H. O.: Opus Caementium – Bautechnik der Römer; 4th ed., Verlag Bau + Technik, Düssefdorf 1993
p. 12 (1.6), p. 13 (1.8), p. 14 (1.9, 1.10, 1.11, 1.12), p. 16 (1.17, 1.18, 1.19, 1.21) Haegermann, G. et al: Vom Caementum zum Spannbeton; vol. 1. Bauverlag GmbH, Wiesbaden/Berlin 1964
p. 16 (1.17), p. 17 (1.20), p. 21 (1.35) Billington, D.P.: Robert Maillart und die Kunst des Stahlbetonbaus; Verlag für Architektur Artetuis, Zurich/Munich 1990
p. 18 (1.22, 1.23) Niebelschütz, W. von: Züblin-Bau; Stuttgart 1958
p. 18 (1.24) Ackermann, K.: Industriebau; Deutsche Verlagsanstalt, Stuttgart 1984
p. 19 (1.27), p. 20 (1.28, 1.29) Perret. L'Architecture d'Aujourd'hui, October 1932
p. 20 (1.30), p. 21 (1.31), p. 33 (1.74), p. 38 (1.91) Frampton, K.: Grundlagen der Architektur-Studien zur Kultur des Tektonischen (Studies in Tectonic Culture); Oktagon Verlag, Munich/Stuttgart 1993
p. 21 (1.32), p. 27 (1.55, 1.56) Riley, T.: Frank Lloyd Wright, Architect; The Museum of Modern Art, New York 1994
p. 22 (1.38, 1.39) Hersel, O.: Die Jahrhunderthalle zu Breslau; Beton- und Stahlbetonbau 12/1987
p. 23 (1.40) Pausen, A.: Eisenbeton 1850-1950; Manz Verlag, Vienna 1994
p. 23 (1.43), p. 32 (1.73) Gössel, P.; Leuthäuser G.: Architektur des 20. Jahrhunderts; Benedikt Taschen Verlag, Cologne 1990
p. 24 (1.46), p. 25 (1.48, 1.49) Lampugnani, V. M., Schneider, R. (ed.): Moderne Architektur in Deutschland 1900-1950 – Expressionismus und Neue Sachlichkeit; Verlag Gerd Hatje, Stuttgart 1994
p. 25 (1.47) Garnier, T.: Die ideale Industriestadt; Verlag Ernst Wasmuth, Tübingen 1989
p. 25 (1.50) Joedicke, J.: Architekturgeschichte des 20. Jahrhunderts; Edition Krämer, Stuttgart/Zurich 1990
p. 28 (1.60, 1.61) Le Corbusier, Studio Paperback; Verlag für Architektur Artetuis, Zurich 1972
p. 29 (1.62) Le Corbusier: Feststellungen zu Architektur und Städtebau; Ullstein Verlag, Frankfurt a.M./Berlin 1964
p. 30 (1.65, 1.66), p. 155 (3.2.33, 3.2.34, 3.2.35, 3.2.37, 3.2.41 ), p. 158 (3.2.50, 3.2.59, 3.2.60) Joedicke, J.: Dokumente der modernen Architektur – Schalenbau; Karl Krämer Verlag, Stuttgart 1962
p. 31 (1.68, 1.70), p. 155 (3.2.32), p. 165 (3.2.77) Siegel, C.: Strukturformen der Modernen Architektur; 3rd ed., Verlag Georg D.W. Callway, Munich 1970
p. 32 (1.71) Paul Rudolph, Bauten und Projekte; Verlag G. Hatje, Stuttgart 1970
p. 35 (1.82) Nestler, P.; Bode, P.M.: Deutsche Kunst seit 1960 – Architektur; Verlag F. Bruckmann, Munich 1976
p. 39 (1.92, 1.93) Bomhard, H.: Konstruktion und Bau der Paketumschlaghalle in München; pub. by Deutscher Beton-Verein, Wiesbaden 1967
p. 39 (1.94) Klotz, H., (ed.): Vision der Moderne, Das Prinzip Konstruktion; Prestel Verlag, Munich 1986
p. 40 (1.100) Domenig, G.: Werkbuch; Residenz Verlag, Salzburg 1991
p. 69 (2.2.11) Zement-Merkblatt Schalung für Beton; Bauberatung Zement (pub.), Verlag Bau + Technik, Düsseldorf 1988
p. 76 (2.2.30, 2.2.31, 2.2.32, 2.2.33, 2.2.34, 2.2.35) Huberty, J. M.: Fassaden in der Witterung; Verlag Bau + Technik, Düsseldorf 1983
p 78 (2.3.2) Frank, W.: Raumklima und thermische Behaglichkeit – Bericht aus der Bauforschung; issue 104; Wilhelm Ernst & Sohn, Berlin 1975
p. 95 (2.3.40) Karl Schwanzer, Architektur aus Leidenschaft; modulverlag GmbH, Vienna/Munich 1973
p. 139 (3.1.85, 3.1.90, 3.1.91, 3.1.92) Zukowsky, J. (ed.): Chicago – Architecture and Design; The Art Institute of Chicago and Prestel-Verlag, Munich 1993
p. 139 (3.1.86) Klotz, H. (ed.): Murphy/Jahn – Messeturm Frankfurt; Oktagon Verlag, Munich/Stuttgart 1991 (Schriften zur Architektur der Gegenwart)
p. 139 (3.1.89) Klotz, H.; Saban, L. (ed.): New York Architecture 1970-1990; Prestel-Verlag, Munich 1989
p. 141 (3.1.97) 75 Jahre Welt des Betons; Deutscher Beton-Verein e.V., Wiesbaden 1973
p. 158 (3.2.51, 3.2.52) Le Corbusier: Feststellungen zu Architektur und Städtebau; Ullstein Verlag, Frankfurt/Berlin 1964.
p. 165 (3.2.78) Conrad, U.; Sperlich, H.G.: Phantastische Architektur; Verlag G.Hatje, Stuttgart 1960
p. 165 (3.2.83, 3.2.84) "Hochtief Nachrichten" (company brochure), Essen
p. 212 L'Architecture d'Aujourd'hui, September 1992

### Full page plates

|  | |
|---|---|
| Page 7 | Formwork to a church dome in Tbilisi, Georgia, 1903, architects: Zarambianz and Aknazaron |
| Page 45 | Pulpit, Notre-Dame-du-Haut pilgrimage church, Ronchamp, 1950-54, architect: Le Corbusier |
| Page 105 | Vaulting, St Antonius Church, Basel, 1927, architect: Karl Moser |
| Page 177 | Spring at the Brion family cemetery, San Vito d'Altivole 1970-72, architect: Carlo Scarpa |
| Page 193 | Inner courtyard of conference pavilion, Weil am Rhein 1993, architect: Tadao Ando |